# Soft Condensed Matter:
# Configurations, Dynamics and Functionality

# NATO Science Series

*A Series presenting the results of activities sponsored by the NATO Science Committee. The Series is published by IOS Press and Kluwer Academic Publishers, in conjunction with the NATO Scientific Affairs Division.*

| | |
|---|---|
| A. Life Sciences | IOS Press |
| B. Physics | Kluwer Academic Publishers |
| C. Mathematical and Physical Sciences | Kluwer Academic Publishers |
| D. Behavioural and Social Sciences | Kluwer Academic Publishers |
| E. Applied Sciences | Kluwer Academic Publishers |
| F. Computer and Systems Sciences | IOS Press |
| | |
| 1. Disarmament Technologies | Kluwer Academic Publishers |
| 2. Environmental Security | Kluwer Academic Publishers |
| 3. High Technology | Kluwer Academic Publishers |
| 4. Science and Technology Policy | IOS Press |
| 5. Computer Networking | IOS Press |

**NATO-PCO-DATA BASE**

The NATO Science Series continues the series of books published formerly in the NATO ASI Series. An electronic index to the NATO ASI Series provides full bibliographical references (with keywords and/or abstracts) to more than 50000 contributions from international scientists published in all sections of the NATO ASI Series.
Access to the NATO-PCO-DATA BASE is possible via CD-ROM "NATO-PCO-DATA BASE" with user-friendly retrieval software in English, French and German (WTV GmbH and DATAWARE Technologies Inc. 1989).

The CD-ROM of the NATO ASI Series can be ordered from: PCO, Overijse, Belgium

**Series C: Mathematical and Physical Sciences – Vol. 552**

# Soft Condensed Matter: Configurations, Dynamics and Functionality

edited by

## A.T. Skjeltorp

Institute for Energy Technology,
Kjeller, Norway and
Department of Physics,
University of Oslo,
Norway

and

## S.F. Edwards

Cavendish Laboratory,
University of Cambridge,
Cambridge, United Kingdom

**Kluwer Academic Publishers**

Dordrecht / Boston / London

Published in cooperation with NATO Scientific Affairs Division

Proceedings of the NATO Advanced Study Institute on
Soft Condensed Matter: Configurations, Dynamics and Functionality
Geilo, Norway
April 6-16, 1999

A C.I.P. Catalogue record for this book is available from the Library of Congress.

ISBN 0-7923-6402-3 (HB)
ISBN 0-7923-6403-1 (PB)

Published by Kluwer Academic Publishers,
P.O. Box 17, 3300 AA Dordrecht, The Netherlands.

Sold and distributed in North, Central and South America
by Kluwer Academic Publishers,
101 Philip Drive, Norwell, MA 02061, U.S.A.

In all other countries, sold and distributed
by Kluwer Academic Publishers,
P.O. Box 322, 3300 AH Dordrecht, The Netherlands.

Printed on acid-free paper

# CONTENTS

# PREFACE

This volume comprises the proceedings of a NATO Advanced Study Institute held at Geilo, Norway, April 6 - 16 1999. The ASI was the fifteenth in a series held biannually on topics related to cooperative phenomena and phase transitions, in this case applied to soft condensed matter and its configurations, dynamics and functionality. It addressed the current experimental and theoretical knowledge of the physical properties of soft condensed matter such as polymers, gels, complex fluids, colloids, granular materials and biomaterials.

The main purpose of the lectures was to obtain basic understanding of important aspects in relating molecular configurations and dynamics to macroscopic properties and biological functionality. To our knowledge, the term Soft Condensed Matter was actually coined and used for the first time in 1989 at Geilo and some selected topics of soft matter were also given at Geilo in 1991, 1993 and 1995. A return to this subject 10 years after its instigation thus allowed a fresh look and a possibility for defining new directions for research.

Soft condensed matter encompasses a wide range of substances which are neither ordinary solids nor ordinary liquids, but have much more complexity and subtlety of character than either as well as having vestiges of each. Systems range from foams and complex fluids to granular materials and biomaterials like protein, DNA and membranes. They exist in a wide variety of structures that are driven by subtle competition between intermolecular interaction energies and entropic forces, both of which are often close to thermal energies at room temperature. These same forces, plus the constraints imposed by the configurations adopted by these systems also have a strong effect on the molecular motions or dynamics. Both the configurations and their dynamical evolution are known to be important in determining a wide variety of mesoscopic and macroscopic properties, including those linked to "function" in the case of biomolecular assemblies.

Many of these so-called "soft materials" display what one can call "adaptive" behaviour – that is strong changes in some physical property that results from a small change in an internal or external driving force. Indeed, such effects are likely a prerequisite for life itself but are only beginning to be understood in physical terms. Among the adaptive phenomena themselves there is a kind of logical progression, from the behaviour of disordered cooperative systems to evolved cooperative systems such as RNA, proteins, and possibly the immune system.

The language needed to discuss these systems are reviewed and basic questions regarding phenomena such as competing ground states, nonlinear feedback and slow dynamics are presented in introductory lectures, with later talks emphasizing subfields in more detail. Granular matter are discussed in regard to segregation of powders, equations of granular materials, granular flow and mechanics as well as the modeling of granular flows. Various aspects of interfaces and confinement are reviewed in relation to nucleation and engineering of crystalline architectures at the air-liquid interface, as well as x-ray and neutron studies of complex confined fluids. Soap films and general properties of the evolution of froth are discussed, both theoretically and experimentally. Broad reviews are given of hierarchical protein folding and protein evolution in vitro. Related to this, DNA recognition and computation, gene expressions and measurements as well as torsion-induced phase transition in single DNA molecules are also discussed. Fungus growth modeling and cell attachment and spreading are also presented.

The Institute brought together many lecturers, students and active researchers in the field from a wide range of countries, both NATO and NATO partners, and non-NATO. The lectures fulfilled the aim of the Study Institute in creating a learning environment and a forum for discussion on the topics stated above. They were supplemented by a few contributed seminars and a large number of poster presentations. These seminars are included in the proceedings and the posters were collected in extended abstract form and issued as an open report available at the Institute for Energy Technology, Kjeller, Norway (Report IFE/KR/E-99/008).

Financial support was principally from the NATO Scientific Affairs Division, but also from the Institute for Energy Technology and the Research Council of Norway.

The editors are most grateful to A. Hansen, M.H. Jensen, R. Pynn, D. Sherrington and H. Thomas who helped them plan the programme and G. Helgesen for helping with many practical details. Finally, we would like to express our deep gratitude to Mary Byberg of the Institute for Energy Technology, for all her work and care for all the practical organization before, during, and after the school, including the preparation of these proceedings.

June 1999


Arne T. Skjeltorp                                                    Sam Edwards

# LIST OF PARTICIPANTS

**Organizing Committee:**

Skjeltorp, Arne T., director
  Institute for Energy Technology, POB 40, N-2027 Kjeller, Norway

Edwards, Sam, co-director
  Cavendish Laboratory, University of Cambridge, Madingley Road,
  Cambridge CB3 0HE, United Kingdom

Byberg, Mary, secretary
  Institute for Energy Technology, POB 40, N-2027 Kjeller, Norway

**Participants:**

Als-Nielsen, Jens
  H.C. Ørsted Institute, Universitetsparken 5, DK-2100 Copenhagen Ø, Denmark

Alstrøm, Preben
  CATS Niels Bohr Institute, Blegdamsvej 17, DK-2100 Copenhagen Ø, Denmark

Antipeshev, Stefan
  Dept. Phys. Chem. Univ. Sofia, 1. J. Bourchier Ave., 1126 Sofia, Bulgaria

Armagan, Turgay
  Istanbul University, Science Faculty, Physics Department, 34459 Vezneciler-
  Istanbul, Turkey

Avgin, Ibrahim
  EGE University Electrical Eng., 35100 Bornova, Izmir, Turkey

Bensimon, David
  Lab. de Physique Statistique, 24 rue Lhomond, Paris 75005, France

Berre, Bjørn
  Norges landbrukshøgskole, Institutt for tekniske fag, Box 5065, N-1432 Ås, Norway

Bickel, Thomas
  LDFC - 3 rue de l'Université, 67084 Strasbourg, France

Bobarykina, Gueia
Frank Laboratory of Neutron Physics, Joint Institute for Nuclear Research, Dubna,
Moscow Region 141980, Russia

Borg, Jesper
Forhaabningsholmsallé 43, 1.tv., 1904 Frb.C, Denmark

Buchanan, Mark
Rm 4305, JCMB, Kings Buildings, Dept. of Physics, The University of Edinburgh,
Scotland, UK

Calonder, Claudio
Biozentrum, Klingelbergstrasse 70, CH-4056 Basel, Switzerland

Castelnovo, Martin
Institut Charles Sadron, 6 rue Boussingault, 67083 Strasbourg Cedex, France

Cernák, Josef
Diamantova 8, SK-04011 Kosice, Slovak Republic

Charitat, Thierry
Institut Charles Sadron, 6 rue Boussingault, 67083 Strasbourg Cedex, France

Clarysse, Francis
Laboratorium Vaste Stof-Fysica en Magnetisme, Celestijnenlaan 200D,
B-3001 Leuven, Belgium

Deger, Deniz
University of Istanbul, Faculty of Science, Physics Department, 34459-Vezneciler,
Istanbul, Turkey

Elgsaeter, Arnljot
Institutt for fysikk, NTNU, S. Sælandsvei 9, N-7491 Trondheim, Norway

Ertosun, Süheda
Istanbul Tıp Fakultesi, Patoloji Anabilim Dalı, Temel Tıp Bilimleri Binası,
34390 Capa-Istanbul, Turkey

Evangelou, Spiros
Physics Dept., Univ. of Ioannina, P.O.Box 1186, GR-45110 Ioannina, Greece

Flekkøy, Eirik G.
University of Oslo, Department of Physics, P.O.Box 1048 Blindern, 0316 Oslo,
Norway

Fogedby, Hans
    Institute of Physics and Astronomy, University of Aarhus, DK-8000 Aarhus C,
    Denmark

Fossum, Jon Otto
    Institutt for fysikk, Gløshaugen, NTNU, S. Sælandsvei 9, N-7491 Trondheim,
    Norway

Giaever, Ivar
    Institute of Science, Rensselaer Polytechnic Institute, Troy, NY 12180, USA

Gorchkova, Ioulia
    Frank Laboratory of Neutron Physics, Joint Institute for Nuclear Research, Dubna,
    Moscow Region 141980, Russia

Grinev, Dmitri
    Department of Physics, Cavendish Laboratory, University of Cambridge,
    Madingley    Road, Cambridge CB3 0HE, UK

Guyon, Etienne
    Ecole Normale Superieur, 45, rue d'Ulm, 75230 Paris Cedex 5, France

Habdas, Piotr
    Institute of Physics, University of Silesia, ul. Uniwersytecka 4, 40-007 Katowice,
    Poland

Habib, Khaled
    Materials Application Dept., KISR, P.O.Box 24885 SAFAT, 13109 Kuwait

Hansen, Alex
    Institutt for fysikk, NTNU Gløshaugen, N-7491 Trondheim, Norway

Hauback, Bjørn C.
    Institute for Energy Technology, P.O.Box 40, N-2027 Kjeller, Norway

Hedin, Niklas
    Physical Chemistry, Royal Institute of Technology, S-100 44 Stockholm, Sweden

Helgesen, Geir
    Institute for Energy Technology, P.O.Box 40, N-2027 Kjeller, Norway

Imer, Filiz
Yıldız Teknik Üniversitesi, Fen-Edebiyat Fakültesi, Kimya Bölümü, 80270 Sisli,
Istanbul, Turkey

Jensen, Mogens Høgh
Niels Bohr Institute, Blegdamsvej 17, DK-2100 Copenhagen, Denmark

Kalkan, Nevin
University of Istanbul, Faculty of Science, Physics Department, 34459 Vezneciler,
Istanbul, Turkey

Kihlman, Sofia
Department of Applied Physics, Chalmers University of Technology,
S-412 96 Göteborg, Sweden

Koukiou, Flora
Laboratoire de Physique Theorique et de Modelisation Universite de Cergy-Pontoise
BP 222, 95302 Cergy-Pontoise, France

Lagerwall, Sven T.
Physics Dept., Chalmers University of Technology, S-412 96 Göteborg, Sweden

Leiserowitz, Leslie
The Weismann Institute of Science, Rehovot, 76100, Israel

Libchaber, Albert
Center for Studies in Physics and Biology, The Rockefeller University,
1230 York Avenue, New York, NY 10021, USA

Lin, Min
National Institute of Standards and Technology, 100 Bureau Dr., Build. 235,
Stop 8562, Gaithersburg, MD 20899-8562, USA

Lise, Stefano
Department of Mathematics, Imperial College, 180 Queen's Gate,
London SW7 2BZ, UK

Lombardo, Domenico
LURE, Bat. 209-D, B.P. 34, F-91898 ORSAY cedex, France

Lopes, António
ITQB-UNL, Ap. 127, P-2781-901 Oeiras, Portugal

Luchsinger, Rolf
    Physik Institut Uni Zürich-Irchel, Winterthurerstr. 190, CH-8057 Zürich,
    Switzerland

Maeland, Arnulf J.
    305, Cactus Hill Court, Royal Palm Beach, Fl 33411, USA

Major, András G.
    Institut für Theoretische Physik Teil 3, Universität Stuttgart, Pfaffenwaldring 57,
    D-70550 Stuttgart, Germany

Manificat, Guillaume
    Gruppe for teoretisk fysikk, NTNU Gløshaugen, N-7491 Trondheim, Norway

Masloboeva, Julia
    University of Oslo, Department of Physics, P.O.Box 1048 Blindern, N-0316 Oslo,
    Norway

McCauley, Joseph
    Physics Dept., University of Houston, Houston, TX 77204, USA

Melø, Thor Bernt
    Dept. of Physics, NTNU (Section Lade), N-7491 Trondheim, Norway

Måløy, Knut Jørgen
    University of Oslo, Department of Physics, P.O.Box 1048 Blindern,
    N-0316 Oslo, Norway

Oxaal, Unni C.
    Dept. of Agricultural Engineering, Agricultural University of Norway,
    P.O.Box 5065, N-1432 Ås, Norway

Pynn, Roger
    P.O.Box 1663, MS H845, Los Alamos, NM 87545, USA

Robert, Aymeric J.F.
    European Synchrotron Radiation Facility ESRF, BP 220, 38043 Grenoble, France

Rossi, Andrea
    SISSA,, Via Beirut 2-4, 34014 Grignano (TS), Italy

Sams, Thomas
    DDRE, Ryvangs Alle 1, P.O.Box 2715, DK-2100 Copenhagen Ø, Denmark

Savyak, Mariya
Institute for Problems of Material Science, 3 Kzhizhanovsky Street, Kiev 252 680, Ukraine

Schmalian, Joerg
ISIS Facility, Rutherford Appleton Laboratory, Chilton, Didcot, Oxon OX11 0QX, Oxfordshire, UK

Settanni, Giovanni
SISSA, Via Beirut 2, 34014 Trieste, Italy

Sherrington, David
Theoretical Physics, University of Oxford, 1 Keble Road, Oxford OX1 3NP, UK
Sinha, Sunil
Advanced Photon Source, Argonne National Laboratory, Argonne, IL 60439, USA

Sitnikov, Ruslan
Department of Physics & Chemistry, Royal Institute of Technology, S-100 44 Stockholm, Sweden

Sneppen, Kim
Nordita, Blegdamsvej 17, DK-2100 Copenhagen, Denmark

Sommelius, Ola
Nordita, Blegdamsvej 17, DK-2100 Copenhagen, Denarmk

Stavans, Joel
Dept. of Physics of Complex Systems, The Weizmann Institute of Science, Rehovot 76100, Israel

Steinsvoll, Olav
Institute for Energy Technology, P.O.Box 40, N-2027 Kjeller, Norway

Thomas, Harry
Dept. of Physics, University of Basel, Klingelbergerstrasse 82, CH-4056 Basel, Switzerland

Tiana, Guido
Niels Bohr Institutet, Blegdamsvej 17, DK-2100 Copenhagen, Denmark

Tsekov, Roumen
Department of Physical Chemistry, University of Sofia, 1 James Bourchier Avenue, 1126 Sofia, Bulgaria

Uhomoibhi, James
Department of Pure and Applied Physics, The Queen's University of Belfast,
University Road, Belfast, BT7 1NN, Northern Ireland, UK

Ulutas, Kemal
University of Istanbul, Faculty of Science, Physics Department, 34459 Vezneciler,
Istanbul, Turkey

Weaire, Denis
Department of Pure and Applied Science, Univ. of Dublin, Trinity College,
Dublin 2, Ireland

Yartys, Volodymyr
Institute for Energy Technology, P.O.Box 40, N-2027 Kjeller, Norway

Zapotocky, Martin
Department of Physics and Astronomy, University of Pennsylvania,
209 South 33rd Street, Philadelphia, PA 19104, USA

# WHAT IS SOFT CONDENSED MATTER ?

ETIENNE GUYON
*Ecole normale supérieure*
*45 rue d'Ulm 75005 Paris, France*

## 1. Context

A well identified community of physicists has established itself over the last thirty years, which was present in the 1999 Geilo Institute. Looking back at the themes of the 14 previous institutes, it is easy to recognize constant themes throughout the meetings as well as a drift away from the initial ones. We are dealing with Material Science as analysed by Solid State physicists using the tools of Statistical Physics. The study of electronic properties which was present in the first meetings has progressively been dismissed. From microscopic systems, the studies have been progressively turned towards mesoscopic -supramolecular- ones using analogies with the microscopic case : an example is the numerical study of granular media based on models developed in molecular dynamics studies of liquids and gases. There has been a progressively significant place given to Continuum Mechanics and Rheology, a field not much in fashion for physicists and chemists. Finally, the opening towards Biological systems is a new and strong component of the present institute. In fact, Soft Condensed Matter does not characterize a field of study, rather a spirit!

Consideration of terminology is not of great use to characterize the field of study. It has been long associated with sligthy depreciative names: ""ill condensed", "complex", "dirty", "mou" ("weich"), "tenuous", "fragile". The qualificative "soft" which apparently was introduced first in a Geilo Institute 10 years ago [1] is more appropriate and with a positive connotation *"yielding readily to touch, easily penetrated or changed in shape"* (the word thixotropy involves indeed the notion of touch , θιξισ). The late Tormod Riste, the organiser of the first 12 Institutes, and David Sherrington characterized it in 1989 by *"the weak interactions between polyatomic components, important thermal fluctuations, mechanical softness with emphasis on fundamental collective physics, plus a rich range of behaviours "*, a definition mostly shared by the examples treated in this meeting, except possibly for the role of thermal fluctuations since several of the examples treated here fall beyond the range of systems where Brownian effects play a role. I will myself not take into account this restriction and consider, following Henri Van Damme, that we are just working with "condensed matter".

The notion of "soft" indeed implies that of touch and feeling; there is an amusing developing field in Engineering Sciences called "psycho-rheology" in which the qualitative appreciation of properties of materials used by man is expressed in terms of

*A.T. Skjeltorp and S.F. Edwards (eds.), Soft Condensed Matter: Configurations,*
*Dynamics and Functionality, 1-14.*
© 2000 *Kluwer Academic Publishers. Printed in the Netherlands.*

measurable parameters. Thus, the "hand evaluation" of a fabric implies such feeling properties as stiffness, smoothness, softness, crispiness, which can be objectively defined from a combination of tensile, bending, shearing and compression moduli, and make use of a quantified description of the surface state, of weight and thickness, of acoustical, thermal and permeability properties [2]. The quality of a lipstick or of a shaving cream which involve subtle physico chemical properties require the use of decriptions and characterizations in the spirit of this Institute, as well as a subjective appreciation we can experience everyday!

In France, a wide-open community was formed in the 1970 by the Groups around de Gennes in Orsay working on Liquid Crystals and Polymers, on one hand, and a community which he and I both initiated a few years later around MIAM (MIlieux Aléatoires Macroscopiques) in which one looked for applications of the more recent developments of statistical physics, like *percolation* and *fractals* concepts, to describe properties of supra-atomic species. A first experimental work done in this context was the study of a mixture of geometically identical conducting and insulating grains to check the universality of the critical variation of electrical conductivity above a critical concentration of conducting grains [3]. But beyond these issues, this study led us, in the 1980, to get involved with the geometry of random granular assemblies as well as with the hydrodynamic properties within the pore space.

## 2. Small disorder and its limits

The subject of "hard" composites itself is an adequate example of the main theme of this introduction which focuses on the necessity of various scales of descriptions in the materials concerned in this "soft matter" community, a characteristic of so called "complex" systems. This could be called the 3M (*Micro, Meso, Macro*) principle.

At a *microscale*, one wishes to describe the local heterogeneous environment using a small enough Representative Elementary Volume. A classical example is the treatment by Clausius and Mossotti in the 1850 of the dielectric constant of a mixture of two components, such as a dilute assembly of spheres of constant $\varepsilon_1$ imbedded in a continuum medium ($\varepsilon_2$). The problem could be solved exactly, the extra field contribution being treated as an average over the dilute inclusions. Many treatements have been given since, in the mean field spirit, using self-consistency requirements. Conveniently symmetrized for the two phases, they lead (Bruggeman treatment) to the existence of a threshold in the limiting case of the mixture of a conductor plus an insulator (if one makes use of the classical correspondance between permitivity and conductivity). However, this is still a mean field result with classical exponents for the variation of the electrical conductance near threshold (I refer to a excellent review by R. Landauer for a historical and scientific description of the subject [4]). Quite generally, homogeneization techniques tend to give an exact description at the local level of a cell and determine from it a global behaviour. This can be done mathematically with most rigour if one uses a periodic structure of cells (corresponding to the Wigner Seitz cells in Solid State Physics). Two examples illustrate the success of small disorder approaches.

In porous media, the Darcy law (1856) expresses the proportionality between the pressure gradient applied to a single phase fluid of viscosity filling the medium and the flow rate through it

$$\mathbf{v} = K\,(\mathbf{grad}p)\,/\,\eta$$

where the Darcy permeability K scales like an area (typically that of a single pore) [5]. The Darcy length scale of a few grain sizes is the scale over which a local averaging has to be made. It is to be noted that the above average potential law results from an highly dissipative process due to viscosity!

Other developments were given along this line. In particular, a remarkable correspondance was established by Johnson [6]: in a porous medium filled with a fluid, in addition to the ordinary sound, where the solid and fluid move in phase, there is another mode ( soft Biot mode in ordinary fluids at high frequency; fourth sound in the case of superfluid He) in which the two phases move out of phase. Due to the multiple scattering of the wave by the medium, the sound speed is reduced from what it would be in a medium made of straight channels aligned along the wave vector axis by a scattering index factor n. Another factor of importance to characterize a porous medium is the Formation Factor F which is the ratio of the conductivity of the fluid (one assumes the solid phase to be insulating) to that of the fluid filled porous medium. One finds exactly that FP =n, where the porosity P is the void fraction. The correspondance indeed comes from the potential nature of both problems; this had been anticipated by Lord Rayleigh in the "theory of sound" [7] on the simple case of the comparison between the potential scattering by a single sphere and the variation of conductivity due to a spherical inclusion. Other exact correspondances were established dealing in particular with an estimate of the surface over volume ratio of the pore space ( skin effect).

A third example of a classical homogenisation treatment is the Einstein result for the viscosity $\eta$ of a dilute suspension of grains (volume concentration c) in a fluid of viscosity $\eta_0$. It expresses simply as

$$\eta = \eta_0\,(1+2.5\,c)$$

independently of the geometry. However as soon as the concentration increases and non linear effects in concentration become effective, the viscosity depends on the characteristics of the flow and the nature of particles [8]. A perturbation treatment beyond order 2 in concentration would be very difficult to perform.

Many problems do not fit so nicely with these geometryless treatments to systems to which we attach the expression *"small disorder"* (Mandelbrot [9] uses, in this context, the adjective *"mild"*). The MIAM community rather concerned itself with examples of *"large disorder"* (M calls it *"wild"*) where *Meso*-physics descriptions do not simply rest on local averaging.

This was indeed the case of composites made of a random mixture of conducting and insulating grains near a percolation threshold. It is known that the percolation structure at threshold is fractal and is amenable to scaling descriptions. The percolation threshold itself can be easily measured from a d.c. conductivity experiment, and the critical behaviour observed through the critical exponent for the conductivity exponent $t \cong 1.3$ in 2D and $\cong 1.8$ in 3D, quite independently of the material disorder [10]. A typical problem adressed in the recent ETOPIM (Electrical Transport and Optical Properties of Inhomogeneous Media) meeetings has been the study of the electromagnetic response of such composite systems [11]. On the one hand, it was found that the absorption has a peak around percolation threshold and, on the other hand, that the optical properties (reflexion, transmission, absorption) are quite independent of wavelength. These properties can be understood by a proper consideration of various length scales which enter the description, well beyond the local scale of the heterogeneities: the wavelength $\lambda$, the percolation coherence length $\xi$ which diverges at threshold, but also a characteristic length $L(\omega)$ function of frequency which gives the range of the electromagnetic response. It has been suggested that it corresponded to an anomalous diffusion of electrons on the fractal lattice existing at scales $< \xi$. The length $L(\omega)$ is much smaller than $\lambda$ and $\xi$. Indeed, we are dealing with a completely heterogeneous problem .

This *Meso* physics has been a factor of common interest for the European and French community of MIAM's scientists. The same concepts of percolation and fractals were used in particular to understand the displacement of one fluid by a second immicible one in a porous medium and the properties of a gel near the gelation threshold which we will meet again later. They were extended to approach the heterogeneous distribution of stresses in a granular medium under compression. We will illustrate briefly these properties using the three examples : granular media, porous media and gels. We will examine what the actual limits of these studies are and where we could go from here.

One limit is that of *"very large disorder"* and of the exploration of the *Macro* world. At large scales, one needs to take into consideration the continuous or discontinuous drift of material properties through space, like the continuous variation of porosity of a sedimentary rock with depth due to the lithostatic pressure above it, or the change of rock material in field studies (like in a layered stratified medium). Here, getting averages over larger scales does not help since it tends to increase the effects of heterogeneities. We are still looking for appropriate descriptions for this case which is often the most important in practice.

There is a also need for a better consideration of specific local effects (particularly interfacial ones) beyond the unified approaches of the mesoworld. The development of nanoscopic observation using tools such as Surface Force Apparatus and Atomic Force Microscopy have contributed a lot to this new studies.

## 3. Some examples of mesophysics

## 3.1 GRANULAR MEDIA

It has been well established from photoelastic studies in 2D and 3D on photoelastic aligned cylinders under compression that stresses propagate through the assemblies using continuous sets of chain forces [12]. The grains present between the chains are "protected" from the stress field by arches but can become compressed if the applied stress increases. The force -F- deformation -D- law is much more non linear than would the local force -f- deformation -d- relation of a Hertz contact,which leads to $f \propto d^{3/2}$. The global law is of the form $F \propto D^x$ with x of the order of 4 instead of 3/2 because of the geometrical effect of the increase of active bonds with pressure [13].

The problem was approached numerically by using a lattice made of diodes randomly distributed such that they would be active only for one sign of bias (corresponding to compression of two grains) and not in the other direction (separation of grains). Recent numerical experiments on 2D and 3D arrays [14] clearly show these lines of forces resisting compression, surrounded by weakly compressed grains with an isotropic force distribution which prevent the lines of presssed grains to buckle. A bimodal distribution of forces is obtained : it has an exponential distribution of strong intergrain forces contrasting with a broad power law distribution of small forces which behaves similarly to a fluid which would surround and compress isotropically the backbone. It is an open question to see if , in the case of an optimal random limit packing as considered in S. Edwards lecture with an array of contacts and equally distribution of forces on the grains [15], such chain forces would persist. The mechanics of an assembly of grains is indeed a highly non local problem which is a function of filling, as discussed by S. Edwards; but the connection with the transport properties of grains has to be done systematically.

An amusing example of large scale effect is the still poorly understood "Branly effect" discovered in 1890 [16]. A dense packing of conducting grains will usually remain insulating (due to bad contacts). However, if one approaches a radiofrequency source from the packing, the assembly will become conducting and remain in that state even after the source has been taken away; this is due to the formation of a weak conducting chain between the electrodes which can be isolated. The phenomenon was used in the first radioreceptors at that period.

Very compact structures can be produced by mixing grains of a large range of sizes down to micron silica fumes mimicing the ideal Appolonian packing made by filling grains in decreasing hierarchical order to just fit in the voids of the previous generation. The "high performance concrete" thus obtained requires little water (5 %) while remaining quite fluid when poured, as likely due to a uniform coating of the individual grains by liquid film in the presence of wetting agents in the preparation [17]! It leads to resistances to compression more than 10 times larger than ordinary concrete . The very small value of the permeability leads to resistance to corrosion. High performance concrete resists traction unlike ordinary cement; this is due to the fact that the grains are reactive; one should not forget that cement is also a problem of chemistry which was first addressed by H. Le Chatelier one century ago.

In this respect, in this problem, like in the other examples I will present, there has been a change of attitude of the physicists. In this example, a *micro* description is needed as well as a macro one if one gets concerned with the realisation of structural materials.

## 3.2. POROUS MEDIA

This is another topic approached by members of this community. Invasion percolation [18] illustrates the need for a mesoscale description. One usually assumes that a porous medium is made of interconnected pores having radii $r_i$ distributed at random. If the material is initially filled with a fluid which wets its walls, a non wetting (N.W.) fluid will be capable of displacing it and occupy a given pore of radii $r_i$ if it is connected to the entry by a continuous path of N.W. fluid and if the pressure applied to the meniscus separating both fluids at this pore exceeds a value

$$\Delta p_i = 2\gamma / r_i$$

where $\gamma$ is the interfacial tension (this assumes that the flow rate is slow enough in order to be able to neglect the viscous pressure drop). The invasion by the NW fluid increases as $\Delta p_i$ increases. For a critical value $\Delta p_c$ corresponding to a value of radius $r_c$, the invasion cluster extends across the medium. The problem can be expressed rigorously in percolation terms and is clearly a non-local one. Variants of it have been found as a function of flow rate and fluid properties (expressed in terms of a capillary number $Ca = v \, \eta / \gamma$, and of the ratio of the viscosities of the two fluids). In particular, when the viscosity of the injected fluid is low and the flow rate large, a fingering structure is obtained which corresponds precisely, to the classical Laplacian growth (DLA) [19]. This results from the potential nature of the average flow Darcy equation.

Percolation had been originally introduced [20] to describe the onset of permeability at a critical concentration of open bonds in a porous medium (actually the clogging of a filter). This notion can be extended to the case of a well connected media with a large distribution of pore sizes as proposed by us [21] and by Katz & Thomson [22]. Because of the important local variation of the Poiseuille flow resistance with pore size (as $1 / r^3$ or $1/ r^4$ in fractures or circular pores), the flow rate is determined by the intermediate pore size (values around $r_c$), not by the large ones which contribute little to the pressure drop, or by the small ones which do not contribute much to the flow rate. The last authors propose a relation

$$KF = (1/ 56) \, r_c^2$$

which is verified within a factor of 2 for a large range of permeabilities. The expression implies on one hand that the permeability and conductivity of the fluid phase behave in the same way controlled by a critical path geometry, and on the other one, that the correct length scale in the dimensional form of the permeability has to be the critical radius.

Other hydrodynamic properties, such as the spreading of markers injected in a flow (dispersion), have been analysed [23]. It is expected that the dispersion of a marker can be expressed in terms of a diffusion like coefficient which is of the order of a product of the average flow rate times a typical pore length (this can be understood simply by analogy with a random walk motion which would be superimposed on an average drift). However field data (spreading of pollutants in hydrological water-bearing bed) show usually much larger spreads, scaling as the distance instead of the square root of it (as would be expected for Gaussian diffusion). Quite generaly, the *Meso* treatments are not always very useful to describe the *Macroworld* for realistic hydrogeological or oil extraction problems. New tools have to be looked for, by reference to certain characteristic structures (e. g. layered geological fields) or by making use of the knowledge of 3D maps of fields which is obtained to day accurately in particular from acoustic field inversion data encountered in the field. Similarly, the *meso* description often fails to consider in sufficient detail local flow characteristics and the local conditions of wetting and film formation (mixed wettability). Local techniques, such as pulsed gradient NMR [24], provide an interesting description of the geometry and transport properties at the level of the grains. They can get involved in logging tool in order to characterise better the nature of oil reservoirs and to improve oil recovery.

3.3 GELATION

The formation of gels is the result of cross bonding of linear polymeric chains. This can be achieved by chemical means (polycondensation or polyaddition) or physically (e. g. the entanglement of helices in gelatine when the temperature is decreased below the helix-coil transition). The transition from a sol (viscous) to a gel (elastic) has been understood since the classical work of Flory using a simple mean field treatment. Later, de Gennes [25] & Stauffer [26] showed that percolation would explain the sol-gel transition while including implicitely effects of excluded volume and loops. Indeed the first appearance of elastic moduli is due to a continuous set of chains spanning the full sample, much beyond the length of an individual chain. The model was tested succesfully on the critical behaviour at the sol-gel transition such as in the divergence of viscosity below threshold or the continuous appearance of elasticity above it (the critical exponent of it turns out to be the electrical conductivity exponent as stressed by Alexander [27] although a vector percolation model appropiate to the elasticity of bars would predict a larger value (typically $\tau = t + 2\ v$ where $v$ is the correlation length exponent expressing the bending effect of bonds of length equal to the correlation length [28]). Deviations are found from the percolation scaling behaviour in particular when kinetic effects are important. In such a case, the problem looks more like an aggregation process, a problem which was much studied for fractal aggregates. Presently, the research on gels has somewhat been displaced from this scaling approach near threshold. It is more important to understand the rheological properties of well formed gels and also to see how the gel formation is conditionned by external conditions such as flow maintained during the gelation process, since most operations involve strong gels well beyond gelation point.

## 4. Rheology of soft matter

Since some general features characterize the rheological properties of a heterogeneous flow, we will spend some time on this problem well in the sprit of "softness". The factors that can affect the rheological behavior are of a different nature : they are flow alignment of undeformable species, deformation under shear, and control of association and dissociation of aggregates by flow.

We consider first the case of a dilute solution of rigid brownian rods of length L and diameter d (d<<L) (in xanthane L is of the order of 1 micron, d of 1 nanometer). In the absence of flow, the rods experience a Brownian motion whose frequency is obtained by stating that it should be of the inverse of the diffusive time over its length L. Thus

$$\tau \sim (kT/6\pi\eta r).\ 1/r^2 = kT/6\pi\eta r^3$$

Let us consider the alignment of these fibers in a simple shear flow. This flow is the addition of a pure elongation at 45° with the flow plus a rotation. Due to the first term, the average alignment of the fluctuating fibers is also at 45° with the flow, if the shear is small enough, as can be checked using ellipsometric techniques in a Couette flow cell where the laser light is along the cylinder axis [30]. However, when the shear increases there is a tendency for the particles to align along the flow characterised by the product $s\tau$. This dimensionless combination compares the efficiency of the shear s (the shear rate is the inverse of a time) and the random rotation induced by Brownian forces measured by the characteristic time $\tau$.

The result applies only to suspensions which are dilute in terms of an influence volume $L^3$ built over the large length L. The semidilute limit where entanglements effects are present is obtained when influence volumes $L^2d$ overlap. In this case, $1/L^3 < c < 1/(L^2d)$, the rotation diffusion coefficient is strongly reduced (by a factor $(cL^2d)^2$ ); alignment effects are then largely suppressed [31]. It is amusing to note that the upper concentration limit would correspond in a suspension to the order of magnitude of a percolation threshold for entangled fibers but also, depending on the preparation conditions, to that of a liquid crystal phase of rods!

The ambivalent role of "simple" shear ( elongation plus rotation) is also illustred by the behaviour of drops. It was observed long ago by Taylor [32] that drops would not break easily even for large shears in simple shear flows whereas moderate elongational flows (pure shear), as obtained between four counter rotating rolls, would easily break bubbles. This is due to the fact that, in the first case, the fluid particles within the drops keep on rotating, and undergo successive phases of extension and compression at +/- 45° from the flow lines. From this analysis one gets that drops of low viscosity fluid (with respect to the suspending one) which feel this effect most strongly as well as highly viscous ones which resist deformation are hard to break. However, it appears that for more concentrated suspensions of droplets like in emulsions, the effect is much more reduced probably due to screening effects. Once more, the consideration of dilute problems does not help us much to deal with concentrated ones.

We examine next the effect of aggregation as modified by shear. Instead of Brownian instantaneous collisions, the shear applied to two close particles will force them to remain in contact over a time of the order of $s^{-1}$ (in the dilute case, $s^{-1}$ is the hydrodynamic time for the rotation of a two particles cluster by $\pi$) [33]. By increasing the shear rate, one increases both the rate of aggregation and that of separation of clusters. However, it is noted that sheared suspensions usually display shear thinning effects; this is due to the increased rotation mobility of smaller clusters. It was suggested long ago by de Gennes [34] that the replacement of Brownian contacts by shear induced ones should lead to a particular kind of dynamic percolation with shear induced formation of clusters. Experiments using non brownian particles floating at the surface of a fluid under shear [35] have shown the increase of size of durable clusters as the concentration of particles increases independently of the shear value. This is the origin of a divergence of the viscosity, like for a gel, when a dynamic cluster links the surfaces between which the shear is applied. But the result apparently has not yet been confirmed from 3D experiments.

Deformation and alignment effects are also present in polymers. It is possible to analyze it as for the case of non deformable particles. The relaxation time for deformation of a dilute suspension of polymers can be expressed in the simple form $(kT / \eta R^3)^{-1}$ where R is the gyration radius. Here again the competition between shear induced deformation and shear relaxation is the cause of the alignment effect. However, for a larger concentration there is competition between the deformation of chains which tends to increase viscosity and the alignment which decreases it. It is clearly of interest to study the effect of flow in polymer solutions using different kinds of prespecified flow fields.

The behavior of polymer suspensions in flows is connected with the important but poorly understood problem of drag reduction in turbulent flows [36]. The concentration needs not to be large (the ppm range, which corresponds to the dilute range of concentration if the chains are moderately extended). Clearly, the effect of elongation under the shearing part of the turbulent flow field should play a dominent role. But it is not clear if it reduces the dissipation at small scales of turbulence or rather if it is an elastic stretching effect. What makes it particularly confusing is the fact that the presence of polymer does not seem to affect the small scales of turbulence. A classical interpretation of the drag reduction is based on the quenching of turbulence in the boundary layer near walls starting from small scales. However, this quenching effect is also observed in the bulk of turbulent flow fields where the energy cascades from large scales. In particular, we were able to check this using contactless measurements of turbulent diffusion using forced Rayleigh scattering [37] : one observes a strong decrease of thermal or mass diffusion in bulk turbulent flows by just adding to water a few ppm of polyoxyethylene. Much remains to be done on the subject in particular using local optical studies.

Another physical aspect in the characterization of "softness" is through friction. This is a subject with an old scientific history initiated by L. da Vinci, Ammontons and Coulomb which has left his name to the law of proportionality between the shear force applied to a solid sliding on a flat surface and the normal force applied between it

$$F = \mu \, F_n$$

where $\mu$ is a material dependent coefficient of friction [38]. It was pointed out that one should consider that the area of contact $A_r$ is much lower than the apparent one and varies proportionally to the normal force (this is the reason why a parallelipipedic solid has the same limit angle of dry friction when resting on any of its faces on an inclined plane). Based on this representation of a large random distribution of small multicontacts critically sheared, there have been recent theoretical and experimental progresses on the dynamic aspect of the field of dry lubrication, including ageing effects, dynamic hysteretic response, and adhesive effects [39]. A further field of interest is the role of a lubricant interfacial layer or that of pollutants who can undergo phase changes under the very large local friction encountered. We are clearly in a range of problems where a microdescription is an important step for a rheological description.

## 5. Polymers

I have deliberately deliberately talking about the basic developments of polymer science, since I have not been involved personally in research on this field... which, by the way, is also largely absent from the present Institute. A jubilee meeting, in honour of P-G de Gennes, a most important initiator and contributor to the field, was held in Les Houches a few months ago with the title *"soft conduced matter; what is new after 30 years?"*. This could have equally been the title of the present talk but on different systems.

The development of polymeric studies, described in that unpublished les Houches meeting, in France (to limit my historical scope) was clearly initiated around de Gennes from the impact in the 70's of a well established community on polymers present in Strasbourg on a Solid State Physics group in Orsay which was at that time also involved in Liquid Crystals. The ingredient of the progress made was in the systematic use of scaling effects, leading to multiple scale descriptions coupled, with simple -sometimes apparently naive- geometrical representations, as well as to the impact of experimental tools in physics such as light scattering and neutronics (involving variable contrast experiments) or resonance techniques. I will not attempt to give any summary of them, but I would like to emphasize the efficiency of minimal models such as the concept of *reptation* where individual molecules interact in suspension with other ones as if they were in rigid tubes where they can undergo translational and sideway movements [40]. This turns out to be simpler than a detailed consideration of entanglements and loops, although this last notion is of importance in the study of polymer melts [41].

The representation of *polymer blobs* has also been very fruitful. It emphasizes the presence of different scales in semidilute polymer solutions : at the local scale, the polymer behaves freely like in a Brownian chain (random walk); at an intermediate one one has to consider electrostatic repulsion effects; at larger scales, blobs of polymers are linked together again like in a polymeric chain of beads.

A third example to which multiple scale effects apply is that of polymers near surfaces; they can be grafted to the substrate, attracted to it or repelled, depending on whether the polymer "prefers" the environment of the substrate or the solvent. The images attached to these situations have helped clarifying the role of polymers in the stabilisation of suspensions, in the understanding of friction and adhesion mechanisms : images of *brushes* attached to walls to describe grafting of polymer chains, of *self similar grids* to take into account the attraction of polymers by wall, or the existence a repulsion length of the order of a blob size for the depletion layer near a repulsive wall [42]. All these images have shown themselves remarkably efficient coupled to scaling descriptions despite their simplicity and have been confirmed by experiments.

Polymers can also be used as constitutive elements of larger scale assemblies. Two such classes of examples are block copolymers and large micelles. *Block copolymers* use two or more elements which have different affinities between themselves and with the solvent in an heteropolymeric chain [43]. In the case of two elements, layered systems as well as micelles are commonly obtained depending on the volume fraction and relative affinities. Three component block polymers A-B-C can give rise to a very rich zoology of structures in particular if frustration is introduced in the relation between elements as illustrated by the poor syllogism proposed by lud.wik Leibler, a major contributor in the field *"A likes C better than B; B likes A better than C"... So what?* In this case, the interface between two phases A and C can become decorated by blobs of the intermediate third one B [44] , but other results can be obtained depending on the relative lengths of the components.

Macroscopic *micellar objects* can form in the presence of certain surfactants and take the form of long tubular objects. They can organise in various polymers-like phases such as lyotropic liquid crystal phases or entangled random chains ( note however that the structure are not permanent; they grow and break due to thermal fluctuations). Their conformation and size result from a competition between the scission energy that favors growth and entropy of mixing which would favor small micelles. This "zoo" of micellar objects displays a vast richess nearly as varied as their applications in many field of applied science to which soft matter physics adresses itself [45]. It would be tedious to try to describe them and to see how scientific approachs have enriched a pragmatic and, often ancient, sets of recipes used in the various fields where micellar objects are used.

## 6. Experimental

Emphasis on characterisation methods has been, in the recent years, given to reciprocal scale descriptions using scattering techniques (neutrons; synchrotron radiation light). It is being partially replaced today by direct imaging in the range of 10 to 100 nm, with incoming techniques such as soft X rays in the wavelength of window of water. Many images are only done on two dimensional samples, and there is a large need of tomographic techniques. For transparent media when the index of various phases can be matched, the "cut" by a laser plane of light is much used. Confocal microscopy uses the sharp focusing of light in a small region of a sample. Diffusing media can also be studied in some instances if one has a strong light source using the few beam rays which are propagating in straight lines [46]. We can also now use a large set of tomographic techniques first derived for medical applications. Other techniques have been developed which make use of the contrast available in an heterogeneous system such as inversion of the measurement of the complex dielectric constant, index of refraction or acoustic absorption of a material.

But experiments can also be used for pedagogical purposes as many experiments in soft condensed matter can be made or reproduced at home or in a school without too much sophistication. This has been a subject of an evening session in Geilo. As an illustration, let us just consider the field of *physical and molecular gastronomy*. This is both a respectable and enjoyable one (again a possible field of application of psychorheology!) [47]. The case of stability of sauces appears to be understood today in relation with colloidal physics. Solids and liquid foams are also the object of studies with new recipes, analysis of taste and consistency. There is a lot yet to be done on the problems of rheology of these phases as well as on the structure and mechanics of weak edible solids (the "craquant" of biscuit). Transformation and degradation under heating which play an important role in the texture and taste are also analysed. This subject which can be taken very seriously can also be used in elementary education in connection with an introduction to experimenting and appreciation (ateliers du goût). This chapter of the use of soft matter physics in the kitchen can be made a remarkable tool for an inductive and inventive approach of physicochemical science using observation and simple experiments.

## 7. Acknowledgments

I have benefitted from suggestions and informations along the preparation of this review from Jean Candau, Madeleine Djabouroff, John Hinch, Jean Pierre Hulin, Roland Lenormand , Hervé This, M. Valdes and Henri Van Damme. The work here presented was largely inspired by the research in our group *"Physique des Milieux Héterogènes Complexes"*.

## 8. References

1. Riste, T. and Sherrington, D. (1989) *Phase transition in soft condensed matter*, Plenum Press, New York.
2. Kawabata, S. (1987) Application of fabric objective evaluation of fabric hand and quality, *J. Text. Mech. Soc.* **40**, 10-20.
3 Ottavi, H., Clerc, J., Giraud, G., Rousserq, J., Mitescu C.D., and Guyon, E. (1978) Electrical conductivity of a mixture of conducting and insulating spheres, *J. of Phys.* *C* **11**, 1311.
4. Landauer, R. (1978), in Garland J.C. and Tanner D.B. (eds.*)*, *ETOPIM 1. AIP conf. Proc.* AIP, New York. 1.
5. Dullien, F.A.L. (1979) *Porous media; fluid transport and pore structure*, Acad. Press, New York.
6. Johnson, D.L. (1980) *App. Phys.* Lett. **37**, 1065.
7. Lord Rayleigh *The Theory of sound*, Dover, New York.
8. Batchelor, G.K. and Green, J.T. (1972) the determination of the bulk stress in a suspension of spherical particles to order 2 *J. Fl. Mech.* **56**, 375.
9. Mandelbrot, B.B. (1999) *Multifractals and 1/f noise*, Springer, New York.
10. Stauffer, D. (1985) *Introduction to percolation theory*, Taylor and Francis, London.
11. Yagil, L., Gadenne, P., Julien, C. and Deutscher, G. Optical properties of a semi continuous film on a substrate. Does an effective dielectric constant exist?, *Physica* **207**, 228-233. The proceedings of ETOPIM 3 and 4 have been edited by Physica A **207** and **241**.
12. Dantu, P. (1957) *Proc. 4th int. Con. Scien. Mechanics*, Butterworths, London.
13. Roux, S. and Guyon, E. (1985) Mechanical Percolation, *J. de Phys. Lett.* **46 L**, 999.
14. Radjai, F., Wolf, D., Roux, S., Jean, M. and Moreau, J.J. *Force networks in granular media.*
15. Edwards, S. these proceedings.
16. *L'onde électrique*, special issue (1991) **71**, 5-53
17. Richard P. and Cheyrezy M. Composition of Reactive Powder Composites (1995), *Cement and Concrete research* **7** 1501-1511
18. de Gennes, P.G. and Guyon, E. (1978) Lois générales pour l'injection d'un fluide dans un milieu poreux aléatoire, *J. de mécanique* **39**, 1280.
19. Paterson, L. (1984) *Phys.Rev.Lett.* **52**, 1621
20. Hammersley, J.M. and Welsh, D.J.A. (1980) *Contemp. Phys.* **21**, 593.
21. Charlaix, E., Roux, S. and Guyon, E. Critical effects in the permeability of heterogeneous porous media, Englman R. and Jaeger Z. (eds*) Fragmentation, form and flow in factured media,* Am. Inst. Phys., New York pp. 316-325.
22. Katz A. J. and Thompson A. H. (1985*) Phys. Rev. Lett.* **54**, 1325.
23. Groupe poreux P.C. (1987) Tracer dispersion, a new characteristic length scale in porous media in R. Pynn and T. Riste (eds*), Time dependent effects in disordered materials*, Plenum, New York.
24. Lebon, L. Leblond, J. and Hulin, J.P. (1997) Experimental measurement of dispersion processes at short times using a pulsed field gradient NMR technique, *Phys. of Fluids* **9** 481-490
25. de Gennes, P.G. (1976), *J. de Phys.* (Paris) **37** L, 1.

14

26. Stauffer, D. (1976) J. *Chem. Soc. Faraday Trans.* **II 72**, 1354.
27. Alexander, S. (1984) *J. de Phys.* (Paris) **45**, 1939.
28. Roux, S. , *J. de Phys.* **A. 19** L 351.
29. Djabouroff, M. , Capron,I.., Costeux S. and Kané M. (1999) Physical network formation under shear , to appear in *The Wiley Polymer network Group Review*, Vol. 2.
30. Fuller,G.G. (1995) *Optical Rheometry of complex fluids*, Oxford Univers. Press
31 Edwards,S.F. and Doi M. (1978) *J. chem. Soc. Farad. Trans.II*, **7** 9184
32. Taylor, G.I. (1934) The formation of emulsions in definable fields of flow, *Proc. Roy. Soc. Lond.* **A 146**, 501-523.
33. Arp, P.A. and Mason, S.G. (1971) *Annual Rewiew Fluid Mecanics* **3**, 291.
34. de Gennes, P.G. (1979) *J. de Phys.* **40**, 783.
35. Guyon, E. and Blanc, R. (1983) Transport properties of an assembly of spheres, in G. Deutscher, R. Zallen and J. Adler (eds.), *Percolation structures and processes*, Am. Inst. Phys., New York, pp. 229-249.
36. Hinch, E.J. (1977), Mechanical models of dilute polymer solutions in strong flows, *Phys. Fluids* **20**, 22-30.
    Szaso, P., Rallison, J.M. , and Hinch, E.J. (1997), Startup of flow of a F.E.N.E. fluid through a constriction, *J. Non Newt. fluid mechanics* **72**, 73-86.
37. Ambari, A., Jenffer, P. and Guyon E. (1986) Réduction de la diffusion turbulente par adjonction de polymères, *CRAS* **303 II** 161.
38. Bowden, T.P. and Tabor, D. (1950) *Friction and lubrication of solids*, Clarendon, Oxford.
39. Baumberger, T. and Caroli, C. (1998) Multicontact solid friction, *MRS bulletin juin 1998*, 41-46.
40. de Gennes, P.G. (1971) Reptation of a polymer chain in the presence of fixed obstacles, *J.Chem. Phys.*. See de Gennes P.G. *Scaling concepts in polymer physics,* Cornell Univ. Press, New York
41. Doi, M. and Edwards, S. (1978) Dynamics of concentrated polymer systems, *J. Chem. Soc. Faraday Trans.* II **74**.
42. de Gennes, P.G. (1987) Polymers at an interface : a simplified view, *Adv. in Coll. and Interface Science* **27**, 189-209.
43. Hamley, I. W. (1999) *Block copolymers*, Oxford U. P., England.
44. Bates, F. S and Fredrickson, G. H. (1999) Block Copolymers Designer Soft Materials, *Physics Today* **52**, 32-38.
45. Lequeux,F. and Candau S. J. Structural properties of worm like micelles, T. Mc Leich (ed.) in *Theoretical challenges in the dynamics of complex fluids* 181-190
46. Yodth,A. and Chang,B. (1994) *Phys. Today* **48** 34
47. There has been a series of international workshops on *molecular gastronomy* held in Erice (Italy). The last one in april 1999 was directed by H. This and T. Blake and held in honor of the recently disappeared Low Temperature physicist N. Kurti who did so much to promote the field.

# A COCKTAIL OF SOFT CONDENSED MATTER

JOEL STAVANS
*Department of Physics of Complex Systems,*
*Weizmann Institute of Science*
*Rehovot 76100, Israel*

## 1. Introduction

Soft condensed matter physics has evolved within the last two decades into a thriving field of inquiry. Arguably, it is unique amongst all other fields in physics in the diversity of the topics it encompasses. You just have to have a look at the subjects to be covered in this School to convince yourself.

In this set of lectures I will review a selection of topics on which I have been personally active. As you will see, they are very different from one another, and I hope this will reflect the huge breadth and flavour of the field. Before I start, an apology is in order: since the number of topics I cover is large, the tutorials you will find below are by necessity both sketchy and brief. The point is just to give you the gist, and then refer you to more extensive reviews in the list of references. Cheers!

## 2. Evolution of two-dimensional non-equilibrium systems

### 2.1. CELLULAR STRUCTURES

Consider quenching a sample of a metal alloy from a high temperature molten state, into a sufficiently low temperature, as illustrated in the phase diagram depicted in Fig. 1. If the final temperature is below the solidus line, crystallites with different crystallographic orientations begin to form all over the sample and grow, until their boundaries meet. A polycrystalline material consisting of polyhedral domains is thus formed. Near room temperature, this would be for all intents and purposes the end state of the system. However, if we would be patient and wait for aeons, we would discover that this final state is really not the true equilibrium state of our system. The reason is that the mismatch between the crystallographic orientations at the grain boundaries costs energy. The system would therefore prefer to

*A.T. Skjeltorp and S.F. Edwards (eds.), Soft Condensed Matter: Configurations,*
*Dynamics and Functionality*, 15-36.
© 2000 *Kluwer Academic Publishers. Printed in the Netherlands.*

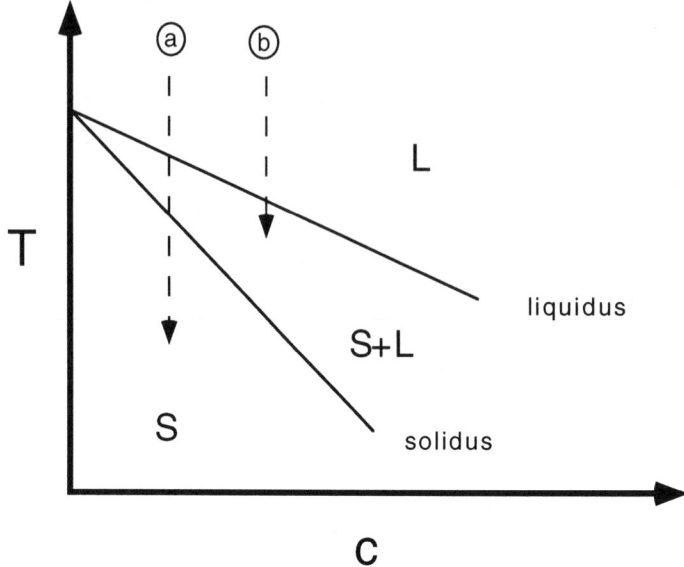

*Figure 1.* One side of a temperature-concentration (T, P respectively) phase diagram of an alloy, exhibiting solid (S), liquid (L) and S+L coexistence regions. Two quenches are shown: (a) below the solidus line, and (b) into the S+L region.

minimize as much as possible the total grain boundary area. Driven by this tendency, the system therefore evolves v..v..e..e..r..r..y..y slowly. Actually we can speed up things considerably, and instead of waiting hundreds if not thousands of years, we can increase the temperature close, but still below the solidus line. Close scrutiny of our sample as it evolves, reveals that some grains shrink and disappear, while others grow, so that the total number of grains and the grain boundary area decreases in time. Metallurgists call this process *recrystallization*.

Why should this matter to us? Well, the ability of a metal (say that of an airplane wing) to withstand stresses depends on the grain size, so youd better know how to control the process to achieve the desired characteristics. From a more fundamental point of view, this system provides an example of the evolution of a system out of equilibrium and we want to understand why it evolves the way it does. One can make considerable headway in the problem by studying two dimensional systems. This makes life easier for an experimentalist since one does not have to make cuts in a sample, and infer from these the statistical characteristics of the system at each point in time. Theoreticians can also cut some ice since the geometric properties of three dimensional polyhedral tilings are much more complicated than two-dimensional polygonal patterns for which exact mathematical results

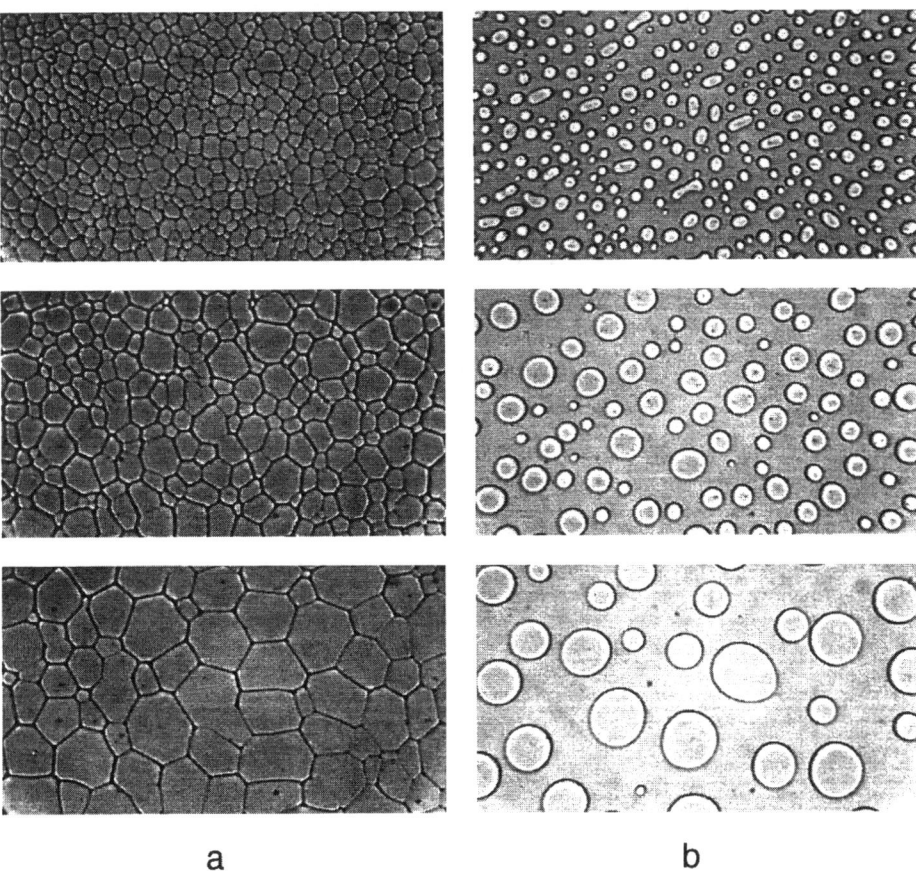

a                                    b

*Figure 2.* Snapshots of a coarsening thin film of succinonitrile (a) in the cellular regime ($\phi = 1$), and (b) in the Ostwald ripening regime with $\phi = 0.3$. For details see [32].

exist.

So let us look at Fig. 2a, where I show three snapshots taken from a thin film ($\sim 20\ \mu m$) of succinonitrile, an organic material, at three times after a quench. The succinonitrile film has been crystallized and the grain boundaries stained to enable their visualization. Crystallites of polygonal shape, which tile the plane are readily observed. Notice that the average crystal size grows with time, just as I described before. Note furthermore that the size difference aside, the patterns look very similar.

In the fifties, C. S. Smith [1], a metallurgist at the University of Chicago had the original insight of using soap froths between two glass plates to

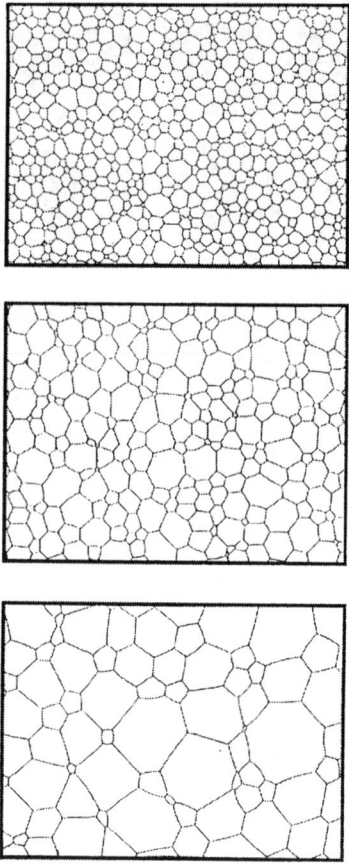

*Figure 3.* Snapshots of the evolution of a two-dimensional foam in the asymptotic scaling state. The run spanned a few days.

model the evolution of polycrystalline materials. Indeed, the structures look very similar as can be seen by comparing Fig. 3 with Fig. 2a, even though the microscopic mechanisms of evolution of both systems are very different. In contrast to grain boundaries which evolve by atom migration along the boundary, froths evolve by the transfer of gas across soap films, driven by pressure differences between neighbouring bubbles.

Polycrystalline films and froths are examples of materials having a *cellular structure* (for reviews see [2, 3, 4]). Other systems include polycrystalline metallic films [5], monolayers of lipids on water [6, 7], and magnetic domains [8]. Foams are ideal in the sense that all the boundaries between bubbles have identical properties. The same is not true for crystals: the energy of a grain boundary depends on the mismatch between the crystallographic orientations of the adjacent grains. A glance at Figs. 1 and 2 illustrates this nicely: note first that at each of the vertexes in the boundary network only three boundaries meet. A vertex with four or more boundaries would decay into a number of vertices with three boundaries each, since this minimizes total boundary length. For the vertices to be at mechanical equilibrium (the driving processes are much slower than the mechanical relaxation time of the boundaries), films must meet at angles of 120° at the vertices as a balance of forces shows. Since grain boundaries have different energies per unit length, this does not apply in the case of polycrystalline materials.

If one follows in detail the evolution of particular cells, one quickly comes across a remarkable finding: those cells that shrink and disappear have five sides or less. This fact is the experimental manifestation of a theorem proved by John von Neumann [9]:

$$\frac{\mathrm{d}a_n}{\mathrm{d}t} = \kappa(n-6) \qquad (1)$$

where $a_n$ is the area of an n-sided bubble and $\kappa$ a constant which depends on the permeability of the films to the gas filling the bubbles [9]. The physical idea behind the von Neumann theorem is that on average, bubbles with less than six sides look convex, while those with more than six sides look concave. A bubble with convex boundaries has a larger inside pressure than its neighbours, and therefore the gas inside will tend to flow out and the bubble will shrink. Just think of a spherical bubble: the pressure inside must be larger to balance both the outside pressure and the surface tension (Laplace's theorem) The opposite argument applies to bubbles with concave sides. Similar arguments can be applied to polycrystalline materials [10].

This does not mean however that if we wait long enough, no cells with less than six sides will be found. The reason is that as cells cells shrink and disappear, they change the number of sides of some of their neighbours. In this way new cells with five sides or less are formed. Incidentally, this also takes care of complying with a theorem by Leonhard Euler who showed that the average number of sides per cell in a graph in which all vertexes have three sides is six.

Two dimensional froths have been the focus of intensive study within the last two decades [11, 12, 13, 14]. These have revealed that the system reaches an asymptotic regime of evolution in which its statistical characteristics do not change with time. Thus for example, within this regime one always finds the same proportion of bubbles with three four five sides etc. In other words, the distribution of number sides $f(n)$ does not change with time in spite of the continuous change in the average characteristic scale. This latter grows like $t^{1/2}$ since the average area grows linearly with time. The same is true with other statistical distributions. Experiments in disparately

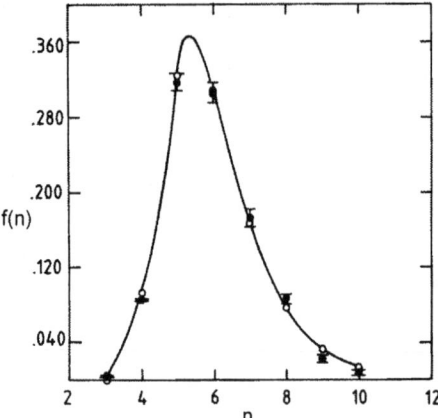

*Figure 4.* Distribution of number of sides $f(n)$ of a two-dimensional soap froth in the asymptotic scaling state: experiment (full circles) and mean-field model (empty circles). The solid line is a guide to the eye [16].

diferent systems with an inherent cellular structure have revealed another no less surprising fact: within statistical error, the distributions obtained in these systems are very much the same. Hence the evolution is pretty much system independent or *universal*. It depends only on the fact that the inherent structure of the system is cellular and two dimensional, irrespective of the detailed kinetics in the system.

There have been many theoretical approaches, aiming at reproducing the experimentally observed distribution with varying degrees of success. Analytical calculations include the use of the maximum entropy formalism [15], and mean-field theories [16, 17] coupled evolution equations for the concentration of bubbles with n sides are constructed. Computer simulations using vertex models have also been performed [18]. A surprising finding emerges from the mean field calculations: one does not get a unique distribution as a fixed-point solution, but rather a one-parameter family. The question then is which member of the family is the one observed in experiments. A dynamical selection principle has been proposed [16]. According to this principle, since a real system is finite, any initial distribution will be such that for sufficiently large $n$, $f(n) = 0$. Therefore, as the system evolves towards the asymptotic regime, the distribution will evolve towards the member of the family "closest" to it, namely the distribution whose tail decays the fastest with increasing n. This is then the distribution observed among all the members of the family.

Statistical characterizations of cellular structures such as distributions do not provide any information about the spatial arrangement of cells. Close scrutiny of cellular structure patterns reveals that this arrangement is not random: there are actual correlations between neighbouring bubbles. On average, bubbles with less than six sides (shrinking ones) are surrounded by bubbles with more than six sides (growing ones). This fact is described mathematically by the Weaire-Aboav law [19, 20] which gives the average number of sides $m(n)$ of the bubbles adjacent to an n-sided bubble:

$$m(n) = a + \frac{b}{n} \qquad (2)$$

Experimental data are very well fitted with by this equation when $a = 5$ and $b = 6 + \mu_2$, where $\mu_2$ is the second moment of the distribution of the number of sides. Furthermore, it has been observed that $< a_n >$ grows monotonically with $n$ (Lewis's law) [21, 22]. Note that the Weaire-Aboav and Lewis's laws together suggest that small cells are surrounded on average by large ones.

## 2.2. OSTWALD RIPENING

Let us go back now to Fig. 1. Up to now I have discussed the evolution of cellular structures in which the coarsening domains fill the plane. What happens if instead of choosing the final quench temperature below the solidus line, we choose it within the solid-liquid coexistence region? In this case crystals do not grow to fill the whole plane. Instead, a number of them nucleate and grow until the ratio of the sum of their areas relative to the total area, the area fraction $\phi$, reaches a finite value different from one. But again, as in the cellular structure case this is not the end of the story. Some crystals shrink while others grow at their expense, keeping the value of $\phi$ constant. The driving force behind this evolution, known as *Ostwald ripening*, [23] is the tendency of the system to minimize the contact between crystals and liquid. The contact involves an energy cost due to the surface tension of the liquid-solid interface. Fig. 2b illustrates this process, which represents the last stage of a phase separation process in a large number of systems such as binary mixtures of polymers or of liquids.

In the forties, Lifshitz and Slyozov [24], and Wagner [25] proposed a model for the evolution (the *LSW* model), appropriate for small volume fractions of solid. They assumed diffusion and curvature as the driving forces in the system. Furthermore, they assumed the domains to evolve independently (since the volume fraction is small, domains are far away from one another). The model, which could be solved exactly, predicted that: (*i*) the average size of the domains grows as the third power of time $< r > = K t^{1/3}$ with K a universal constant, and (*ii*) that the system reaches a scaling regime where the distribution of sizes becomes scale invariant. They were able to calculate this distribution in closed form. The LSW model has been generalized to two dimensions and similar conclusions have been reached. [26, 27] While the 1/3 exponent has been observed in a large number of experiments in various systems, it was found that (i) K is volume-fraction dependent, and (ii) the size distributions are broader than the LSW prediction. It was later realized that domains do not evolve independently of one another, but rather interact via their diffusion fields. Correlations then develop, which perturbative calculations in $\sqrt{\phi}$ [28, 29, 30, 31] showed to be of two types: (i) direct correlations between domain sizes: large domains are on average surrounded by small ones, and (ii) correlations between rates of growth: shrinking domains are on average surrounded by growing ones and viceversa. These correlations extend within a finite neighbourhood whose scale is determined by the volume fraction. The models make an analogy with electrostatics: the diffusion equation for the concentration field of one of the components has to be solved in the presence of sources (shrinking domains) and sinks (growing domains):

$$\nabla^2 c = 4\pi \sum_i Q_i(t)\delta(r - r_i) \tag{3}$$

This is nothing more than the Poisson equation in which the charges $Q_i(t)$ (at positions $r_i$) are given by the rate of change of the domain area. Therefore, shrinking domains correspond to negative charges while growing ones to positive charges. Global charge neutrality means that the area fraction $\phi$ is conserved, while the correlation between rates of growth corresponds to charge screening in this Coulomb gas language.

The existence of these correlations has been explicitly demonstrated recently [32] in the case of the two-dimensional system shown in Fig.2b. As an illustration of the rate of change correlations, one can divide crystals into two classes, according to whether they grow or shrink. One can then calculate the pair correlation functions which measure the likelihood of finding two crystals either within the same class, or in different classes, as function of distance (normalized by the average size). Experiment and theory are shown in Fig. 5. Clearly, for sufficiently small distances, shrinking crystals are preferably surrounded by growing ones. At large distances however, (for $r \geq 6$ in the case of Fig. 5) correlations are lost.

Notice the close similarity between the correlation of growth rates and the Weaire-Aboav law for cellular structures. In both cases shrinking domains are "screened" by growing ones.

It has been claimed recently that the solution of the equations in the Lifshitz-Slyozov model may not be unique [33]. Further theoretical support for this contention has been provided recently [34]. This again is analogous to the family of solutions of the mean-field equations describing cellular structures. However, no selection mechanism has been proposed, and the

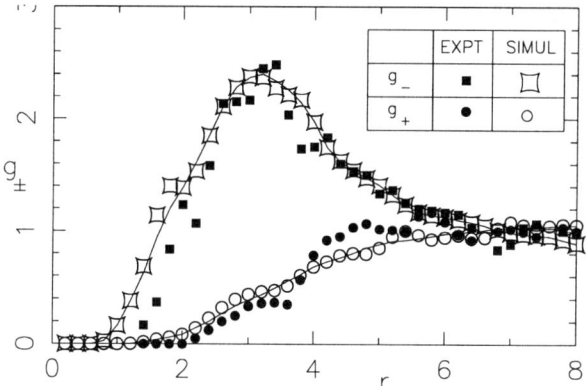

*Figure 5.* Charge-charge pair correlation functions as function of normalized distance for (a) crystals both growing or both shrinking (full circles: experiment; empty circles: theory); (b) crystals one growing and the other shrinking (full squares: experiment; empty squares: theory). The area fraction is $\phi = 0.13$. From [31].

connection between physical distributions and those belonging to the family of solutions has not been established.

## 3. Entropy-driven phase separation: the depletion effect

Imagine an idealized system consisting of a closed box containing two species of spheres, identical except for their different radii $r_s$ and $r_l$. Imagine further that: ($i$) the spheres are immersed in a fluid of the same density, so that gravity is neutralized. Besides kicking the spheres in brownian motion, the fluid plays no other role. ($ii$) The only interaction between the spheres is a contact repulsion. In short we have a binary mixture of hard spheres, which has been considered as a primitive model of binary fluid mixtures. We then ask the following question: is the homogeneous, well-mixed state of the system stable? For a long time practitioners of statistical mechanics and liquid state theory thought this to be so [35]. After all, one is used to think of a phase separation process as involving the competition of the enthalpic and entropic contributions to the free energy, and there is no enthalpic contribution in the system under consideration. Surprisingly the answer to the above question is negative: if the size difference is large enough, in certain regimes of total and relative concentrations the system separates into two phases, each one rich in one of the two species. This has been shown both at the theoretical [36, 37, 38] and experimental level [39, 40].

To understand the mechanism driving this phase separation process, consider one large and one small sphere touching (see Fig. 6 top). Since

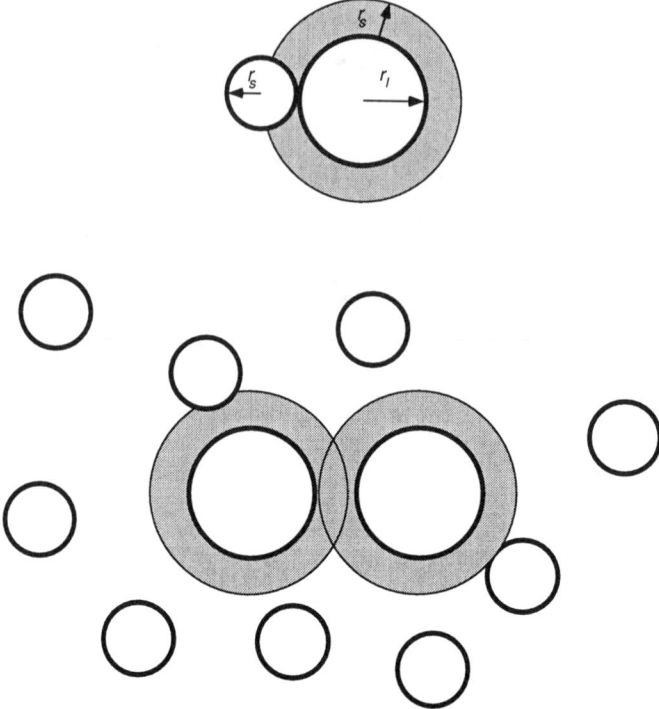

*Figure 6.* The depletion effect. Top: the center of a small sphere of radius $r_s$ cannot penetrate into a shell of this same thickness surrounding a large spere of radius $r_l$. Bottom: the shells of two nearby large spheres overlap, increasing the volume accesible to the small ones.

the spheres are considered as impenetrable, a shell of thickness $r_s$ about the large one is inaccesible to the center of mass of the small sphere. If the large spheres are few in number and far away from each other, the volume comprised of all the shells around them is inaccesible to the small spheres. Now comes the crucial step: if two or more large spheres approach (Fig. 6 bottom) so that the distance between their surfaces is less than $2r_s$, the shells overlap. This increases the volume accesible to the small spheres, and thereby their contribution to the total entropy. This is a state that the system thermodynamically favours, and as a result the large spheres cluster and phase separation ensues.

Note that the effective *depletion* attraction is of purely entropic origin. Furthermore, while hard spheres were considered in our example, the effect is quite general and occurs also when the large objects are softer and squishier as the droplets in an emulsion, and the small objects are say polymer coils or micelles. In fact, the depletion effect was discovered by Asakura and Oosawa [41] and independently by Vrij [42] when they considered the interaction induced between two walls (spheres of infinite radius) by a solution of non-adsorbing polymer coils. It is the exclusion of the coils when the walls are sufficiently near that leads to an imbalance in osmotic pressure and therefore to an effective attraction in this case.

One can readily observe depletion-induced phase separation in the laboratory. Experimentalists have mapped phase diagrams in the case of mixtures of solid particles [39], emulsions consisting of oil droplets of two different sizes [40], colloid-polymer [43, 44], and colloid-surfactant [45, 46] mixtures. The folowing picture emerges from these studies: for small values of the size ratio $\xi = r_s/r_l$, say around $\xi \sim 0.1$, one observes a phase diagram exhibiting fluid, crystal and fluid-crystal coexistence regions. For larger values of $\xi$, phase diagrams become more complicated, and coexistence regions with gas and liquid phases appear. The phase diagrams agree semi-quantitatively with predictions from statistical mechanical calculations [47] and computer simulations [48] in the case of colloid polymer systems (in these works, polymers are treated as point particles as far as their interaction with themselves to make the problem tractable), and with calculations of hard-sphere mixtures in the case of dispersions of spheres [37, 38]. The experiments also reveal that for large concentrations, kinetically-arrested phases such as gels are formed in the case of colloid-polymer and colloid-surfactant mixtures [49, 44], and glasses in the case of binary mixtures of spheres [39, 40]. The viscoelastic properties of these fragile gels have recently been studied by light scattering techniques [50]. The depletion attraction induced by small beads between a large sphere and a wall has recently been measured directly [51].

## 4. Stratification or the "quantum soap effect" in soap films

I turn now to soap films, the basic building elements of froths and foams. Looking back at Fig. 3, you will notice that the snapshots shown were taken over a period of days. In fact, typical experimental runs could span even weeks. Considering the fact that films can achieve very tiny thicknesses, (of the order of 40 Å) this is really surprising. How can films last for so long being apparently so flimsy? What are the interactions which stabilize them? While questions of stability were not the main concern of either Isaac Newton [52] or Robert Hooke [53], these scientists noted the bright interference colors and black spots soap films exhibit as they drain. It was until this century that most of the interactions relevant for soap films have been understood. Important questions remain as you will see, and research is still going on.

In an idealized model, a soap film can be regarded as consisting of two monolayers of surfactant molecules separated by a water core. In the case of ionic surfactants, the monolayers are charged, releasing counterions into the water. The state of a film is then determined by the competition of attractive Van der Waals and repulsive screened electrostatic forces (together, these two comprise the DLVO interaction [54, 55]). When the amount of ions in solution is low, the competition results in the existence of a minimum (known as the *common black film*) in the interaction potential, and a barrier that prevents the film from falling into a state of much lower thickness, known as the *Newton black film*. Within this very simple picture, a film drawn out of a solution with a *small* amount of surfactant drains slowly due to capillary forces and gravity, until it reaches the common black film state. If one adds enough salt to the solution from which the film is drawn, the barrier separating the common and Newton film states becomes smaller, and the film can reach the Newton black film state.

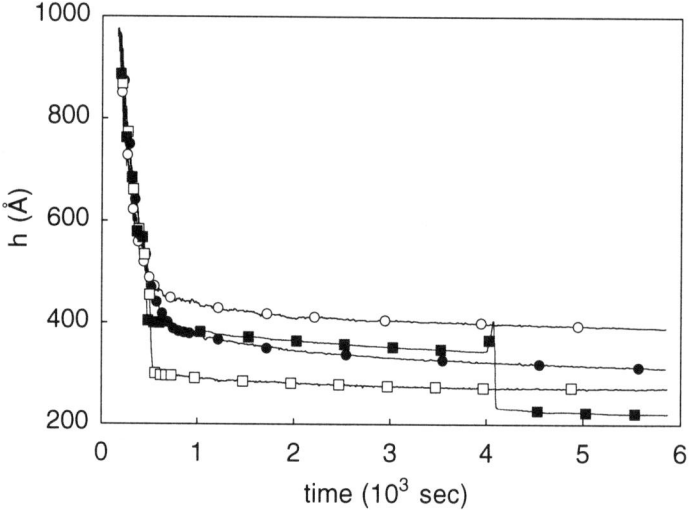

*Figure 7.* Thickness $h$ as function of time of draining soap films of SDS. Solution surfactant concentrations (in weight percent) : 0.15 (empty circles), 0.25 (full circles), 0.40 (empty squares) and 0.8 (full squares). From [58].

One can monitor the thickness of a film by measuring the intensity of laser light it reflects. Such simple experiments reveal [56, 57, 58] that the picture sketched above is not too far off the mark, provided the concentration of soap in the solution is sufficiently small. Still, quantitative differences exist, some which I illustrate in Fig. 7. Note that as the surfactant concentration increases, the equilibrium thickness becomes smaller. This effect cannot be attributed only to increased screening by the extra surfactant, but to depletion as well [58]. Furthermore, above a threshold concentration

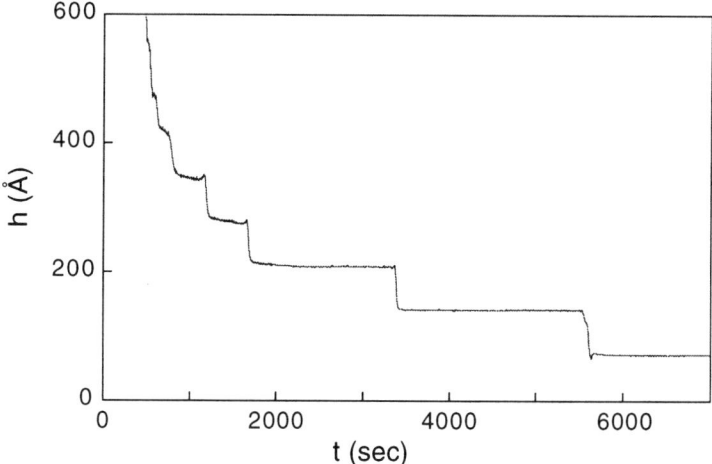

*Figure 8.* Thickness $h$ as function of time of a draining soap film, drawn out of an SDS solution with surfactant concentration of 12.0 (in weight percent). From [58].

there is one abrupt change in thickness. More striking discrepancies emerge when a film is drawn from a solution having a large surfactant concentration [57, 58, 59, 60, 61]. I illustrate this in Fig. 8, where I show a plot of thickness versus time for a film drawn from a solution of sodium dodecyl sulfate (SDS) at a concentration of 12 percent by weight. Instead of reaching smoothly the equilibrium thickness, the film goes through a series of equal height steps of about 100 $\mathring{A}$ each. These steps cannot be accounted for by the DLVO model, and must correspond to minima in the interaction potential $V(h)$, besides those corresponding to the common and Newton black film states. The multiple states have been visualized in microscope [62]. What is the origin of these steps? To understand this behavior, imagine now pouring a collection of marbles between two glass panes separated by a large distance (say hundred times the marbles' diameter). While well inside the marble sample one obtains a random closed packed arrangement, adjacent to the panes the marbles form a layer with nearly-ordered hexagonal order. Naturally, this nearly-ordered layer also induces a certain degree of layering order in the second layer, and so on with the third and next layers. But eventually all traces of an ordered arrangement are lost, and the random closed packed arrangement is recovered. Now let us start bringing the panes near to one another. Nothing will change as long as the distance between them is still large. However, when this distance is of the order of ten marble diameters or less, all the marbles between the panes will be arranged in layers parallel to the panes. Were we to measure the pressure to bring the panes together, we would then see oscillatory behavior, with each period corresponding to the removal of one layer.

What plays the role of the marbles in the case of soap films? Recall that above a certain concentration of surfactant, soap molecules form quasispherical aggregates called *micelles*, in order to hide their hydrophobic tails from the surrounding water. In the case of SDS [63] the hard core diameter of these micelles is about 30 Å. However, since micelles are charged (remember, SDS is an ionic surfactant), they repel each other and hence their effective diameter is larger.

Careful light scattering experiments [57, 58] as well as direct measurements of film disjoining pressure [59, 60] yield a wealth information about the interaction potentials responsible for this stratification phenomenon. These studies have revealed the following features about *structural* forces: the amplitude

1. at the highest concentrations studied not more than 7-8 steps are observed.
2. the amplitude of the oscillatory potential decreases exponentially.
3. the step height decreases with soap concentration.

The first two features are reminiscent of the behavior of confined fluids. Indeed, surface force apparatus measurements of the pressure needed to squeeze a liquid between two surfaces show oscillatory behavior with a period of the size of the molecule, about the same number of steps and their exponentially decreasing amplitude [64]. One can therefore borrow models devised to describe confined fluids and test them in the case of soap films. A simple density functional model by Tarazona and Vicente [65] predicts that away from a wall the fluid density $\rho$ behaves as:

$$\rho(z) = \rho_0 + Ce^{-ik_0 z}e^{-z/\xi} \tag{4}$$

where C is a constant, $k_0$ the value at which the structure factor and the decay length $\xi$ is given by:

$$\xi^{-2} = -\frac{S(k_0)}{S''(k_0)} \tag{5}$$

$\xi$ is the characteristic lengthscale over which micellar order prevails as we move away from the wall. The experiments show that $\xi$ increases as the micellar concentration increases [57, 58]. In other words, micellar stratification into layers increases with the micellar concentration. In Fig. 9 I show the ratio of this lengthscale over the step height $\delta$ as a function of soap concentration. Also shown are the predictions of the model having as input the micellar charge, size and density (the structure factor was calculated using the RMSA approximation which describes well ionic fluids). The agreement

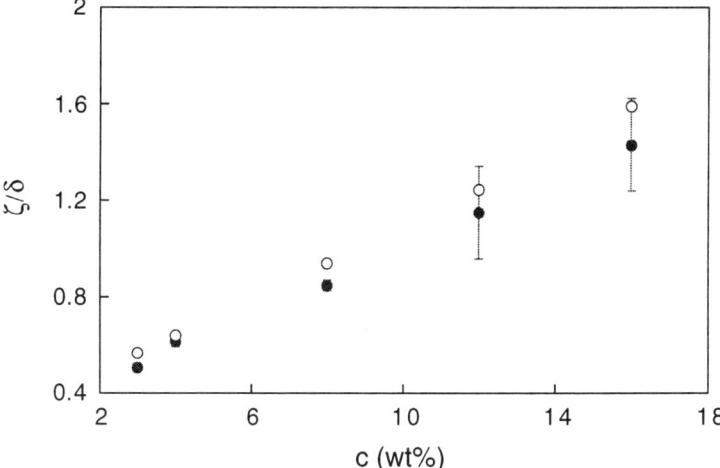

*Figure 9.* Ratio of penetration length over step size $\xi/\delta$ as function of surfactant concentration [58].

between the predictions and the observation is really good. At large surfactant concentrations, when intermicellar interactions are reasonably well screened, the micelles form a more concentrated supermolecular fluid, and thus one may expect their behavior to resemble that of ordinary fluids, whose main structure is determined by their hard core. Indeed this is the case: the ratio $\xi/\delta$ falls in the same range as that of liquids [64]. Thus we can describe both with the same theoretical framework. It is nice that we can vary the density of our supermolecular fluid of micelles in such a wide range whereas with ordinary liquids we are stuck with one value! Both the effects of depletion and structural forces have been seen in surface force measurements of micellar solutions [66, 67].

## 5. Segregation of Granular Matter

Anyone who eats cereal in the mornings has certainly found that the raisins are invariantly at the bottom of the box. This mundane yet annoying observation illustrates a very general phenomenon of granular materials : their tendency to remain segregated according to size, in spite of being vibrated, jostled or sheared (for a general review of granular matter see [68]).

One configuration in which segregation in granular matter can be studied controllably consists of a horizontal tube filled to about half its volume by a mixture of two powders (for a review see [69]). The slope formed as a result of the rotation gives rise to a downflow of the powders [70] in a thin

*Figure 10.* Four stages in the segregation of a mixture of glass beads and sand grains inside a tube rotating with an angular velocity 15.5 rpm, starting from a well-mixed state. From [77].

layer which has a characteristic s-shaped profile. When such a tube is rotated slowly, one finds under certain conditions that grains segregate into bands of two types along the tube axis, each type having different concentrations of the two constituent species. This is illustrated in Fig. 10.

This *axial segregation* was first documented by Oyama [71], sixty years ago. In the sixties, Donald and Roseman [72] proposed a criterion for axial segregation to occur. They claimed that the two species must have not only different size, but also *angles of repose*. Roughly speaking, the angle of repose is the angle of the slope you observe when you form a mound by droping grains from a source above (in reality, the angle at which avalanches fall is larger than the angle of the slope after an avalanche has fallen. Furthermore the angle of repose of two powders may vary differently with frequency, allowing for demixing remixing transitions [73]). According to the criterion of Donald and Roseman, the angle of repose of the small species must be larger than that of the large species. It turns out that this criterion provides a good rule of thumb as to what one may expect in an experiment. An intuitive explanation of its plausibility goes as follows: imagine an initially nearly-homogeneous mixture. As the mixture is rotated, fluctuations in the local concentration of the species develop. If we accept the criterion as valid, to some approximation regions with a higher concentration of small particles will have a steeper slope than adjacent regions in which the concentration of small particles is smaller. An axial slope gradient results which grains can follow, in addition to rolling down. Now, larger grains have higher mobility since their motion is not hindered by the roughness of the slope as much as the smaller grains. Consequently, larger grains are preferentially driven out of steeper sloped regions. This mechanism thereby provides a positive feedback which amplifies concentrations fluctuations.

Contrary to what Donald and Roseman conjectured, recent experiments in long tubes (their length is much larger than the diameter) show that band formation occurs all along the tube, and is not triggered by the tube ends [74, 75, 76, 77]. Furthermore, bands evolve slowly: some bands shrink and disappear while others grow at their expense. This can most easily be seen by recording the evolution with video, and selecting out of each frame a

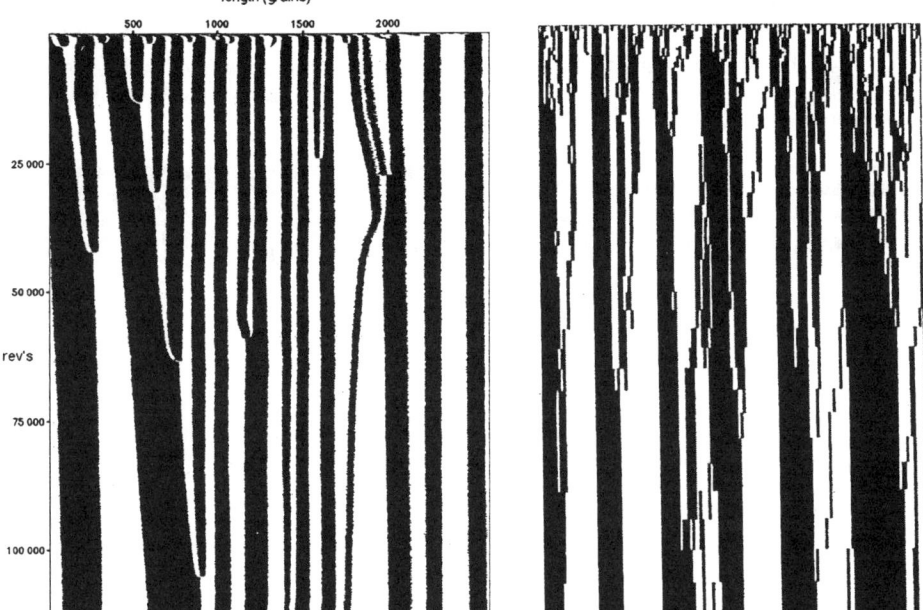

*Figure 11.* Left: space(horizontal)-time(vertical) diagram corresponding to the experiment shown in Fig. 10 [77]. Right: space-time diagram obtained from the computer simulations of Levitan [78].

line of pixels parallel to the tube axis. Stacking these lines one on top of the other, one generates a space-time diagram such as the one shown in the left side of Fig. 11. I also show in Fig. 11 the results of a space-time diagram generated by Levitan [78], which captures many features observed in the experiments. He did this by solving numerically an equation for the deviation of the local relative concentration of glass beads from the average [75], including a noise term.

In this experiment black sand (small and irregular grains, having a large angle of repose) and white glass beads (large and smooth, with a smaller angle of repose) are well mixed before the experimental run. The experimental conditions are such that glass beads flow smoothly, whereas the sand flow is punctuated by avalanches. Note that some regions are quiescent, whereas others show band growth and shrinkage. As a result, the average band size grows with time, but ever more slowly (almost logarithmically [78]). The question is, does it go on growing forever? Since the growth becomes so slow and it is hard to give an answer to this question with this

32

type of experiment. However, one can concoct initial conditions in which the powders are completely segregated, each occupying one half of the tube, and ask whether this state is stable. The answer is negative!. Glass beads find their way along the sand and form bands there as shown in Fig. 12.

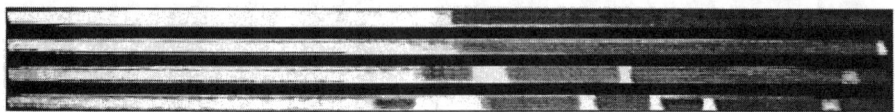

*Figure 12.* Four stages in the segregation of a mixure of glass beads and sand grains inside a tube rotating with an angular velocity 15.5 rpm starting from a fully segregated state. From [77].

Notice also the peculiar order in which these bands form. Were the basic transport mechanism a purely diffusive process, one would expect the nucleation of bands to occur progressively as the distance from the initial interface between the two types of grains increases. This is clearly not the case. Fig. 12 also provides a clue as to what transport mechanism may be at play. Note that while glass bands form within the sand, sand bands do not form within the glass.

Recall now that the glass beads flow smoothly whereas the sand flow is punctuated by avalanches which propagate axially. Thus avalanches may carry glass beads. Careful scrutiny of both the avalanches in a sand band, and the change in the size of the flanking glass bands reveals that this is indeed so [77].

What happens when the inequality of Donald and Roseman is strongly violated? It turns out that a nearly-homogeneous mixture is also unstable in this case. A radial core of small grains is formed along the axis of the tube, a phenomenon dubbed *radial segregation* (in fact some experiments reveal that even when the inequality *is* obeyed, there is a certain degree of radial segregation *within* bands [79, 80]). Last but not least, one may also wonder about cases when the angles of repose are very similar. This can be achieved when the experiment is repeated with smooth glass beads of two different sizes. As expected no bands are formed in accordance with our heuristic criterion (see Fig. 13a). However one can observe that there is a certain degree of segregation near the curved end of the tube, while no segregations is observed near the flat end wall. This suggested that curvature modulations in the tube may promote segregation. This expectations were confirmed by the experiments: peristaltic axial modulations of the tube induce band formation locked with the tube modulation (Fig. 13b), in the same mixture which did not segregate in the ordinary tube. Furthermore, helical modulations (Fig. 13c) induce complete segregation! The spheres

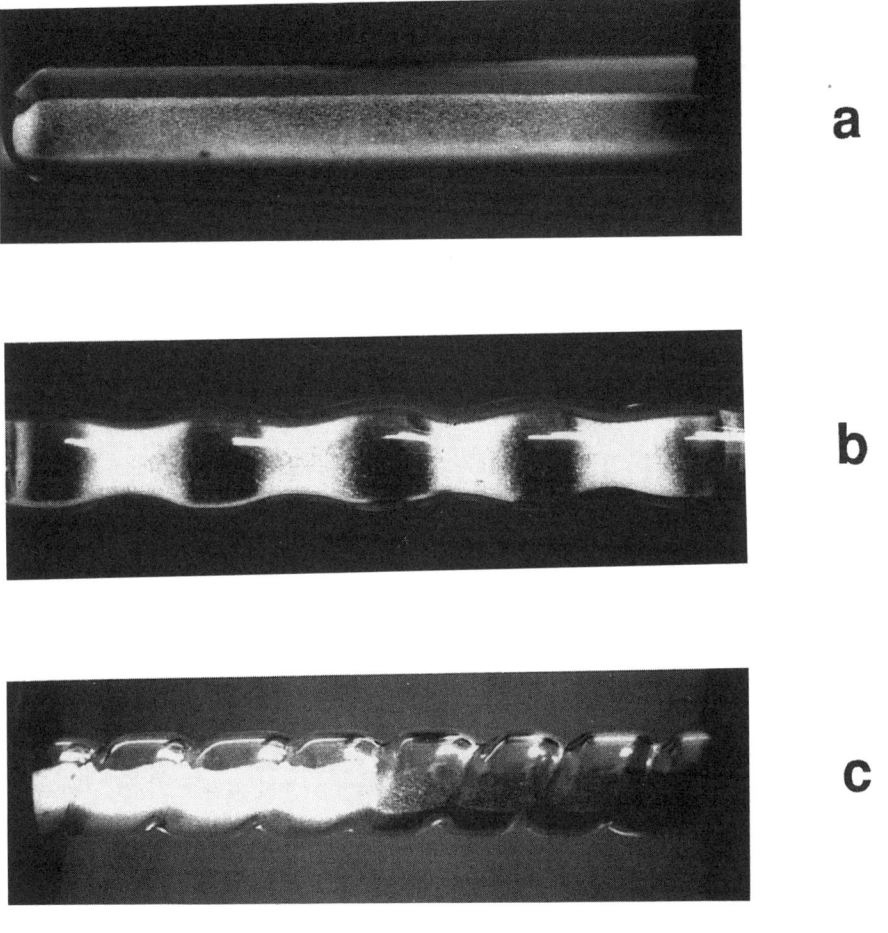

*Figure 13.* Segregation of small (dark) and large (white) glass beads of in an ordinary test tube (a), tube with a peristaltic modulation (b) and tube with a helical modulation.

change sides upon a change in the tube helicity or in the direction of rotation. The explanation of this effect is based on the asymmetric s-shaped nature of the flowing layer profile [75].

### References

1. Smith, C. S. (1952) Grain shapes and other metallurgical applications of topology *Metal Interfaces*, American Society for Metals, Cleveland, OH
2. Stavans, J. (1993) Evolution of cellular structures, *Rep. Prog. Phys.* , **Vol. no. 56**, pp. 733–789.
3. Stavans, J. (1993) Evolution of two-dimensional cellular structures: the soap froth *Phys-*

34

*ica A*, **Vol. no. 194**, pp. 307–314.

4. Weaire, D. and Rivier, N. (1984) Soap cells and statistics-random patterns in two-dimensions *Contemp. Physics*, **Vol. no. 25** , pp. 59–99.

5. Fradkov, V. E., Kravchenko, A. S. and Shvindlerman, L. S. (1985) Experimental investigation of normal grain growth in terms of area and topological class*Scripta Metall.*, **Vol. no. 19**, pp. 1291–1296.

6. Stine, K. J., Rauseo, S. A., Moore, B. G., Wise, J. A. and Knobler, C. M. (1990) Evolution of foam structures in Langmuir monolayers of pentadecanoic acid *Phys. Rev. A*, **Vol. no. 41** , pp. 6884–6892.

7. Berge, B. Simon, A. J. and Libchaber A. (1990) Dynamics of gas bubbles in monolayers *Phys. Rev. A*, **Vol. no. 41** , pp. 6893–6900.

8. Weaire, D., Bolton. F., Molho, P. and Glazier, J. A. (1991) Investigation of an elementary model for magnetic froth *J Phys.: Condens. Mat.* , **Vol. no. 3**, pp. 2101–2114.

9. von Neumann, J. (1952) Discussion: shape of metal grains *Metal Interfaces* , American Society for Metals, Cleveland, OH.

10. Mullins W. W. (1956) Two-dimensional motion of idealized grain boundaries *J. Appl. Phys*, **Vol. no. 27**, pp. 900–904.

11. Glazier, J. A., Gross, S. P. and Stavans, J. (1987) Dynamics of two-dimensional soap froths *Phys. Rev. A*, **Vol. no. 36** , pp. 306–312.

12. Stavans, J. and Glazier, J. A. (1989) Soap froth revisited: dynamic scaling in the two-dimensional froth *Phys. Rev. Lett.* , **Vol. no. 62** , pp. 1318–1321.

13. Stavans, J. (1990) Temporal evolution of two-dimensional drained soap froths *Phys. Rev. A*, **Vol. no. 42**, pp. 5049-5051.

14. Aste, T., Szeto, K. Y. and Tam, W. Y. (1996) Statistical properties and shell analysis in random cellular structures *Phys. Rev. E*, **Vol. no. 54**, pp. 5482-5492.

15. Iglesias, J. R. and de Almeida, R. M. C. (1991) Statistical thermodynamics of a two-dimensional cellular system *Phys. Rev. A*, **Vol. no. 43**, pp. 2763–2770.

16. Stavans, J., Domany, E. and Mukamel, D. (1991) Universality and pattern selection in two-dimensional cellular structures *Europhysics Lett*, **Vol. no. 15**, pp. 479–484.

17. Flyvbjerg, H. (1993) Model for coarsening froths and foams *Phys. Rev. E* , **Vol. no. 47**, pp. 4037–4054.

18. Nagai, T., Ohta, S., Kawasaki, K., and Okuzono, T. (1990) Computer simulation of cellular pattern growth in two and three dimensions, *Phase Transitions* **Vol. no. 28**, pp. 177–211.

19. Aboav, D. A. (1970) The qrrangement of grains in a polycrystal, *Metallography* , **Vol. no. 3** , pp. 383-390.

20. Lambert, C. J. and Weaire, D. L. (1981) The arrangement of cells in a network *Metallography* , **Vol. no. 14** , pp. 307-318.

21. Lewis, F. T (1928) The correlation between cell division and the shapes and sizes of prismatic cells in the epidermis of cucumis *Anat. Rec.* , **Vol. no. 38** , pp. 341–362.

22. Rivier, N. (1982) On the correlation between sizes and shapes of cells in epithelial mosaics *J. Phys. A*, **Vol. no. 15** , pp. L143–L148.

23. Langer, J.S. (1992) An introduction to the kinetics of first order phase transitions, in *Solids Far From Equilibrium*, Cambridge University Press, Cambridge.

24. Lifshitz, I.M. and Slyozov, V.V. (1961) The kinetics of precipitation from supersaturated solid solutions,*J. Phys. Chem. Solids* , **Vol. no. 19**, pp. 35–50.

25. Wagner, C. (1961) *Z. Elektrochem.* , **Vol. no. 65**, pp. 581–587.

26. Rogers, T. M. and Desai, R. C., (1989) Numerical study of late-stage coarsening for off-critical quenches in the Cahn-Hilliard equation of phase separation *Phys. Rev. A*, **Vol. no. 39**, pp. 4848–4853.

27. Marqusee J. A. Dynamics of late stage phase separation in two dimensions *J. Chem. Phys.*, **Vol. no. 81**, pp. 976-981.

28. Tokuyama, M. and Kawasaki, K. (1984) Statistical mechanical theory of coarsening of spherical droplets*Physica A*, **Vol. no. 123**, pp. 386–411.

29. Marder, M. (1985) Correlations and droplet growth *Phys. Rev. Lett.* , **Vol. no. 55**,

pp. 2953-2956.

30. Zheng, Q.and Gunton J.D. (1989) Theory of Ostwald ripening for two-dimensional systems *Phys. Rev. A*, **Vol. no. 39**, pp. 4848–4853.

31. Levitan, B. and Domany, E. (1998) Ostwald ripening in two dimensions: correlations and scaling beyond mean field *Phys. Rev. E* **Vol. no. 57** 1895-1911; Ostwald ripening in two dimensions: treatment with pairwise interactions *J. Stat. Phys.* **Vol. no. 93** 501-510.

32. Krichevsky, O. and Stavans, J. (1993) Correlated Ostwald ripening in two dimensions *Phys. Rev. Lett.* , **Vol. no. 70** , pp. 1473–1476; (1995) Ostwald Ripening in a two-dimensional system: correlation effects *Phys. Rev. E* , **Vol. no. 52** , pp. 1818–1827.

33. Brown, L. C. (1989) A new examination of classical Coarsening Theory *Acta metall.* , **Vol. no. 37**, pp. 71–77.

34. Meerson, B. and Sasorov, P. V. (1996) Domain stability, competition, growth, and selection in globally constrained bistable systems *Phys. Rev. E* , **Vol. no. 53** , pp. 3491–3494.

35. Lebowitz, J. L. and Rowlinson, J. S. (1964) *J. Chem. Phys.* **Vol. no. 41** 133-xxx.

36. Biben, T. and Hansen, J. P. (1991) Phase separation of asymmetric binary hard-sphere fluids *Phys. Rev. Lett.* **Vol. no. 66** 2215-2218.

37. Rosenfeld, Y. (1994) Phase separation of asymmetric binary hard-sphere fluids: self-consistent density functional theory *Phys. Rev. Lett.* **Vol. no. 72** 3831-3835.

38. Poon, W. C. K. and Warren, P. B. (1994) Phase behavior of hard-sphere mixtures *Europhys. Lett.* **Vol. no. 28** 513-518.

39. Imhof, A., and Dhont, J. K. G., (1995) Experimental Phase Diagram of a Binary Colloidal Hard-Sphere Mixture with a Large Size Ratio *Phys. Rev. Lett.* **Vol. no. 75** 1662-1665.

40. Steiner, U.; Meller A.; Stavans, J. (1995) Entropy driven phase separation in binary emulsions *Phys. Rev. Lett.* **Vol. no. 74** 4750-4754.

41. Asakura, S. and Oosawa, J. (1958) On interaction between two bodies immersed in a solution of macromolecules *J. Polymer Sci.* **Vol. no. 32** 183-184.

42. Vrij, A. (1976) Polymers at interfaces and the interactions in colloidal dispersions *Pure Appl. Chem.* **Vol. no. 48** 471-483.

43. Ilett, S. M., Orrock, A., Poon, W. C. K., and A. D. Pusey, (1995), Phase behavior of a model colloid-polymer mixture, *Phys. Rev. E* **Vol. no. 51** 1344-1352.

44. Meller, A. Stavans, J. (1996) Stability of Emulsions with Non-Adsorbing Polymers *Langmuir* **Vol. no. 12** 301-304.

45. Piazza, R.; Di Pietro, G.(1994) Phase Separation and Gel-like Structures in Mixtures of Colloids and Surfactant, *Europhys. Lett.* **Vol. no. 28** 445-450.

46. Bibette, J.; Roux D.; Pouligny, B. (1992) Creaming of Emulsions: the Role of Depletion Forces Induced by Surfactant *J. Phys. II France*, **Vol. no. 2** 401-424.

47. Lekkerkerker H. N. W., Poon, W.C.K., Pusey, P. N., Stroobants A. and Warren, P. B. (1992) Phase Behavior of Colloid+Polymer Mixtures. *Europhys. Lett.* **Vol. no. 20** 559-564.

48. Meijer, E. J. and Frenkel D. (1994) Colloids dispersed in polymer solutions. A computer simulation study *J. Chem. Phys.* **Vol. no. 100** 6873-6886.

49. Poon, W. C. K., Pirie, A. D., and Pusey, P. N., (1995), Gelation in colloid-polymer mixtures.*Faraday Discuss.* **Vol. no. 101** 65-76.

50. Meller, A. Gisler, T. Weitz, D. and Stavans, J. (1999) Viscoelasticity of Depletion-Induced Gels in Emulsion-Polymer Systems to be published *Langmuir*.

51. Kaplan, P. D. Faucheux, L. P. and Libchaber, (1994) A. Direct observation of the entropic potential in a binary suspension *Phys. Rev. Lett.* **Vol. no. 73** 2793-2797.

52. Newton, I. Optiks (Smith and Walford, London, 1704), R. Hooke, in The History of the Royal Society of London (T. Birch, London, 1757).

53. Hooke, R., (1672) On holes (black film) in soap bubbles *Comm. Royal Soc.* march 28.

54. Derjaguin, B. V. and Landau, L.D., (1941) *Acta Physicochim. URSS* **Vol. no. 14** 633.

55. Verwey, E. J. W. and Overbeek, J. Th. G. (1948) *Theory of the Stability of Lyophobic*

36

*Colloids* Elsevier, Amsterdam.

56. Lyklema, J., and Mysels, K. J., (1965) A study of double layer repulsion and van der Waals attraction in soap films *Jour. Am. Chem. Soc.* **Vol. no. 87** 2539-2546.

57. Krichevsky, O. and Stavans, J. (1995) Micellar stratification in soap films: a light scattering study *Phys. Rev. Lett.* **Vol. no. 74** 2752-2756.

58. Krichevsky, O. and Stavans, J. (1997) Confined fluid between two walls: the case of micelles inside a soap film *Phys. Rev. E.* **Vol. no. 55** 7260-7266.

59. Bergeron, V. and Radke, C. J. (1992) Equilibrium measurements of oscillatory disjoining pressures in aqueous foam films *Langmuir* **Vol. no. 8** 3020-3026.

60. Nikolov, A. D. and Wasan, D. T. (1992) *Langmuir* **Vol. no. 8** 2985.

61. Nikolov, A. D. and Wasan, D. T. (1989) Ordered micelle structuring in thin films formed from anionic surfactant solutions *J. Colloid Interface Sci.* **Vol. no. 133** 1-12.

62. Sonin, A. and Langevin D. (1993) Stratification dynamics of thin films made from aqueous micellar solutions *Europhys. Lett.* **Vol. no. 22** 271-277.

63. Corti, M. and Degiorgio, V. (1978) Laser light scattering investigation on the size, shape and polydispersity of ionic micelles *Ann. de Physique* **Vol. no. 3**, 303-308.

64. J. Israelachvili, *Intermolecular and Surface Forces* Academic Press, London.

65. Tarazona, P. and Vicente, L., (1985) A model for density oscillations in liquids between walls *Mol. Phys.* **Vol. no. 56** 557-572 .

66. Richetti, P. and Kékicheff, P. (1992) Direct measurement of depletion and structural forces in a mIcellar system *Phys. Rev. Lett.* **Vol. no. 68** 1951-1955 .

67. J.L. Parker, P. Richetti, P. Kékicheff, P. and Sarman, S.(1992) Direct measurement of structural forces in a supermolecular fluid *Phys. Rev. Lett.* **Vol. no. 68** 1955-1958.

68. Jaeger, H. M. and Nagel, S. R. (1992) Physics of the granular state, *Science* **Vol. no. 255** 1523-1531.

69. Stavans, J. (1998) Axial segregation of powders in a horizontal rotating tube *J. Stat. Phys.* **Vol. no. 93** 467-475.

70. Boutreux, T. and de Gennes, P. G. (1996) Surface flows of granular mixtures: I. general principles and minimal model, *Jour. Phys. I France* **Vol. no. 6** 1295-1304.

71. Oyama, Y. (1939) *Bull. Inst. Phys. Chem. Res. Japan* **Vol. no. Rep. 18**, 600 in Japanese.

72. Donald, M. B.and Roseman, B. (1962) Mechanisms in a horizontal drum mixer, *British Chem. Eng.* **Vol. no. 7** 749-753.

73. Hill, K. M. and Kakalios, J. (1994) Reversible axial segregation of granular materials, *Phys. Rev. E* **Vol. no. 49** R3610-R3613.

74. Das Gupta, S. Khakhar D. V. and Bhatia, S K. (1991) Axial segregation of particles in a horizontal rotating cylinder, *Chem. Eng. Sci.* **Vol. no. 46** 1513-1517.

75. Zik, O. Levine, D. Lipson, S. G. Shtrikman, S. and Stavans, J. (1994) Rotationally induced segregation of granular materials, *Phys. Rev. Lett.* **Vol. no. 73** 644-648.

76. Choo, K., Molteno, T. C. A. and Morris, S. W. (1997) Traveling granular segregation patterns in a long drum mixer, *Phys. Rev. Lett.***Vol. no. 79** 2975.

77. Frette V. and Stavans, J. (1997) Avalanche mediated transport in a rotated granular mixture *Phys. Rev. E* **56** 6981-6990.

78. Levitan, B. (1998) Long time limit of rotational segregation of granular media, *Phys. Rev. E* **Vol. no. 58** 2061-2064.

79. Hill, K. M., Caprihan, A. and Kakalios, J. (1997) Bulk segregation in rotated granular material measured by magnetic resonance imaging, *Phys. Rev. Lett.* **Vol. no. 78** 50-54.

80. Metcalfe, G. Shinbrot, T. McCarthy J. J. and Ottino, J. M. (1995) Avalanche mixing of granular solids *Nature* **Vol. no. 374** 39-41.

# MATERIALS-DRIVEN SCIENCE:
# FROM HIGH T$_C$ TO COMPLEX ADAPTIVE MATTER

JÖRG SCHMALIAN, DAVID PINES AND BRANKO STOJKOVIC
*ISIS Facility, Rutherford Appleton Laboratory Chilton,*
*Oxfordshire, U K and*
*LANSCE Division, Center for Materials Science, and Center*
*for Nonlinear Science, Los Alamos Laboratory, Los Alamos,*
*USA*

**Abstract.** The past decade has seen a paradigm shift from the study of solids as a collection of weakly interacting elementary excitations to a materials-driven exploration of systems in chemistry, physics, and biology that display unexpected emergent behavior because they contain collections of agents (electrons, atoms, molecules) that couple nonlinearly with each other in an enviroment which both influences agent behavior and is influenced by it. Experimental and theoretical investigations of the high T$_c$ cuprate superconductors and related members of the strongly correlated branch of the hard matter family (heavy electron systems, organic superconductors, manganites) have raised profound questions in physics. Because these questions have deep intellectual and practical connections with the behavior of soft matter and biological matter, the exploration of these connections can open new directions for further progress.

To make explicit the study of these potential connections we have introduced the term, complex adaptive matter. Complex adaptive matter thus denotes materials that display intrinsic nonlinear behavior and typically must choose between competing ground states. Such materials change their properties dramatically (adapt) in response to small changes in external parameters such as temperature, pressure, doping level, applied fields. For strongly correlated electron systems, such as the cuprate superconductors, intrinsic nonlinear behavior arises because the dominant interaction between quasiparticles (here largely confined to a plane) is electronic; that interaction is both determined by the quasiparticles and determines their behavior. As a result nonlinear feedback plays a crucial role in determining systems behavior. here that feedback is negative, it keeps the system close to a man field behavior; where it becomes positive, as in the case

A.T. Skjeltorp and S.F. Edwards (eds.), Soft Condensed Matter: Configurations,
Dynamics and Functionality, 37-70.
© 2000 Kluwer Academic Publishers. Printed in the Netherlands.

of magnetically underdoped cuprate superconductors, it leads to nascent spin density waves, dramatic changes in quasiparticle behavior, quantum critical behavior, and three distinct normal state phases before the system undergoes a superconducting transition.

In this lecture, we will present a complex adaptive matter perspective on the cuprate superconductors and organic superconductors with an emphasis on the organizing principles at work in these systems, principles that may play a role in other members of the complex adaptive matter family.


## 1. INTRODUCTION

The exploration of increasingly more complex materials has been a central issue in hard condensed matter research during the last decade. Such systems display an extremely rich spectrum of dynamic behavior, including competing ground states, new kinds of symmetry breaking, strong sensitivity with respect to disorder and external field etc., posing many new experimental and theoretical challenges. Besides being complicated, a *complex* material is intrinsically determined by the interference of excitations and ordering on different time and length scales and by nonlinear feedback. In the theoretical description of a complex material it is not possible to decouple characteristic degrees of freedom and to apply mean field type theories for each of these separately.

Cuprate superconductors are arguably the best studied examples for such kind of behavior, yet a number of mysteries remain. The cuprates are strongly correlated nearly two dimensional systems, in which the dominant interaction between the quasiparticles is of electronic origin and caused by a collective mode of the quasiparticles under consideration.

In this context it is helpful to first summarize the nature of *hard matter* physics in the past. In the early days of the quantum theory of solids and in many successful contemporary approaches , the concept of very weakly interacting constituents forming a solid was and still is fruitfully exploited. The free electron gas theory and the description of lattice degrees of freedom in terms of noninteracting phonons are famous examples. Even though the electrons in a metal or semiconductor do interact strongly, the effective quasiparticles are exposed only to a strongly reduced Coulomb interaction due to screening, an effect which also causes enormous modifications of the lattice degrees of freedom, but eventually leads to an almost free Bose gas of phonons. Even though modern electronic structure calculations such as density functional theory capture an impressive amount of details in the description of a real system, they are, in principle, based on the same assumption of weakly interacting agents. Interesting consequences of this type of physics arise often due to a weak coupling between different degrees of freedom. Examples are the theory of conventional superconductivity due to Bardeen, Cooper and Schrieffer [1] or the theory of charge density waves [2] in quasi one-dimensional solids. If, in a next step, the interaction between the different excitations increases, another type of phenomena occurs, such as the Kondo effect [3], where the strong interaction of localized magnetic impurities causes anomalies in the transport and magnetic properties of an otherwise noninteracting electron gas. In this case or in the similar case of polarons, where one has a strong electron lattice interaction, the behavior of the system is drastically changed due to strong correlations. However, nonlinear feedback, i.e. qualitative changes of all parts of the system due to their mutual interaction, does not occur.

Feedback behavior implies that the system under consideration must choose between competing ground states and is expected to occur in situations where the interaction between quasiparticles is strong and mediated by their own collective excitations. Depending on the physical system under consideration, these collective modes can be for example plasma oscillations, charge or spin excitations or collective pairing degrees of freedom. In this lecture we will investigate strongly correlated electrons in cuprate superconductors and organic superconductors, where the collective modes under consideration are antiferromagnetic spin fluctuations. We will in particular emphasize the organizing principles at work that may play a role in other members of the complex adaptive matter family.

## 2. CUPRATE SUPERCONDUCTORS: AN OVERVIEW

Following a decade of work, there is now an experimental and theoretical consensus that the behavior of the elementary excitations in the Cu-O planes provides the key to understanding the normal state properties of these cuprate superconductors, and that essentially no normal state property (save one) resembles that found in the Landau Fermi liquid description of the normal state of a conventional, low $T_c$, superconductor. As may be seen in Table 1, both the spin response (measured in static susceptibility, nuclear magnetic resonance (NMR) experiments and inelastic neutron scattering (INS) experiments) and the charge response (measured in transport and optical experiments) of the high $T_c$ materials are dramatically different from their low $T_c$ counterparts, as is the single particle spectral density measured in angle-resolved photoemission studies (ARPES). Moreover, essentially no property of the superconducting state is that of a conventional superconductor, in which BCS pairing takes place in a singlet s-wave state, and the quasiparticle energy gap at low temperatures is finite and isotropic

TABLE 1. Some ways in which the normal state of high $T_c$ materials is anomalous.

| Physical quantity | Conventional | High $T_c$ |
|---|---|---|
| Uniform susceptibility, $\chi_0(T)$ | Flat | Varies with temperature, possesses a maximum at $T^{cr} > T_c$ for optimally doped and underdoped systems |
| Low frequency spin excitation spectrum | Nearly wave vector independent | Sharply peaked close to $\mathbf{Q} = (\pi/a, \pi/a)$ |
| Maximum strength of spin excitations | $\approx 1\,\mathrm{state/eV}$ | $\approx 20 - 300\,\mathrm{states/eV}$ |
| Characteristic spin excitation energy, $\omega_{\mathrm{sf}}$ | $\approx \varepsilon_F$ | $\omega_{\mathrm{sf}} < T \ll \varepsilon_F$ |
| Antiferromagnetic correlations | None | Strong and temperature dependent, with $\xi > 2a$ for optimally doped systems |
| Resistivity | $\rho \propto T^2$ | $\rho \propto T$ (for optimally doped systems) |
| Quasiparticle scattering rate | $aT^2 + b\omega^2$ | $a'T + b'\omega$ |
| Quasiparticle spectral density | Sharp peak at the Fermi surface | Highly ansiotropic, displaying considerable evolution with temperature |

as one moves around the Fermi surface. Despite the fact that something quite new and different is required to understand normal state behavior, there is also an emerging consensus that BCS theory, suitably modified, will provide a satisfactory description of the transition to the superconducting state, and the properties of that state. There is a near consensus as well on the basic building blocks required to understand the high temperature superconductors. These can be summarized as follows:

- The action occurs primarily in the Cu-O planes, so that it suffices, in first approximation, to focus both experimental and theoretical attention on the behavior of the planar excitations, and to focus as well on the two best-studied systems, the 1-2-3 system ($YBa_2Cu_3O_{7-x}$) and the 2-1-4 system ($La_{2-x}Sr_xCuO_4$).

— At zero doping ($YBa_2Cu_3O_6$ ; $La_2CuO_4$) and low temperatures, both systems are antiferromagnetic insulators, with an array of localized $Cu^{2+}$ spins which alternate in sign throughout the lattice.

— One injects holes into the Cu-O planes of the 1-2-3 system by adding oxygen; for the 2-1-4 system this is accomplished by adding strontium. The resulting holes on the planar oxygen sites bond with the nearby $Cu^{2+}$ spins, making it possible for the other $Cu^{2+}$ spins to move, and, in the process, destroying the long range AF correlations found in the insulator.

— If one adds sufficient holes, the system changes its ground state from an insulator to a superconductor.

— In the normal state of the superconducting materials, the itinerant, but nearly localized $Cu^{2+}$ spins form a non-Landau Fermi liquid, with the quasiparticle spins displaying strong AF correlations even for systems at doping levels which exceed that at which Tc is maximum, the so-called overdoped materials.

It is convenient to discuss the temperature and doping dependence of this "uniformly" anomalous behavior in terms of the schematic phase diagram shown in Fig. 1. There one sees that overdoped and underdoped systems may be distinguished by the extent to which these exhibit crossover behavior in the normal state: underdoped systems exhibit two distinct crossovers in normal state behavior before going superconducting, while overdoped systems pass directly from a single class of anomalous normal state behavior to the superconducting state [4].

A phase diagram similar to Fig. 1 was independently derived from studies of the charge response by Hwang, Batlogg, and their collaborators [5] and from an analysis of the low frequency NMR experiments [6] by Barzykin and Pines [7]. The latter authors identified the upper crossover temperature, $T^{cr}$, from measurements of the uniform susceptibility, $\chi_o$ , in Knight shift experiments, which show that for underdoped systems $\chi_o$ possesses a maximum at a temperature $T^{cr}$, which in underdoped systems, increases rapidly from $T_c$ as the doping level is reduced. The fall-off in susceptibility for temperatures below $T^{cr}$ was first studied in detail by Alloul et al. [8]. Barzykin and Pines identified further crossover behavior in this pseudogap regime by examining the behavior of the $^{63}Cu$ nuclear spin-lattice relaxation time, $^{63}T_1$ , as the temperature was reduced below $T^{cr}$. They noted that between $T^{cr}$ and a lower crossover temperature, $T^*$, the product, $^{63}T_1T$ decreases linearly in temperature, while shortly below $T^*$ this product has a minimum, followed by an increase as the temperature is further lowered, an increase which is strongly suggestive of gap-like behavior. They proposed that these two crossovers were accompanied by changes in dynamical scaling behavior which could be measured directly if NMR measurements of

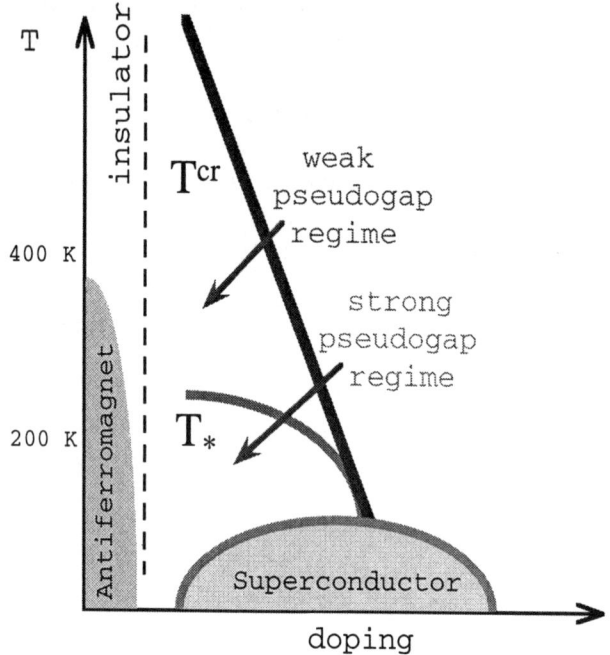

*Figure 1.* Schematic phase diagram of underdoped cuprates. Note the presence of two different crossover temperatures: $T^{cr}$, which characterizes the onset of sizable antiferromagnetic correlations; and T*, which signals the onset of a considerable loss of low energy spectral weight in the quasiparticle spectrum leading to a minimum of the characteristic spin-fluctuation energy. The region between $T^{cr}$ and T* is the weak pseudogap regime discussed in this paper.

$^{63}T_1$ could be accompanied by measurements of the spin-echo decay time, $^{63}T_{2G}$. Above $T^{cr}$ they argued that the ratio, $^{63}T_1T/^{63}T_{2G}^2$, would be independent of temperature, a result equivalent to arguing that the characteristic energy of the spin fluctuations, $\omega_{sf}$, would be proportional to the inverse square of the antiferromagnetic correlation length, $\xi$. Between $T^{cr}$ and T* they proposed that the ratio, $^{63}T_1T/^{63}T_{2G}$ would be independent of temperature, which means that an underdoped system would exhibit $z = 1$ scaling behavior, i.e. $\omega_{sf}$ would be proportional to $\xi^{-1}$ ; below T* they found that the increase in $\omega_{sf}$ would be accompanied by a freezing out of the temperature-dependent antiferromagnetic correlations; i.e. $\xi^{-1}$, which was proportional to $a + bT$ between $T^{cr}$ and T*, would approach a constant. This behavior has recently been confirmed in NMR measurements on $YBa_2Cu_4O_8$ by Curro *et al.* [9] while $z = 1$ pseudoscaling behavior has been found in INS experiments on $La_{1.86}Sr_{0.14}Cu\,O_4$ by Aeppli *et al.* [10].

Because pseudogap behavior of different character is found between $T^{cr}$ and $T^*$, and between $T^*$ and $T_c$, the terms weak pseudogap and strong pseudogap behavior were coined to distinguish between the two regimes [11]. Thus in the weak pseudogap regime one finds $z = 1$ pseudoscaling (because the scaling behavior is not universal) behavior, with both $\omega_{sf}$ and $\xi^{-1}$ exhibiting linear in $T$ behavior, while the rapid increase in $^{63}T_1T$, or what is equivalent, $\omega_{sf}$, found below $T^*$ suggests that strong pseudogap is an appropriate descriptor for this behavior.

An alternative perspective on weak and strong pseudogap behavior comes from ARPES [12, 13] and tunneling experiments [14] , which focus directly on single particle excitations. Above $T^*$, ARPES experiments show that the spectral density of quasiparticles located near the $(\pi, 0)$ part of the Brillouin zone, develops a high energy feature, a result which suggests that the transfer of spectral weight from low energies to high energies for part of the quasiparticle spectrum may be the physical origin of the weak pseudogap behavior seen in NMR experiments. Below $T^*$, ARPES experiments disclose the presence of a leading-edge gap, a momentum-dependent shift of the lowest binding energy relative to the chemical potential by an amount up to $30\,meV$ for quasiparticles near $(\pi, 0)$; it seems natural to associate the strong pseudogap behavior seen in the NMR experiments with this leading edge gap.

Strong pseudogap behavior is also seen in specific heat, d.c. transport, optical experiments, and Raman experiments. Below $T_*$, a reduced scattering rate for frequencies $\omega < \pi T_*$ has been extracted from the optical conductivity using a single band picture [15]. This suggests that excitations in the pseudogap regime are more coherent then expected by extrapolation from higher temperatures. This point of view is supported by recent Raman xperiments [16, 17] which observe in the $B_{1g}$-channel, sensitive to single particle states around $\mathbf{k} = (\pi, 0)$, a suppression of the broad incoherent Raman continuum and a rather sharp structure at about twice the single particle gap of ARPES experiments [16]. Resistivity measurements also show that below $T_*$ the systems becomes more conducting than one would have expected from the linear resistivity at higher temperatures [18].

It is natural to believe that the pseudogap in the spin damping, as observed in $^{63}T_1T$ measurements, the single particle pseudogap of ARPES and tunneling experiments, and the pseudogap of the scattering rate are closely related and must be understood simultaneously. Furthermore, it is essential for any theory of the strong pseudogap to account properly for the already existent anomalies above $T_*$, because they are likely caused by the same underlying effective interactions. As can be seen by inspection of Fig. 1, strong pseudogap behavior in underdoped cuprates only occurs once the system has passed the weak pseudogap state.

## 3. THE SPIN FLUCTUATION MODEL

The nearly antiferromagnetic Fermi liquid (NAFL) model [19, 20] of the cuprates offers a possible explanation for the observed weak and strong pseudogap behavior. It is based on the spin fluctuation model, in which the magnetic interaction between the quasiparticles of the $CuO_2$ planes is responsible for the anomalous normal state properties and the superconducting state with high $T_c$ and $d_{x^2-y^2}$ pairing state [19, 21].

In common with many other approaches, within the spin fluctuation model the planar quasiparticles are assumed to be characterized by a starting spectrum which reflects their barely itinerant character, and which takes into account both nearest neighbor and next nearest neighbor hopping, according to

$$\varepsilon_{\mathbf{k}} = -2t(\cos k_x + \cos k_y) - 4t' \cos k_x \cos k_y - \mu \,, \tag{1}$$

where $t$, the nearest neighbor hopping term, $\sim 0.25$ eV, while the next nearest neighbor hopping term, $t'$, may vary between $t' \approx -0.45t$ for $YBa_2Cu_3O_{6+\delta}$ and $t' \approx -0.25t$ for $La_{2-x}Sr_xCuO_4$.

In distinction to many other models, the spin fluctuation model starts from the ansatz that the highly anisotropic effective planar quasiparticle interaction mirrors the dynamical spin susceptibility [22],

$$\chi_{\mathbf{q}}(\omega) = \frac{\alpha\xi^2}{1 + \xi^2(\mathbf{q} - \mathbf{Q})^2 - i\frac{\omega}{\omega_{sf}}} \,, \tag{2}$$

peaked near $\mathbf{Q} = (\pi.\pi)$, via:

$$V_{\text{eff}}^{\text{NAFL}}(\mathbf{q}, \omega) = g^2 \chi_{\mathbf{q}}(\omega) \,, \tag{3}$$

an ansatz which enables us to construct directly a theory which focuses solely on the relevant low energy degrees of freedom. In Eq. 3, $g$ is the coupling constant characterizing the interaction strength of the planar quasiparticles with their own collective spin excitations. In this model, changes in quasiparticle behavior both reflect and bring about the measured changes in spin dynamics. The dynamic susceptibility, Eq. 2, was introduced by Millis, Monien, and Pines [22] to explain NMR experiments, which can be used to determine the correlation length, $\xi$, the constant scale factor, $\alpha$, and the energy scale $\omega_{sf}$, which characterizes the over-damped nature of the spin excitations. It follows from the experimental data that the static staggered spin susceptibility $\chi_{\mathbf{Q}} = \alpha\xi^2$ is large compared to the uniform spin susceptibility, $\chi_0$, and the relaxational mode energy correspondingly small compared to the planar quasiparticle band width [7, 23].

Of course we have to keep in mind that the dynamical spin susceptibility, Eq. 2, characterizes the magnetic excitation spectrum of the quasiparticles under consideration. Therefore, the quantities $\xi$, $\alpha$ and $\omega_{sf}$ in Eq. 2 are determined by the quasiparticle degrees of freedom. $\alpha$ and $\xi$ characterize the static susceptibility at $\omega = 0$ and are dependent on the behavior of the quasiparticles and their interaction on all energy scales. This makes it difficult to theoretically determine these quantities microscopically. Even if we were able to solve a particular model Hamiltonian exactly, an intention which is hopeless for dimensions larger than one (and smaller than infinity), $\alpha$ and $\xi$ would depend on the unknown details of this Hamiltonian such as extended Coulomb interactions. Therefore it is much more effective and equally appropriate to determine $\alpha$ and $\xi$ from experiment and keep the microscopic details of the Hamiltonian open. The situation is different for the characteristic energy scale $\omega_{sf}$ and the temperature variation of the correlation length, e.g. $\xi(T)^{-1} - \xi(0)^{-1}$ for $\xi(0)^{-1} > 0$. These quantities are determined by the low energy part of the quasiparticle spectrum and should be determined self consistently. Within a weak coupling calculation one then finds $\omega_{sf} \propto \xi^{-2}$ [22]. Introducing a dynamical scaling exponent via

$$\omega_{sf} \propto \xi^{-z} \qquad (4)$$

this leads to $z = 2$. However, using $^{63}T_1 T \propto \omega_{sf}$ and $^{63}T_{2G} \propto \xi^{-1}$, one expects in this weak coupling picture $^{63}T_1 T / ^{63}T_{2G}^2$ to be temperature indepedet, a result which disagrees with the above mentioned NMR findings below $T^{cr}$. Thus, we have to develop an alternative scenario for the spin dynamics and the spin damping $\gamma \equiv 1/(\omega_{sf}\xi^2)$ which goes beyond the simple weak coupling approach. Before we do this, we first investigate the behavior of the quasiparticles strongly interacting with the magnetic degrees of freedom.

Since the dynamical spin susceptibility, $\chi_{\mathbf{q}}(\omega)$, peaks at wave vectors close to $(\pi, \pi)$, two different kinds of quasiparticles emerge [24, 25]: *hot quasiparticles* with

$$|\varepsilon_{\mathbf{k}} - \varepsilon_{\mathbf{k}+\mathbf{Q}}| < v/\xi, \qquad (5)$$

located close to those momentum points on the Fermi surface which can be connected by $\mathbf{Q}$, feel the full effects of the interaction of Eq. (2); *cold quasiparticles* with $|\varepsilon_{\mathbf{k}} - \varepsilon_{\mathbf{k}+\mathbf{Q}}| > v/\xi$, located not far from the diagonals, $|k_x| = |k_y|$, feel a "normal" interaction. In Fig. 2, we show the Fermi surface in the first quarter of the BZ and indicate the evolution with $\xi$ of its hot regions, which fulfill Eq. 5, by a thick line. Note that even for a correlation length $\xi = 1$ a different behavior along the diagonal and away

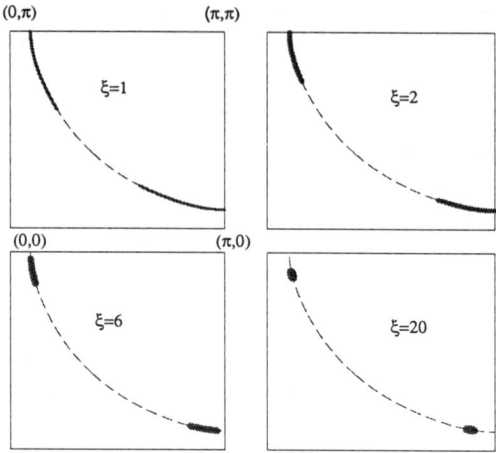

*Figure 2.* A typical bare Fermi surface in the first quarter of the BZ for different AF correlation lengths. The thick sections characterize the hot parts of the Fermi surface.

from it is expected. For larger values of $\xi$, the hot regions become smaller while their effective interaction increases. Close to $T_c$, typical values for $\xi$ of underdoped but superconducting cuprates are $2 < \xi < 8$, depending on doping concentration [7]; $v$ is the magnitude of a typical Fermi velocity in the corresponding momentum regions.

The distinct lifetimes of hot and cold quasiparticles can be obtained from transport experiments: a detailed analysis shows that, due to the almost singular interaction, the behavior of the hot quasiparticles is highly anomalous, while cold quasiparticles may be characterized as a strongly coupled Landau Fermi Liquid [25]. The presence of incommensurate peaks in the spin fluctuation spectrum [10, 26], and hence in the NAFL interaction, although difficult to calculate, may be expected to amplify the role played by hot quasiparticles in the determination of system behavior.

In the spin fluctuation model the anomalous behavior of the cuprates is assumed to originate in a strong interaction between fermionic spins $\mathbf{s_q} = \frac{1}{2} \sum_{\mathbf{k}\sigma\sigma'} c^\dagger_{\mathbf{k+q}\sigma} \sigma_{\sigma\sigma'} c_{\mathbf{k}\sigma'}$ which brings about intermediate range ($\xi > 1$) antiferromagnetic spin correlations and over-damped spin modes. Here, the operator $c^\dagger_{\mathbf{k}\sigma}$ creates a quasiparticle which consists of hybridized copper $3d_{x^2-y^2}$ and oxygen $2p_{x(y)}$ states [27]. The quantity of central physical interest is the dynamical spin susceptibility

$$\chi_\mathbf{q}(\tau - \tau') = \langle T_\tau s^\alpha_\mathbf{q}(\tau) s^\alpha_{-\mathbf{q}}(\tau') \rangle . \tag{6}$$

which after Fourier transformation in frequency space and analytical continuation to the real axis is assumed to take the form, Eq. 2. The intermediate and low energy degrees of freedom are characterized by an effective action [19]

$$
\begin{aligned}
S \;=\; &- \int_0^\beta d\tau \int_0^\beta d\tau' \left( \sum_{\mathbf{k},\sigma} c_{\mathbf{k}\sigma}^\dagger(\tau)\, G_{0\mathbf{k}}^{-1}(\tau - \tau')\, c_{\mathbf{k}\sigma}(\tau') \right. \\
&\left. + g^2 \frac{2}{3} \sum_{\mathbf{q}} \chi_{\mathbf{q}}(\tau - \tau')\, \mathbf{s}_{\mathbf{q}}(\tau) \cdot \mathbf{s}_{-\mathbf{q}}(\tau') \right),
\end{aligned}
\tag{7}
$$

where $G_{0\mathbf{k}}^{-1}(\tau - \tau') = -(\partial_\tau + \varepsilon_{\mathbf{k}})\delta(\tau - \tau')$ is the inverse of the unperturbed single particle Green's function with the bare dispersion, Eq. 1. In using Eq. 7, we implicitly assume that the effect of all other high energy degrees of freedom, which are integrated out to obtain the action $S$, do not affect the Fermi liquid character of the quasiparticles. In Eq. 7, the effective spin-spin interaction is assumed to be fully renormalized; thus it reflects the changes in quasiparticle behavior it brings about, and can be taken from fits to NMR and INS experiments. We will also assume that the spin degrees of freedom are completely isotropic and that all three components of the spin vector are equally active. Finally, another quantities of primary interest to us is the single particle Green's function $G_{\mathbf{k},\sigma}(\tau - \tau') = -\langle T c_{\mathbf{k}\sigma}(\tau) c_{\mathbf{k}\sigma}^\dagger(\tau') \rangle$ which provides information about the quasiparticle spectral density determined in angular resolved photo-emission experiments, the dynamical spin susceptibility itself, and the corresponding charge response functions. How to determine $G_{\mathbf{k},\sigma}(\tau)$ within the spin fluctuation model using Feynman diagrams is discussed in detail in Ref. [28].

## 4. FEEDBACK PHENOMENA: WEAK PSEUDOGAP BEHAVIOR AND QUANTUM CRITICALITY

In the following we first concentrate on the normal state behavior of cuprate systems and demonstrate that the spin fluctuation model can account for anomalies of the quasiparticle and collective mode behavior simultaneously.

For optimally doped and underdoped systems one finds that over a considerable regime of temperatures,

$$
\omega_{\mathrm{sf}} \ll \pi T
\tag{8}
$$

and it is only as $T$ falls below $T_*$ that $\omega_{\mathrm{sf}}$ becomes comparable to and eventually larger than $\pi T$. In detail, between $T_*$ and $T^{\mathrm{cr}}$ one finds: $\omega_{\mathrm{sf}}/(\pi T) \approx 0.17$ for $\mathrm{YBa_2Cu_4O_8}$ and $\omega_{\mathrm{sf}}/(\pi T) \approx 0.14$ for $\mathrm{YBa_2Cu_3O_{6.63}}$ rather independent of $T$ [7]. As a result of this comparatively low characteristic energy found in the weak pseudogap region, the spin system, for $\mathbf{q} \sim \mathbf{Q}$, is thermally excited and behaves quasi-statically [11]; the quasiparticles see a spin system which acts like a static deformation potential, a behavior which is no longer found below $T_*$ where $\omega_{\mathrm{sf}}$ increases rapidly [7] and the lowest energy scale is the temperature itself.

In the quasistatic limit it is possible to sum the whole perturbation series for the single particle Greens function as well as the two particle spin and charge vertex functions, since it suffices to consider only the zeroth bosonic Matsubara frequency in $\chi_{\mathbf{q}}(i\omega_n)$. This follows from Eq. (2) after analytical continuation and is also reflected in ther fact that the second order diagram, which is particularly important for smaller doping concentrations [29], is clearly dominated by the lowest Matsubara frequency if $\pi T \gg \omega_{\mathrm{sf}}$. The remaining momentum summations are evaluated by expanding $\varepsilon_{\mathbf{k}+\mathbf{q}} \approx \varepsilon_{\mathbf{k}+\mathbf{Q}} + \mathbf{v}_{\mathbf{k}+\mathbf{Q}} \cdot (\mathbf{q} - \mathbf{Q})$ for momentum transfers close to $\mathbf{Q}$, where $v_{\mathbf{k}+\mathbf{Q}}^{\alpha} = \partial\varepsilon_{\mathbf{k}+\mathbf{Q}}/\partial k_{\alpha}$. In this limit *all diagrams* can be summed up by generalizing a solution for a one dimensional charge density wave system obtained by Sadovskii [30] to the case of two dimensions, and more importantly, to isotropic spin fluctuations. In detail, the solution consists of two steps: First, it can be shown that for $\omega_{\mathrm{sf}} \ll \pi T$ each crossing diagram of the spin fermion model equals, besides sign and multiplicity, a particular noncrossing diagram. The second step is to determine the total, $\xi$ independent multiplicity of each class of identical diagrams from the straightforward solution in the limit $\xi \to \infty$. Here, care was taken to ensure spin rotation invariance. Details of the calculations can be found in Ref. [28].

We find the following recursion relation for the Green's function $G_{\mathbf{k}}(\omega) \equiv G_{\mathbf{k}}^{(l=0)}(\omega)$, whose imaginary part determines the spectral density $A(\mathbf{k}, \omega)$, seen in ARPES

$$G_{\mathbf{k}}^{(l)}(\omega)^{-1} = g_{\mathbf{k}}^{(l)}(\omega)^{-1} - \kappa_{l+1}\Delta_o^2 G_{\mathbf{k}}^{(l+1)}(\omega). \qquad (9)$$

Here, $\kappa_l = (l+2)/3$ if $l$ is odd, while $\kappa_l = l/3$ if $l$ is even, and $g_{\mathbf{k}}^{(l)}(\omega)$ is the Fourier transform of $-i\Theta(t)e^{-i\varepsilon_{\mathbf{k}+l\mathbf{Q}}t}[K_0(t|\mathbf{v}_{\mathbf{k}+l\mathbf{Q}}|/\xi)/(2\pi)]^l$ with modified Bessel function $K_0(x)$. The recursion relation, Eq. (9), enables us to calculate $A(\mathbf{k}, \omega)$ to arbitrary order in the coupling constant $g$. In the limit $\xi \to \infty$ the Green's function reduces to $G_{\mathbf{k}}(\omega) = \int d\Delta \, p(\Delta) \, G_{\mathbf{k}}^{\mathrm{SDW}}(\Delta)$, where $G_{\mathbf{k}}^{\mathrm{SDW}}(\Delta)$ is the single particle Green's function of the mean field SDW state and $p(\Delta) \sim \Delta^2 \exp(-\frac{3}{2}\Delta^2/\Delta_0^2)$ is the distribution function of a fluctuating SDW gap, centered around $\sqrt{\frac{2}{3}}\Delta_0$.

The quantities we calculate are the single particle spectral density, $A(\mathbf{k}, \omega)$, and the low frequency behavior of the irreducible spin susceptibility $\tilde{\chi}_{\mathbf{q}}(\omega, T)$. In Fig. 1a, we show our results for the product of $A_{\mathbf{k}}(\omega)$ and the Fermi function, $f(\omega)$, which demonstrate the qualitatively different

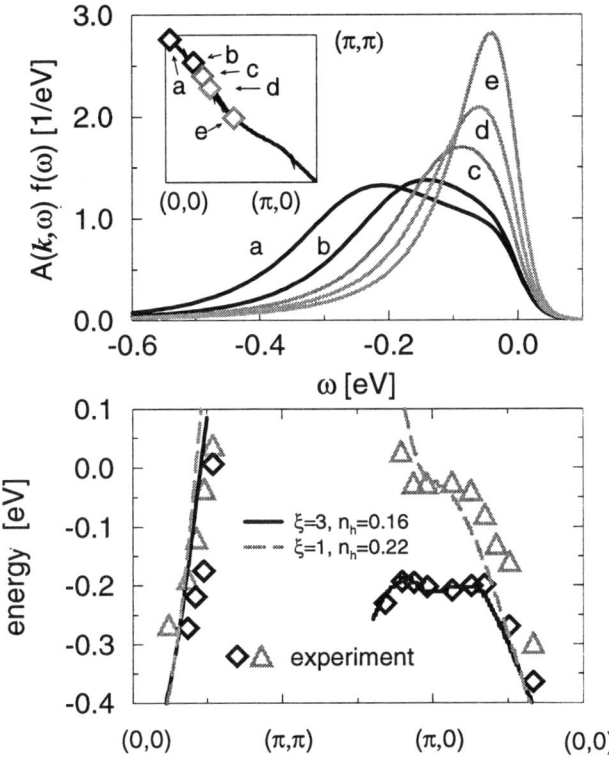

*Figure 3.* (a) Spectral density multiplied with Fermi function on the Fermi surface for $\xi = 3$. The distinct behavior of hot and cold quasiparticles is visible. The inset shows the corresponding Fermi surface. (b) Momentum dependence of local maxima of the spectral density as function of $\xi$ and hole doping concentration $n_h$ is compared with experiments of Ref.[31] for $Bi_2Sr_2Ca_{1-x}Dy_xCu_2O_{8+\delta}$ with $x = 0.01$ (triangles) and $x = 0.175$ (diamonds). Only maxima with relative spectral weight $> 10\%$ are shown.

behavior of hot quasiparticles, close to $(\pi, 0)$ and cold quasiparticles close to the diagonals $|k_x| = |k_y|$. We see that in contrast to the conventional behavior of the cold quasiparticles, whose spectral function is peaked at the Fermi energy, the hot quasi particle density of states develops a pronounced high energy feature (peak) as observed in experiments [12, 31]. Although the hot quasiparticles do not possess a distinct peak at the Fermi level, due to their strong scattering rates, this does not mean that the system has lost pieces of the Fermi surface; rather we find that the Fermi surface, defined by $\varepsilon_{\mathbf{k}} + \mathrm{Re}\Sigma_{\mathbf{k}}(\omega = 0) = 0$, remains large but the coherent peak at the Fermi energy falls below the visibility threshold of present ARPES experiments. Even though these coherent quasiparticles are invisible in the present temperature range, they are, we believe, responsible for the sharp peak for $\mathbf{k} \sim (\pi, 0)$, observed in ARPES below the superconducting transition temperature [12] where the low frequency scattering rate is strongly suppressed. It is only for $\xi > 30$, which is not appropriate for the underdoped cuprates studied here, that the solution of Eq. (9) yields hole pockets, closed around $(\pi/2, \pi/2)$, accompanied by a large piece of a FS closed around $(0,0)$ [29]. If one determines the $\mathbf{k}$-states for which the occupation number $n_{\mathbf{k}} = \frac{1}{2}$, the resulting line in momentum space can hardly be distinguished from the uncorrelated FS, i.e., the total occupied spectral weight of a given momentum state is quite robust with respect to the drastic changes of the line shape of the spectral function. Therefore, high energy features are always more pronounced for the occupied part of the density of states than for its unoccupied part, since states close to $(\pi, 0)$ remain mostly occupied.

In Fig. 3b we compare our results for the momentum dependence of the high energy features with the experimental results of Marshall et al. [31] for two different doping concentrations. The best agreement between theory and experiment was obtained for a next nearest neighbor hopping element $t' = -0.25t$. It can be seen that within our approach one can understand the qualitatively different behavior experimentally seen for underdoped and overdoped systems, as well as the detailed momentum dependence of the high energy features in the underdoped case. While for overdoped cuprates the FS crossing close to $(\pi, 0)$ is clearly visible, no such crossing has been observed in the underdoped systems. In contrast, along the diagonal of the Brillouin zone, there is little difference in the behavior of underdoped and overdoped systems. This detailed agreement between theory and experiment is a strong indication that the high energy features of cuprate superconductors seen in ARPES experiments are indeed due to precursors of a SDW like state.

We sum *all diagrams* of the perturbation series for the electron-spin fluctuation vertex function in similar fashion as the Green's function $G_{\mathbf{k}}(\omega)$ in Eq. (9). We combine the results for $G_{\mathbf{k}}(\omega)$ and the electron-spin fluctuation vertex function and so determine the irreducible spin susceptibility $\tilde{\chi}_{\mathbf{q}}(\omega)$, which gives information about the spin damping $\gamma = \tilde{\chi}''_{\mathbf{Q}}(\omega, T)/\omega\big|_{\omega=0}$ displaying the crossover from $z = 2$ to $z = 1$ at $T^{\mathrm{cr}}$. As may be seen in the inset to Fig.4, $\gamma$ is suppressed below $T^{\mathrm{cr}}$ due to the onset of SDW precursors. Another calculable quantity which can be compared with experiment is $\tilde{\chi}_{\mathbf{Q}}(T)^2/\gamma_{\mathbf{Q}} \equiv \omega_{sf}\chi_{\mathbf{Q}}$, being proportional to the product, ${}^{63}T_1 T/({}^{63}T_{2G})^2$. As may seen in Fig.4, qualitative agreement with the results of Curro et al. [9] for $YBa_2Cu_4O_8$ is found, implying that our theory indeed gives $\omega_{\mathrm{sf}} \approx \xi^{-1}$, i.e. SDW precursors change qualitatively the way spinexcita-

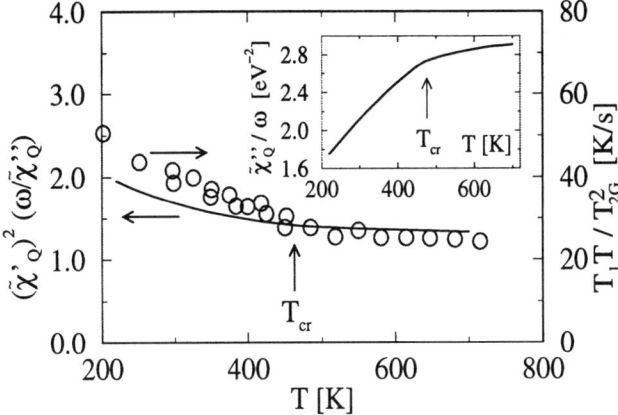

*Figure 4.* Results for $T_1T/T_{2G}^2$ as function of temperature, compared with experimental data of Ref.[9]. The inset shows the crossover in the calculated T-dependence of the spin damping.

tions decay into particle and hole degrees of freedom on the Fermi surface. Here we still had to assume a specific temperature dependence of the correlation length: $\xi^{-1}(T) = \frac{1}{4} + \frac{1}{4}\frac{T-T_*}{T^{cr}-T_*}$ between $T^{cr} = 470\,\mathrm{K}$ and $T_* = 220\,\mathrm{K}$ and $\xi^{-2}(T) = \frac{1}{4} + \frac{1}{7}\frac{T-T^{cr}}{700\,\mathrm{K}-T^{cr}}$ above $T^{cr}$ consistent with the NMR results of Curro *et al.* [9] for $YBa_2Cu_4O_8$. Below we will *calculate* the temperature dependence of $\xi$.

Physically, the most interesting aspect of our results is the appearance of SDW precursor phenomena and, in turn, modified damping, brought about by the strong interaction between the planar quasiparticles, for moderate AF correlation lengths, $\xi > \xi_o \approx 2$, in contrast to earlier calculations, in which SDW precursor behavior was only found in the limit of very large correlation length. Our complete solution of the static problem enables us to access this region of strong coupling. For $\xi > \xi_o$, the hot electron mean free path, $\sim \xi_o^2/\xi$ begins to be small compared to $\xi$, so that the quasi particle can no longer distinguish the actual situation from that of a SDW state; hence we find pseudo-SDW behavior, i.e. SDW behavior without symmetry breaking. The related shift of spectral weight for states close to $(\pi, 0)$ affects mostly the low frequency part of the irreducible spin susceptibility and leads to the calculated crossover behavior.

In the next step we will investigate the consequences of our finding that $\omega_{\mathrm{sf}} \approx \xi^{-1}$ on the critical properties of the collective spin degrees of freedom This will enable us to give for the first time an explicit result for the temperature dependence of the antiferromagnetic correlation length which is in agreement with experiment. Feedback effects due to spin fluctuation induced precursors in the fermionic quasiparticle spectrum will be taken into account in the description of a quantum critical point of itinerant spin systems. Using a one loop renormalization group approach, we obtain a quantitative explanation for the scaling behavior seen in underdoped cuprate superconductors. We eliminate the fermionic degrees of freedom from the spin fermion action by Gaussian integration and expand up to forth order in $\mathbf{S}(q)$; the resulting effective action of the collective spin degrees of freedom is [32, 33]:

$$
S = \frac{1}{2} \int^{\Lambda} dq \, \chi_0^{-1}(q) \, \mathbf{S}(q) \cdot \mathbf{S}(-q)
$$

$$
+ u \int^{\Lambda} dq_1 \cdots \int^{\Lambda} dq_4 \, \delta_{q_1+q_2+q_3+q_4}
$$

$$
\times \mathbf{S}(q_1) \cdot \mathbf{S}(q_2) \, \mathbf{S}(q_3) \cdot \mathbf{S}(q_4) \,. \tag{10}
$$

Here, the coupling constant, $u > 0$ is due to quasiparticle mediated collective mode - collective mode interaction. In Eq. 10 the vector $q = (\mathbf{q}, i\omega_n)$ consists of the 2-dimensional momentum vector $\mathbf{q}$ and the bosonic Matsubara frequency $\omega_n = 2n\pi T$ at temperature, $T$. We write $\int^{\Lambda} dq \cdots = T \sum_n \int_{|\mathbf{q}|<\Lambda} \frac{d^2\mathbf{q}}{(2\pi)^2} \cdots$ and $\delta_{q+q'} = T^{-1}\delta_{n+n'}\delta^{(2)}(\mathbf{q} + \mathbf{q}')$, where $\Lambda$ is the upper momentum cut off.

Usually, $\chi_0(q)$ and $u$ are calculated within a weak coupling approach in which the spin damping due to particle hole excitations is identical to that of a noninteracting electron gas [32, 33]. However, fermionic quasiparticles are strongly affected by their own spin excitations once the system gets close to a magnetic instability [34, 35, 36, 29] and we have to take these modifications of the quasiparticle spectrum in an investigation of the critical behavior into account. Of particular importance are quasiparticles close to the magnetic BZ-boundary, which, on the one hand, determine the spin damping, $\gamma$, and, on the other hand, are mostly affected by the proximity to the ordered state. The propagator of the collective spin modes is then assumed to be:

$$
\chi_0(\mathbf{q}, i\omega_n) = \frac{\alpha}{\xi_0^{-2} + \mathbf{q}^2 + \xi_0^{-1}|\omega_n|/\hat{c} + (\omega_n/c)^2} \,, \tag{11}
$$

with bare correlation length $\xi_0$, spin damping, $\gamma = \hat{c}^{-1}\xi_0^{-1}$ and spin wave velocity $c$. Momenta are measured relative to the ordering vector. For $\hat{c} \to \infty$, i.e. without spin damping, the problem is similar to that investigated by Chakravarty et al. [37] and Chubukov et al. [38]. The alternative limit $\gamma = const.$ and $c \to \infty$ was discussed by Hertz [32] and Millis [33]. The situation $\gamma = const.$ and $c$ finite was discussed by Sachdev et al. [39], who found a $z = 1$ to $z = 2$ crossover for decreasing temperature, in contrast to

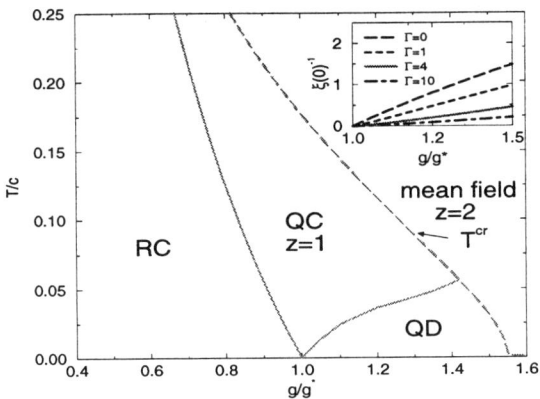

*Figure 5.* Phase diagram of an itinerant AF with z= 1 scaling behavior for $\Gamma = 4$. RC is the renormalized classical regime with exponentially large correlation length QC is the quantum critical regime and QD the quantum disordered regime. In the mean field regime above $T^{cr}$ with correlation length larger 2, no feedback effect due to changed quasiparticle behavior occurs, leading to z=2. The inset shows the inverse zero temperature correlation length in the QD regime for different $\Gamma$.

the experimental finding in the cuprates of a $z = 2$ to $z = 1$ crossover for decreasing $T$ [7, 23, 9].

Using a Kadanoff-Wilson momentum shell renormalization group approach; integrating out states with momenta between $\Lambda e^{-l}$ and $\Lambda$, including the rescaling $T(l) = e^{zl}T$ to reach self-similarity, we obtain the flow equations for the characteristic coupling constant $g \propto u\xi_0^2$. We find a quantum critical point $g^* = 4\pi/(c\Lambda)$ which separates in the usual sense [37] a renormalized classical (RC) regime with exponentially growing correlation length, $\xi(T) = \hat{c}/(2T) \exp\left(2\pi c\Lambda/T(\frac{1}{g^*} - \frac{1}{g})\right)$, for $g(l = 0) < g^*$, from a quantum disordered (QD) state with a finite zero temperature correlation length, $\xi(0)$,. We find for $\Gamma \gg 1$ that $\xi(0)/\log(\xi(0)\Lambda) = \frac{\Gamma g^*}{4\pi\Lambda}(g - g^*)^{-1}$, which is enhanced compared to the situation without damping: $\xi(0) = \frac{g^*}{2\Lambda}(g - g^*)^{-\nu}$ with $\nu = 1$ up to one loop. The results of a numerical evaluation of $\xi(0)^{-1}$ are shown in the inset of Fig. 5.

Modified scaling behavior as a function of temperature can be found in similar fashion. We find up to logarithmic corrections,

$$\xi(T)^{-2} \approx \xi(0)^{-2} + bT^2,\qquad(12)$$

which is the expression anticipated in Ref. [10]. So far no other theory was able to give an explanation for this specific temperature depence. Quantitative agreement between theory and experiment can be reached for $g/g^* = 1.18$. In contrast, in the case of propagating spin excitations a sharp transition to an exponentially weak temperature dependent $\xi$ occurs for $T \approx \Delta$ with spin wave gap $\Delta = c\xi(0)^{-1}$. Thus, due to spin damping the spin wave gap is filled with low energy states and the correlation length continues to grow even for very low temperatures. The crossover between the quantum critical (QC) and QD regime is very gradual. Finally, considering higher temperatures, we have to take into account that $\gamma \propto \xi^{-1}$ occurs only for $\xi \geq \xi_o \approx 2$ [11]. For $\xi < \xi_o$, $\gamma = const.$ leading to a $z = 2$ behavior at a characteristic temperature $T^{cr}$ (see Fig. 5).

Another characteristic phenomenon caused by the proximity to a QCP is the scaling behavior of the frequency dependence of $Im\chi(\mathbf{q},\omega)$. For any point in the $(T,g)$ phase diagram (except $g \leq g^*$ and $T = 0$) $Im\chi_{\mathbf{q}}(\omega)$ increases linearly with $\omega$ due to excitations in the particle hole continuum. In the RC regime however, the spin damping is exponentially suppressed and the spin dynamics is indistinguishable from a system with purely propagating spin waves. On the other hand, in the QC regime $\omega/T$ scaling behavior of $\xi^{-2}Im\chi_{\mathbf{q}=0}(\omega)$ and of the momentum averaged susceptibility $\int d^2\mathbf{q} Im\chi_{\mathbf{q}}(\omega)$ is found and the low frequency slope of $\chi_{\mathbf{q}=0}(\omega)$ behaves like $Im\chi_{\mathbf{q}=0}(\omega)/\omega|_{\omega\to 0} = \alpha\xi^\delta/\hat{c}$ with $\delta = 3$.

In applying these results to cuprate superconductors, we assume that doped systems without long range order are located in the QC and QD regime. Since $g \propto 1/\langle\mathbf{S}(\mathbf{r})^2\rangle$ doping reduces the effective moment, causing the coupling constant to grow until it exceeds $g^*$. For the cuprate material $La_{1.86}Sr_{0.14}CuO_4$ the parameters $\hat{c} \approx 50\,meV$ and $c \approx 220 meV$ are reasonably well known from NMR and INS experiments [7, 23, 10] and it follows $\Gamma \approx 4$. Thus, it suffices for a quantitative understanding of the INS data of Ref [10] to determine the ratio $g/g^*$ from the correlation length at a given temperature. Using $g = 1.18g^*$ as determined above, we can determine the frequency and temperature dependence of the dynamical spin susceptibility. The results for $Im\chi_{\mathbf{q}=0}(\omega)$ with $\alpha = 26\,eV^{-1}$, shown in Fig. 6, are in remarkable agreement with the results of Aeppli $et$ $al.$ [10], who also find $Im\chi_{\mathbf{q}=0}(\omega)/\omega|_{\omega\to 0} \propto \xi^\delta$ with $\delta = 3 \pm 0.3$. This agreement between theory and experiment is a direct consequence of the fact that the experimental data show a $\omega/T$ scaling behavior of $\xi^{-2}Im\chi_{\mathbf{q}=0}(\omega)$ which is a strong indication that the system under consideration is indeed close to a QCP. Note, our results agree quantitatively with the phenomenological description of the $z = 1$ pseudo-scaling regime of Ref. [23] based on NMR experiments [40].

In conclusion, based on previous calculations [11, 34, 35, 36], we have shown that modifications of the particle hole excitation spectrum can change the dynamical scaling behavior of itinerant AF systems close to a QCP by affecting the spin damping. The resulting $z \approx 1$ scaling causes a T-

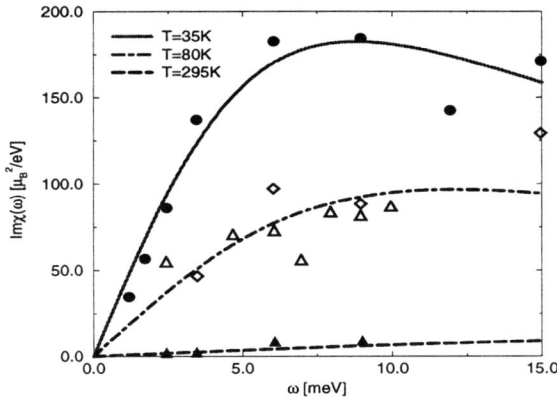

*Figure 6.* Frequency dependence of the dynamical susceptibility at the peak maximum for $\Gamma = 4$ and $g = 1.18g^*$ in comparison with INS experiments from Ref.[3].

dependence of the AF correlation length which is completely different from the usual $z = 2$ case. This new $T$-dependence of the dynamical spin susceptibility is in remarkable agreement with NMR and INS experiments and gives an explanation for the crossover scenario of Refs. [7, 23]. The $z = 1$ scaling and the position of the QCP are the same as those for propagating spin modes in insulating AF. However, due to damping, the correlation length is enhanced and the $T$-dependence of $\xi$ is changed.

We expect similar behavior in the quantum phase transitions of other itinerant systems if the strong interaction between quasiparticles and collective modes changes the dynamics of the collective mode under consideration. Examples are one and two dimensional charge density wave systems, superconductors or saturated ferromagnets. In higher dimensions or for systems with weak quasiparticle-collective mode coupling precursor phenomena are only expected once $\xi$ diverges [36] and will play no role in the quantum disordered regime.

## 5. PATTERN FORMATION IN TRANSITION METAL OXIDES AND OTHER COMPLEX MATERIALS

In the previous section we have seen that critical behavior is expected in a number of systems with competing phases. Near a critical point, this competition can yield spatial inhomogeneities (pattern formation), i.e., a creation of domains in which a competing phase would establish a non-zero order parameter, with or without a long range order. In the former case (so called phase separation) this leads to novel and exciting physics. In this section we show a simple model which yields such pattern formation in transition metal oxides and relate the observed behavior to other complex systems.

As mentioned above INS measurements, which directly probe magnetic properties of a system, have been used to study the AF fluctuations in HTS. While in the insulating phase these measurements yield a sharp peak in $\chi''$ at momentum $(\pi, \pi)$, upon doping the long AF order disappears, and the INS peak broadens and splits into four incommensurate peaks, where the splitting between them is roughly proportional to the doping level. This splitting of the AF fluctuation spectrum in momentum space can be interpreted either as an indication of quasinesting of the FS (see previous sections and Ref. [42]) or as an indication of the ordering of the doped holes in CuO planes: if holes order in an array of stripes, separated by the AF ordered spins with interstripe distance $\ell$ along, say, $x$ direction, then the peak in the spin susceptibility shifts from $(\pi, \pi)$ to $(\pi + \pi/\ell, \pi)$ (and by symmetry to $(\pi - \pi/\ell, \pi)$). Assuming stripe ordering occurs in domains with stripe directions perpendicular to each other, the stripe formation yields the INS spectrum as that seen in Nd doped $La_{2-x}Sr_xCuO_4$, where this type of (long-range) order has been confirmed unambiguously. Without Nd doping, however, the controversy still remains: several authors have proposed that the charge ordering is dynamic, characterized by a specific time scale, with spins ordering antiferromagnetically in between stripes.[43]

On the theoretical side, stripes have been proposed as a result of a competition between short-range attractive interaction between holes from the breaking of AF bonds and the long-range Coulomb interaction [43]. Striped phases have been obtained within mean field approaches to the short-range Hubbard or $t - J$ models[44]. However, few models have obtained stripes as a result of direct numerical simulation, with a notable exception of the recent Density Matrix Renormalization Group simulations (DMRG) [45] of White and Scalapino. More importantly, to date no unbiased calculation has produced a striped phase with *metallic* conductivity.

## 5.1. MODEL

The spin density wave (SDW) picture of the layered transition metal oxides has been very successful in describing the insulating AF phase of these systems at low $T$ [48, 49]. In this picture the electrons move with hopping energy $t$ in the self-consistent staggered field of its spin. In the limit $t \ll J$ it can be shown that the holes interact via two different mechanisms:[1] a short-

---

[1]Note that this limit is poorly satisfied for cuprates not far from optimal doping, mentioned in previous sections. Hence the present model is more applicable to, e.g., Ni based compounds than the cuprates.

range attractive force due to AF bond breaking and a long-range magnetic dipolar interaction due to the spiral distortion of the AF background [49, 50]. The magnetic dipole moment associated with each hole is due to the virtual hopping of holes between neighboring sites and scales with the AF magnetic energy. The magnetic dipolar interaction between two holes with moments $\mathbf{d}_{1,2}$ at distance $\mathbf{r}$ apart has the form:

$$U_{dip} = \frac{1}{r^2}\left[(\mathbf{d}_1 \cdot \mathbf{d}_2) - \frac{2}{r^2}(\mathbf{d}_1 \cdot \mathbf{r})(\mathbf{d}_2 \cdot \mathbf{r})\right]\left(\sigma_1^\dagger \sigma_2^- + \sigma_1^- \sigma_2^\dagger\right), \qquad (13)$$

where $\sigma_i^{(\dagger}, -)$, are the Pauli matrices. The spin part of the many body Hamiltonian can be described by a two dimensional (2D) non-linear $\sigma$ model in the long wavelength limit [50, 51]. In 2D at finite $T$ any long-range ordering in a system with short-range interactions is prohibited[52], i.e., the dipolar phase is short-ranged and the system is magnetically disordered and characterized by a finite magnetic correlation length, $\xi$ [53]. In turn, the magnetic dipolar interaction between the holes, mediated by the AF background, is also short-ranged. However, besides the AF interactions, at very low doping the holes also feel the long-range Coulomb interaction. This is clear if we consider that $r_s = r_0/a_0$ (where $r_0$ is the mean inter-particle distance and $a_0$ is the Bohr radius) can be very large in,e e.g., extremely underdoped cuprates ($r_s \approx 8$) in which case the interaction energy between the holes, which behaves like $e^2/(a_0 r_s)$, is certainly more important than the kinetic term ($\approx e^2/(a_0 r_s^2)$) at low densities. This implies that the interaction terms should be treated first and the kinetic energy as a perturbation. Finally, each hole carries a spin degree of freedom as well, but it is possible to show that the overall spin energy is minimized in the spin anti-symmetric channel, as we assume here. Thus, we are left with only the charge channel with an effective (magnetic in origin) interaction between two holes, 1 and 2, of the form (see Eq. (13))

$$V(\mathbf{r}) = \frac{q^2}{r} - A\,e^{-r/a} - B\cos(2\theta - \phi_1 - \phi_2)e^{-r/\xi}, \qquad (14)$$

where $q$ is the hole charge, $\theta$ is the angle made between $\mathbf{r}$ and a fixed axis and $\phi_{1,2}$ are the angles of the magnetic dipoles relative to the same fixed axis. $A$ is the strength of the short-range uniform interaction ($A \approx U$ when $t \gg U$ and $A \approx 4t^2/U$ when $U \gg t$ [54]) and $B$ is the strength of the magnetic dipolar interaction ($B \approx A/(2\pi\xi^2)$), which we will assume to be independent variables. Both $A$ and $B$ terms include the short-range bond-breaking attractive interaction between holes.

In general, the many-body problem of holes in an AF background is extremely complicated, involving many-particle interaction terms. However,

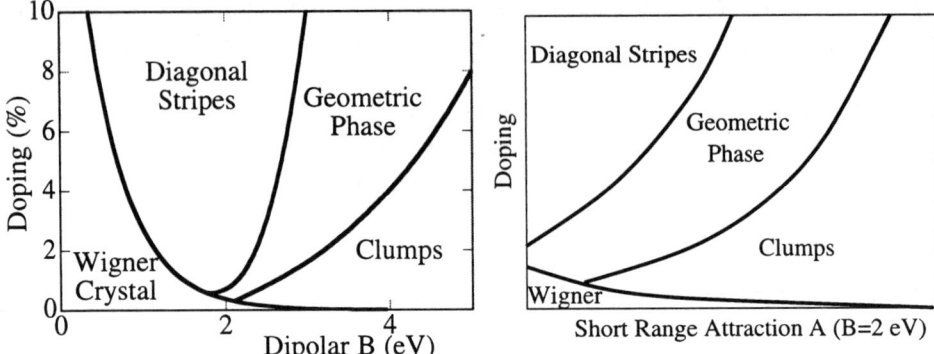

*Figure 7.* Calculated phase diagrams as a function of doping and interaction parameters, $A$ and $B$ (see Eq. 14). The left panel corresponds to $A = 0$, and the right panel corresponds to the fixed value of $B = 2$ eV.

at low densities it is reasonable to assume that the interaction of any two holes is weakly perturbed by other holes within distance $\xi$, and the total potential energy can be expressed in terms of two-particle energies. Therefore, in our numerical calculations we study the physics of a finite density of holes (i.e., $N$ holes per rectangular computational box of size $L_x \times L_y$ with $L_x$, $L_y$ up to 100 unit cells in a $CuO_2$ plane), interacting via $V(\mathbf{r})$ as given in (14).

To account for the finite density of holes we imply the periodic boundary conditions. We handle the long-range Coulomb interaction properly by summing interactions of particles with all of their images residing in cells obtained by translation from the original computational cell[46, 47]. On making integral transformations, Coulomb interactions are computed by summing over fast-convergent Bessel functions with great accuracy. We study the ground state and the dynamics of this system using three different methods: MC method, Langevin MD and a hybrid MC-MD method[55]. All three methods yield essentially the same results.

## 5.2. RESULTS

In the absence of disorder we find four phases depending on the density of holes and the characteristic AF energy scales: a Wigner crystal, diagonal stripes, horizontal-vertical stripes (tiles) and a glassy-clumped phase. The order parameter for charge ordering is the Fourier transform of the hole

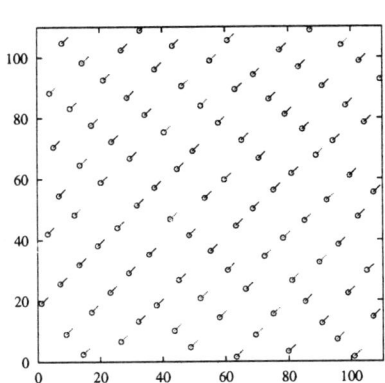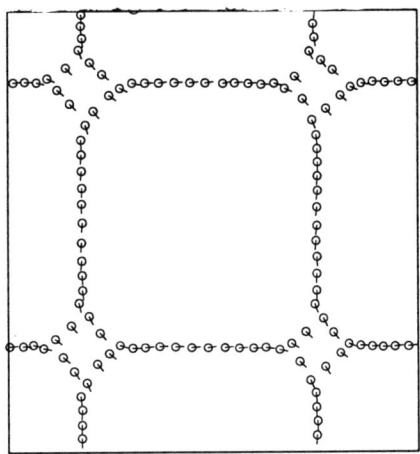

*Figure 8.* The ferro-dipolar phase (a) and the checkerboard phase (geometric) phase (b). Note the presence of domains of stripes in (a), due to large number of metastable states, which are very close in energy to the single domain striped ground state.

density:

$$\rho(\mathbf{q}) = \frac{1}{N} \sum_{i=1}^{N} e^{i\mathbf{q} \cdot \mathbf{r}_i}, \tag{15}$$

where $\mathbf{r}_i$ is the position of the $i^{th}$ hole and $N$ is the total number of holes. A peak in $\rho(\mathbf{q})$ at some wave-vector $\mathbf{q} = \mathbf{K}$ indicates ordering.

For $B = 0$ we find the Wigner crystal with small distortions to be the state of lowest energy, as expected [56]. Increasing $A$ while retaining $B = 0$ reduces the lattice constant of the Wigner crystal until a critical value is reached where holes group together. For $A = 0$ and finite $B$ the situation is quite different. At small $B$ and larger densities the Wigner crystal is unstable and a new phase with diagonal stripes is formed. This phase is characterized by ferro-dipolar order (see Fig. 8b). The situation here is very similar to that observed in $La_{2-x}Sr_xNiO_{4+y}$ [57]. As shown in Fig. 8b, at larger values of $B$ line stripes are formed, which, with increasing density tend to close into a checkerboard pattern. More importantly, the tile formation is accompanied by magnetic dipole orientation along the straight portion of a loop with gradual rotation by $\pi/2$ at each corner [58]. The size of the inter-hole distance within a line is determined by the ratio of $B$ and the Coulomb energy; the loop sizes are determined by the hole density alone. These results appear to be consistent with the DMRG solution of the $t$-$J$ model [45]. If $B$ is increased further the magnetic dipolar interaction becomes dominant over the average Coulomb interaction; the well-defined pattern disappears and one observes star shaped clumps of holes, which can, at sufficiently high density, form another geometric structure (e.g., a

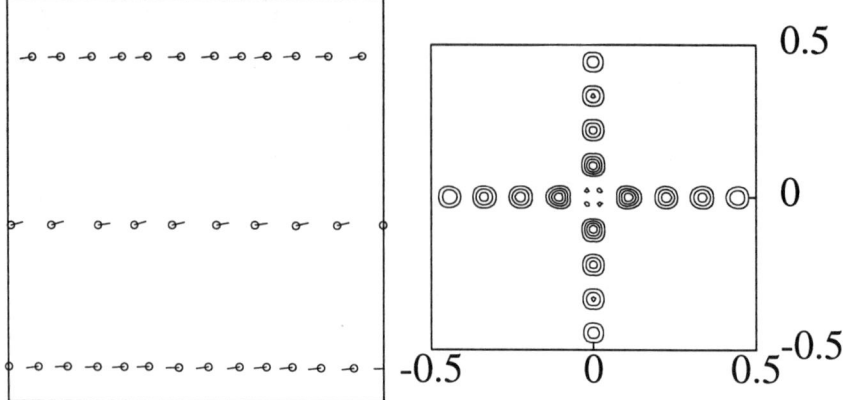

*Figure 9.* (a) If small amount of anisotropy is included into the calculation (∼ 20%), the checkerboard phase is lost and one observes a perfect stripe formation, corresponding to a modulated dipolar phase. Note the direction of dipoles in neighbouring stripes. (b) Assuming twinning the striped phase yields the same $\rho_{\mathbf{k}}$ as the checkerboard phase, shown in Fig. 8.

Wigner crystal of clumps). We remark that in all phases a non-vanishing value of $A$ leads to a decrease in the effective value of $B$ at which the transitions occur (Fig. 7); the isotropic term $A$ alone *never* produces any non-trivial geometric phase (e.g., stripes), even with inclusion of lattice effects. We find that the transition between the ferro-dipolar and the stripe phase is first order, while other transitions appear to be of second order [59].

In the cases presented above we have assumed uniform magnetic dipolar interaction. It is well known that there are orthorhombic and tetragonal distortions in practically all transition metal oxides. In order to study the influence of the anisotropy we assume that the magnetic dipole sizes along $x$ and $y$ directions have anisotropy $\alpha$ ($\alpha = 1$ corresponds to the isotropic case). Figure 9a shows our solution for $\alpha = 0.8$: the symmetry is broken and a stripe superlattice is formed, with a charge ordering vector $\mathbf{K} = (\pi/\ell)\mathbf{x}$, where $\ell$ is the inter-stripe distance. As mentioned above, a peak at $\mathbf{K}$ in (15) leads to a peak at $\mathbf{Q} \pm \mathbf{K}$ in $S(\mathbf{q})$. Thus our results yield a neutron peak at $(\pi/a \pm \pi/\ell, \pi/a)$. Assuming twinning, this would imply neutron peaks at $(\pi/a \pm \pi/\ell, \pi/a \pm \pi/\ell)$ in agreement with experiment[60] and as obtained in the checkerboard phase (see Fig. 8b). If one includes the kinetic energy[59], instead of static stripe formation one obtains dynamical stripes like those believed to exist in $La_2Sr_xCuO_4$. The kinetic energy may also destroys the ferro-dipolar phase and certainly the Wigner crystal phase.

On increasing of the size of the computational box, in a typical calculation the checkerboard pattern often acquires point or line defects, as the temperature is lowered. Moreover, we find that there is an infinite number of geometric patterns with energies close the the ground state energy, which ultimately yields a "glassy" dynamics of the system at higher temperatures [59] where the stripe phases melt. Second, in a finite system with appropriate nutralizing charge background, the holes do not form geometric phases, although they still form stripes; however, in this case even a very small anisotropy ($\alpha \sim 0.95$) again leads to geometric stripe formation, similar to that seen in in Fig. 9. Finally, we have also studied impurity effects (from defects or charged counter-ions). We find that at all values of $A$ and $B$ the phases are unstable towards stripe formation: the geometric (stripe) pattern is destroyed at relatively moderate inpurity concentrations. On the other hand, the magnetic dipole interaction is sufficient to retain the main line orientation, which leads us to conjecture that with the addition of the kinetic energy the holes can move in *stripe segments* in an orientation given basically by the phase diagram of the clean system[59]. These string segments are kept together by the dipolar interaction (i.e., string tension).

On a final note: clearly, the pattern formation presented here is a result of the competing (AF dipolar and the Coulomb) interactions, which dominate behavior at short and long ranges, respectively and a finite value of the density of holes. Otherwise, the calculation is quasiclassical and the behavior is completely generic. In other words, it is plausible that other systems with competing interactions should yield similar pattern formation. This has been shown recently[62] for Langmuir films of lipids and cholesterol, where the short-range interaction is of Lennard-Jones type and the long-range interaction is electrostatic (Coulomb), active only between lipid molecules. In this case stripes occur at particular lipid-cholesterol mixing ratio, which is equivalent to the hole doping level in transition metal oxides. A pattern formation should be expected quite generally for any complex system with competing interaction or competing phases.

## 6. PAIRING DUE TO SPIN FLUCTUATIONS IN CUPRATES AND ORGANICS

The spin-fluctuation approach to cuprates is particularly appealing since it was argued that the exchange of spin fluctuations peaked at or near antiferromagnetic momentum $\mathbf{Q}$ gives rise to an attractive interaction in a $d_{x^2-y^2}-$ pairing channel which, as most researchers now believe, is the dominant gap symmetry in cuprates. This argument is based on the assumption that spin-fluctuations in high $T_c$ materials play the same role as phonons in ordinary superconductors, i.e., one can write an analog of weak coupling BSC theory, and write the gap equation in the singlet channel in the form

$$\Delta_{\mathbf{k}} = -\frac{3}{2}g^2 \int \chi(\mathbf{k} - \mathbf{k}')\Delta_{\mathbf{k}'} \frac{\tanh(\epsilon_{\mathbf{k}'}/2T)}{2\epsilon_{\mathbf{k}'}} d^2\mathbf{k}'. \qquad (16)$$

An important difference with the case of phonons is that the r.h.s. of this equation has a minus sign caused by a summation over spin indices whereas a conventional spin-independent interaction yields the term in brackets without $-3/2$ factor. As a result, if $\chi(\mathbf{q})$ was peaked at zero momentum transfer, Eq. (16) would have no nontrivial solutions. However, when $\chi(\mathbf{q})$ is peaked near $\mathbf{Q}$, one can neutralize the overall minus sign by using an ansatz $\Delta_{\mathbf{k}} = -\Delta_{\mathbf{k}+\mathbf{Q}}$. This last condition is satisfied if the gap has a $d_{x^2-y^2}$ symmetry, i.e., $\Delta \propto (\cos k_x - \cos k_y)$. However, our argumentation applies equally well for the higher $s$−wave harmonic $\Delta \propto (\cos k_x + \cos k_y)$. For cuprates, in all range of doping concentrations where superconductivity is observed, the Fermi surface is rather close to the magnetic Brillouin zone boundary along which $\cos k_x + \cos k_y = 0$. Thus, almost no superconducting condensation energy resulting from states close to the Fermi surface can be gained in the s-wave channel. For this reason the $d_{x^2-y^2}$ channel is a dominant one.

The above solution is based on weak coupling, BSC-like approach. As sown above, in cuprates, a weak coupling approach is applicable only deep in the overdoped regime. Clearly, the weak coupling approach has to be modified in the underdoped regime. Monthoux and Pines [41] addressed this issue by self-consistently including self-energy corrections to fermionic propagators, i.e., performing the same kind of calculations as Eliashberg did for conventional superconductors. They found that in fact the occurence of strong quasiparticle scattering causes a negative feedback in the sense that it $T_c$ is suppressed due to life time effects compared to the naive weak coupling theory above, leading to $T_c \approx 100\,\mathrm{K}$ in agreement with the experimental values. This result is not consistent as the energy scales involved in the problem are much higher than $T_c$. For example, the analog of a Debye frequency in thej spin-fluctuation scenario is the exchange coupling $J$ which in cuprates is $\sim 1500K$. The theory of Ref. [41] does not take weak or strong pseudogap effects into account. A universally applicable theory for the superconducting state for all doping values is still missing.

In addition to the copper-oxide based high-temperature superconductors, heavy fermion superconductors like UPt$_3$ and UBe$_{13}$, and the ruthenate compound Sr$_2$RuO$_4$, organic superconductors are candidates for a pairing mechanism which originates in strong electronic correlations in the normal state.

The organic molecular crystals $\kappa$-(BEDT-TTF)$_2$X (in the following abbreviated as $\kappa$-(ET)$_2$X), are very anisotropic quasi two-dimensional superconductors with transition temperatures up to around $10\,\mathrm{K}$, depending on pressure and the ion X, which can, for example, be $I_3$, $Cu[N(CN)_2]Br$ or $Cu(SCN)_2$ [63]. Shubnikov - de Haas experiments have established the existence of a well defined Fermi surface (FS) in these materials [64, 65], demonstrating the Fermi liquid character of the low energy quasiparticles. Below $T_c$, the low temperature $^{13}$C-NMR spin-lattice relaxation rate varies as $T^3$ [66, 67, 68], the electronic specific heat coefficient exhibits a magnetic field dependence $\gamma \propto B^{1/2}$ [69] and the thermal conductivity $\kappa$ is linear in $T$ [70]; evidences for nodes of the SC gap function on the FS. All these results suggest that superconductivity is caused by a highly anisotropic and likely non-phononic pairing interaction.

It has been pointed out earlier that many anomalous normal state properties of layered organics strongly suggest the important role of electronic correlations [71, 72, 73] and that some, but not all, of these anomalies are in striking similarity to cuprate superconductors [74]. Particularly magneto-transport, the thermoelectric power, and the uniform magnetic susceptibility display similar (pseudogap) behavior. Another resemblance is their proximity to an antiferromagnetic (AF) state, even though, in organics, frustration and strong in-plane anisotropies demonstrate differences between both systems. Particularly, due to the, by now, established, dominant $d_{x^2-y^2}$ order parameter component [75], high-temperature superconductivity in cuprates is believed to be dominated or entirely caused by an electronic mechanism. The spin fluctuation scenario [4, 76], which predicted the $d_{x^2-y^2}$ state, is a very promising approach, because it also offers an explanation for a variety of anomalous normal state properties of cuprates. The unfrustrated AF nearest neighbor coupling and the specific shape of the FS are important prerequisites for a spin-fluctuation induced SC state with $d_{x^2-y^2}$ symmetry in cuprates. In contrast, in organics, the magnetic coupling is frustrated, due to the underlying anisotropic triangular lattice [74]. In addition, their FS consists of two disconnected pieces [64, 65]. Therefore, it is not at all obvious whether, despite these numerous similarities to cuprates, spin fluctuations can bring about superconductivity in the organics.

Within a given conducting layer, the unit cell of $\kappa$-$(ET)_2X$ is occupied with six electrons in the outer electronic shells and consists of four ET- molecules, which can be considered as single electronic degrees of freedom [77]. The four molecules are arranged in two dimers (see Fig. 10a). Due to the large intra-dimer transfer integral, $t_1 \approx 250\,meV$, the two bonding bands are completely occupied and shifted to binding energies $\sim 2t_1$ below the Fermi energy. Only the two remaining anti-bonding bands cross the FS; those are half filled with the remaining two electrons. Unfortunately, we do not have sufficient information about the momentum dependence of the magnetic susceptibility in the organic materials and cannot use a straight-

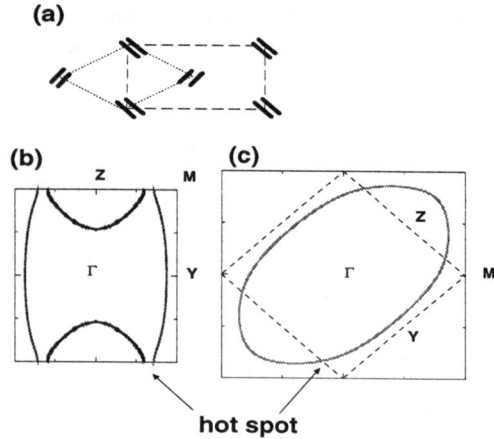

**(a)**

**(b)**  Z  M  **(c)**

hot spot

*Figure 10.* (a) Spatial arrangement of the ET molecules within the planes. The two dimer types are indicated by the different tilting of the molecule pairs, which are, after dimerization, considered as a single electronic degree of freedom. The dashed (dotted) lines indicate the unit cell for $t_3$ ¿ $t_{3'}$ ($t_3 = t_{3'}$). (b) Fermi surface for the case $t_3$ larger $t_{3'}$, consisting of a quasi 1-D band and a hole pocket closed around the Z-point. (c) Fermi surface for the case $t_3 = t_{3'}$. Note that the hot spots due to nearest neighbor AF correlations correspond to the gaps between both Fermi surface parts in (b).

forward phenomenological approach similar to the cuprates. Instead, we have to use a microscopic Hamiltonian to determine the magnetic excitation spectrum selfconsistently.

In what follows [80], we consider only the two anti-bonding bands near the Fermi surface (within a hole picture) and neglect the bonding bands. Due to the strong intra-molecular Coulomb repulsion ($U_{ET} \geq 1\,\mathrm{eV}$), a doubly occupied dimer will likely distribute the two holes on both molecules of the dimer, causing an effective correlation energy $U \sim 2t_1$ determined by the bonding-anti-bonding band splitting [74]. This leads to the following two band Hubbard Hamiltonian

$$H = \sum_{ij,l,\sigma} t_{ij} c_{il\sigma}^\dagger c_{jl\sigma} + U \sum_{i,l} n_{il\uparrow} n_{il\downarrow}$$
$$+ \sum_{ij,\sigma} \tilde{t}_{ij} \left( c_{i1\sigma}^\dagger c_{j2\sigma} + c_{j2\sigma}^\dagger c_{i1\sigma} \right). \qquad (17)$$

Here, $c_{il\sigma}^\dagger$ is the creation operator of a hole with spin $\sigma$ in the anti-bonding band of the $l$-th dimer ($l \in \{1,2\}$) in the $i$-th unit cell and $n_{il\sigma} = c_{il\sigma}^\dagger c_{il\sigma}$. $t_{ij}$ and $\tilde{t}_{ij}$ are inter-dimer hopping elements between dimers of the same and of different type, respectively. Their Fourier transforms $t_{\mathbf{k}}$ and $\tilde{t}_{\mathbf{k}}$ with $\mathbf{k} = (k_y, k_z)$ determine the band-structure of the two bands $\varepsilon_{\mathbf{k},\pm} = t_{\mathbf{k}} \pm \tilde{t}_{\mathbf{k}}$.

Various different tight binding parameterizations have been proposed for $t_\mathbf{k}$ and $\tilde{t}_\mathbf{k}$ [64, 77, 78, 79]. I will use the dispersion relation $t_\mathbf{k} = 2t_2 \cos(k_y)$ and $\tilde{t}_\mathbf{k} = 2\cos(k_z/2)\sqrt{t_3^2 + t_{3'}^2 + 2t_3 t_{3'} \cos(k_y)}$ [64, 79]. Due to the non-symmetrical position of the dimers and the ET-molecule structure itself, the three hopping elements are slightly different. The FS, for $t_2 = 45\,\text{meV}$, $t_3 = 60\,\text{meV}$ and $t_{3'} = 65\,\text{meV}$, shown in Fig. 10b, reproduces de Haas - van Alphen measurements [65] and electronic structure calculations [77]. Note, the two pieces of the FS are disconnected since $t_3 \neq t_{3'}$.

For the investigation of an electronic pairing state, one has to determine the momentum and frequency dependence of the pairing interaction which is responsible for the magnitude, $|\Delta_{ll'}^{ij}|$, of the coordinate space components of the gap function $\Delta_{ll'}^{ij} \propto \langle c_{il\uparrow} c_{jl'\downarrow} \rangle$ and the relative phase relations the $\Delta_{ll'}^{ij}$ establish. The latter effect is determined by the maximum gain of condensation energy on the FS. In order to have a quantitative account for this interplay of magnetic correlations, FS shape and inter- and intra-band coupling, we use a self-consistent summation of bubble and ladder diagrams (fluctuation-exchange approximation [81, 82]) within the Nambu-Gorkov description of the SC state. Both the pairing interaction and the SC gap function will be determined self consistently. Whether superconductivity occurs and whether the gap function possesses nodes on the FS will therefore be the results of our calculation.

The fluctuation-exchange approximation has been successfully used for the investigation of superconductivity in one band models relevant for high temperature superconductors [81, 82, 83]. Our numerical results for the anomalous self energy $\Phi_{\mathbf{k},\pm}(\omega = 0) \propto \Delta_{\mathbf{k},\pm}$ are shown in Fig. 11. We find a stable SC solution, caused by AF spin fluctuations at a temperature $T = 8\,\text{K}$ which exhibits zeroth of the gap-function along the lines $k_{y,z} = 0$. The leading harmonics of the gap-function are given by $\Delta_{\mathbf{k},\pm} \approx \Delta_\pm^0 \sin(k_y/2) \sin(k_z/2)$ with $\Delta_+^0 = -0.9\,\text{meV}$ and $\Delta_-^0 = 1.5\,\text{meV}$. This gives gap nodes for both pieces of the FS. The symmetry of this superconducting state is $\text{d}_{yz}$ which, in addition, is out of phase between the two parts of the FS. The transition temperature $T_c = 13\,\text{K}$ is in fair agreement with the experiment. Recent magneto oscillation experiments by Singleton et al.[84] were able to determine the gap amplitude as function of angle. Their results are in very good agreement with our predictions as far as the position of the gap nodes is concerned.

The physical origin of this pairing state and the close resemblance to the $\text{d}_{x^2-y^2}$ state in cuprates can be understood if, for illustrating purposes, one makes the assumption $t_3 = t_{3'}$ for the hopping elements between different dimer types. Now, the two dimer types, which differ in their orientation of the molecule pairs, are indistinguishable and the unit cell reduces to that indicated by dotted lines in Fig. 10a. Assuming furthermore that the

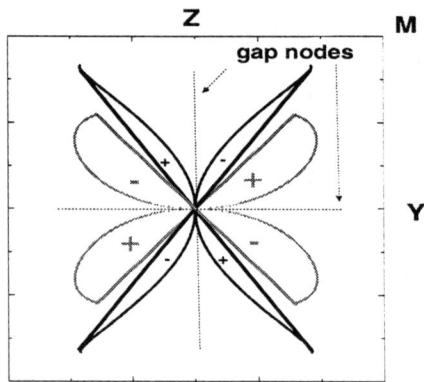

*Figure 11.* Polar plot of the superconducting gap functions $\Delta_{\mathbf{k},\pm}$ of the two bands along the FS. For a given angle the amplitude of the gap corresponds to the distance from the origin.

dimers are arranged on a square lattice, the BZ is doubled and rotated by $\pi/4$. The resulting FS is shown in Fig. 10c. One easily recognizes that the two branches of the FS of Fig. 10b correspond, after down-folding into the reduced BZ, to states separated by the dashed line in Fig. 10c. Since in coordinate space the dominating AF coupling is between nearest neighbors, it causes peaks in the magnetic susceptibility around $(\pm\pi, \pm\pi)$ of the extended BZ, causing anomalies for *hot* quasiparticles, located on FS segments close the corresponding magnetic zone boundary (dashed line in Fig. 10c) [24, 11], which lead to the inter-band coupling in the original BZ of Fig. 10b. As discussed above (see also Refs. [4, 76]), a momentum transfer $(\pm\pi, \pm\pi)$ causes a superconducting state with gap function $\Delta_{\mathbf{k}} \propto \cos k_y - \cos k_z$. Rotation by $\pi/4$ and down folding into the reduced BZ of Fig. 10b yields $\Delta_{\mathbf{k},\pm} \propto \pm \sin(k_y/2)\sin(k_z/2)$ which covers the general behavior of the results presented in Fig. 11. Thus, subject to the modifications brought about by the gap between the two branches of the FS and the additional anisotropy of the underlying lattice, the origin of the pairing state in cuprates and organics is very similar. Due to the important role played by *hot* momentum states for various normal state anomalies of cuprates [24, 11], it is tempting to conclude that the occurrence of two distinct classes of quasiparticles, *hot* and *cold*, depending on the strength of the effective interaction are essential for the unconventional behavior in organics as well.

# 7. SUMMARY AND OUTLOOK

In summary, we have demonstrated that generic features of cuprate super-conductors and organic superconductors can only be understood if takes into account the mutual interaction of the involved degrees of freedom which cannot be treated as being independent and behaving in a mean field type manner. Spin fluctuations change quasiparticle behavior quali-tatively and, in turn, this leads to a qualitative change of the spin fluctu-ations themselves, a kind of behavior which necessarily has to be investi-gated in a self consistent way. Based on these observations we expect that such a system is extremely sensitive to external perturbations or disorder. Phenomena like interaction- induced intrinsic inhomogeneities, droplet or magnetic stripe formation are possible. Here, the inability of the system to find the correct ground state might lead to spatially segregated regions with different local excitation spectra. Because of the enormous amount of experimental data available and because of the many competing inter-actions involved, cuprate superconductors and collosal magneto-resistant manganites are very appealing candidates for an investigation of intrinsic inhomogeneities. Often it is argued that these materials are affected by the formation of charge stripes. However, a simple rigid stripe picture can only be a crude caricature for the actual situation in these systems. In order to understand the rather sharp INS line shape of $La_{2-x}Sr_xCuO_4$, rather rigid charge stripes need to be assumed, a picture which seem to be rather ar-tificial, given that NMR and NQR observation demonstrate a much richer behavior (cluster spin glass regime, charge glass behavior, weak pseudogap behavior). We believe that a coherent motion of spin clusters, accompanied by a rapid charge motion can account much better for the experimental data. Besides cuprates and manganites, other correlated systems which are close to a quantum critical point and which are affected by strong lat-tice coupling, long range interactions or disorder are likely to bring about intrinsic inhomogeneities and non-exponential relaxation. In this context we believe that many concepts, originally developed and successfully ap-plied by the soft condensed matter community, such as the self similarity of the energy landscape or the decsription of competing ground states using replica techniques and replica symmetry,= breaking, will be essential for a further exploitation of the fascinating behavior of strongly correlated hard condensed matter systems as well.

The material presented in these lecture notes has partly been published in Refs. [11, 28, 80] or is taken from a forthcoming publication to appear in Reviews of Modern Physics. C.

## 8. Acknowledgments

This work has been supported in part by the Science and Technology Center for Superconductivity through NSF-grant DMR91-20000, by the Center for Nonlinear Studies and Center for Materials Science at Los Alamos National Laboratory, and by the Deutsche Forschungsgemeinschaft (J.S.). It is our pleasure to thank our colleages A. Balatsky, V. Barzykin, A. Chubukov, P. C. Hammel, A. J. Millis, D. Morr, and P. Monthoux for stimulating discussions. One of us (BPS) wants to acknowlegde a collaboration with Z.G. Yu, A. R. Bishop, A.H. Castro Neto and N. Grønbech-Jensen on reserach presented in part of the ms.

## References

1. J. Bardeen, L. N. Cooper, and J. R. Schrieffer, Phys. Rev. **108**, 1175 (1957).
2. P. A. Lee, T. M. Rice and P. W. Anderson, Phys. Rev. Lett. **31**, 462 (1973).
3. K. G. Wilson, Rev. Mod. Physics **47**, 773 (1975).
4. D. Pines, Z. Phys. B**103**, 129 (1997); Proc. of the NATO ASI on The Gap Symmetry and Fluctuations in High $T_c$ Superconductors, J. Bok and G. Deutscher, eds., Plenum Pub. (1998).
5. H. Y. Hwang et al, Phys. Rev. Lett. **72**, 2636 (1994).
6. C. P. Slichter, in Strongly Correlated Electron Systems, ed. K. S. Bedell et al. (Addison-Wesley, Reading, MA,1994).
7. V. Barzykin and D. Pines, Phys. Rev. B **52**, 13585 (1995).
8. H. Alloul, P. Mendels, G. Collin, and Ph. Monod, Physica C **156**, 355 (1988).
9. N. J. Curro, T. Imai, C. P. Slichter, and B. Dabrowski, Phys. Rev. B **56**, 877 (1997).
10. G. Aeppli, T. E. Mason, S. M. Hayden, H. A. Mook, J. Kulda, Science **278**, 1432 (1997).
11. J. Schmalian, D. Pines, and B. Stojković, Phys. Rev. Lett. **80**, 3839 (1998).
12. A. G. Loeser, et al., Science, **273**, 325 (1996).
13. H. Ding, et al., Nature, **382**, 51 (1996).
14. C. Renner, O. Fischer to appear in Phys. Rev. Letters.
15. A. V. Puchkov, D. N. Basov, and T. Timusk, J. Phys. Condens. Matter 8, 10049 (1996), and references therein.
16. G. Blumberg, M. Kang, M. V. Klein, K. Kadowaki, C. Kendziora, Science **278**, 1374 (1997).
17. R. Nemetschek, M. Opel, C. Hoffmann, P. Müller, R. Hackl, H. Berger, L. Forro, A. Erb, E. Walker, Phys. Rev. Lett. **78**, 4837 (1997).
18. B. Bucher et al., Phys. Rev. Lett. **70**, 2012 (1993).
19. P. Monthoux and D. Pines, Phys. Rev. B **47**, 6069 (1993); ibid **48**, 4261 (1994).
20. P. Monthoux, A. Balatsky, and D. Pines, Phys. Rev. Lett. **67**, 3448 (1993); Phys. Rev. B **46**, 14803 (1992).
21. D. Pines, P. Monthoux, J. Phys. Chem. Solids, **56**, 1651 (1995).
22. A. Millis, H. Monien, and D. Pines, Phys. Rev. B **42**, 1671 (1990).
23. Y. Zha, V. Barzykin, and D. Pines, Rev. B **54**, 2561 (1996).
24. R. Hlubina and T.M. Rice, Phys. Rev. B **51**, 9253 (1995); ibid **52**, 13043 (1995).
25. B. P. Stojković and D. Pines, Phys. Rev. Lett. **76**, 811 (1996); Phys. Rev. B **55**, 8576 (1997).
26. Pengcheng Dai, H. A. Mook, and F. Dogan, Phys. Rev. Lett. **80**, 1738, (1998).
27. F. C. Zhang and T. M. Rice, Phys. Rev. B **37**, 3759 (1988).

28. J. Schmalian, D. Pines, and B. Stojković, to appear in Phys Rev. B.
29. A. Chubukov, D. Morr, and K. A. Shakhnovich, Phil. Mag. B **74**, 563 (1994).
30. M. V. Sadovskii, Sov. Phys. JETP **50**, 989 (1979); J. Moscow Phys. Soc. **1**, 391 (1991). see also R. H. McKenzie and D. Scarratt, Phys Rev. B **54**, R12709 (1996).
31. D. S. Marshall et al., Phys. Rev. Lett. **76**, 4841 (1996).
32. J. A. Hertz, Phys. Rev. B **14**, 1165 (1976).
33. A. J.Millis, Phys. Rev. B **48**, 7183 (1993).
34. A.P. Kampf and J.R. Schrieffer, Phys. Rev. B **42**, 7967 (1990); E. Dagotto, A. Nazarenko and A. Moreo, Phys. Rev. Lett **74**, 310 (1995); M. Langer, J. Schmalian, S Grabowski, and K. H. Bennemann, Phys. Rev. Lett. **75**, 4508 (1995); A.V. Chubukov and D.K. Morr, Phys. Rep., **288**, 355 (1997); R. Preuss, W. Hanke, C. Gröber, and H. G. Evertz, Phys. Rev. Lett. **79** 1122 (1997).
35. J. R. Schrieffer, J. Low Temp. Phys. **99**, 397 (1995).
36. Y. M Vilk and A. M. S. Tremblay, J. Phys. I (France) **7**, 1309 (1997).
37. S. Chakravarty, B. I. Halperin, and D. R. Nelson, Phys. Rev. Lett. **60**, 1057 (1988); ibid. Pys. Rev. B **39**, 2344 (1988).
38. A. V. Chubukov and S. Sachdev, Phys. Rev. Lett. **71**, (1993); A. V. Chubukov, S. Sachdev, and J. Ye, Phys. Rev. B **49**, 11919 (1994).
39. S. Sachdev, A. V. Chubukov, and A. Sokol, Phys. Rev. B **51**, 14874 (1995)
40. S. Ohsugi et al., J. Phys. Soc. Jpn. **63**, 700 (1994).
41. P. Monthoux, and D. Pines, Phys. Rev. Lett. **69**, 961 (1992); Phys. Rev. B **49**, 4261 (1992).
42. Q. Si et al, Phys. Rev. B (1993).
43. V. J. Emery and S. A. Kivelson, Physica C **209**, 597 (1993).
44. H. J. Schulz, J.Phys. *France* **50**, 2833 (1989); J. Zaanen and J. Gunnarsson, Phys. Rev. B40, 7391 (1989); A. R. Bishop *et al.*, Europhysics Letters, **14** ,157 (1991).
45. S. R. White and D. J. Scalapino, cond-mat/9801274.
46. J. Lekner, Physica (Amsterdam) **176A**, 485 (1991).
47. N. Grønbech-Jensen *et al.*, Molec. Phys. **92**, 941 (1997).
48. J. C. Slater, Phys. Rev. **82**, 538 (1951).
49. J. R. Schrieffer, X. G. Wen and S. C. Zhang, Phys. Rev. B **39**, 11663 (1989).
50. B. I. Shraiman and E. D. Siggia, Phys. Rev. B **40**, 9162 (1989).
51. T. Dombre, J. Phys. France **51**, 847 (1990).
52. N. D. Mermin and H. Wagner, Phys. Rev. Lett. **17** 1133 (1966).
53. S. Chakravarty *et al.*, Phys. Rev. Lett. **60**, 1057 (1988).
54. D. M. Frenkel and W. Hanke, Phys. Rev. B **42**, 6711 (1990).
55. J. Bonča and J. E. Gubernatis, Phys. Rev. E **53**, 6504 (1996).
56. E. P. Wigner, Phys. Rev. **46**, 1002 (1934).
57. J. M. Tranquada *et al.*, Phys. Rev. Lett. **70**, 445 (1993).
58. This is a highly degenerate configuration which can be mapped into a classical six vertex model [59]. See, for instance, R. J. Baxter in *Exactly Solved Models in Statistical Mechanics* (Academic Press, London, 1982), pg. 127.
59. B. P. Stojković et al, to appear in Phys. Rev. Letters.
60. G. Aeppli *et al.*, Science **278**, 1432 (1997).
61. J. Bardeen, Phys. Rev. Lett. **45**, 1978 (1980); D. S. Fisher, Phys. Rev. B **31**, 1396 (1985).
62. B. P. Stojković and Niels Grønbech-Jensen, in preparation.
63. M. Lang, Superconductivity Review, **2**,1 (1996).
64. J. Caulfield et al., J. Phys.: Cond. Mat. **6**, 2911 (1994).
65. C. H. Mielke, et al., Phys. Rev. B **38**, R4309 (1997).
66. S. M. De Soto, C. P. Slichter, A. M. Kini, H. H. Wang, U. Geiser, and J. M. Williams, Phys. Rev. B **52**, 10364, (1995).
67. H. Mayaffre, et al., Phys. Rev. Lett. **75**, 4122 (1995).
68. K. Kanoda, et al., Phys. Rev. B **54**, 76, (1996).

69. Y. Nakazawa and K. Kanoda, Phys. Rev. B **55**, R8670 (1997).
70. S. Belin, K. Behnia, and A. Deluzet, preprint, cond-mat/9805354.
71. L. N. Bulaevskii, Adv. Phys. **37**, 443 (1988).
72. H. Kino and H. Fukuyama, J. Phys. Soc. Jpn. **65**, 2158 (1996).
73. K. Kanoda, Physica C **282**, 299 (1997).
74. R. H. McKenzie, Science **278**, 820 (1997); preprint, cond-mat/9802198.
75. D.J. Van Harlingen, Rev. Mod. Phys. **67**, 515 (1995).
76. D. J. Scalapino, Phys. Rep. **250**, 329 (1995).
77. K. Oshima, T. Mori, H. Inokuchi, H. Urayama H. Yamochi, and G. Saito, Phys. Rev. B **38**, 938 (1988).
78. M. Tamura, et al. J. Phys. Soc. Jpn. **60**, 3861 (1991).
79. V. Ivanov, et al., Physica C **275**, 26 (1997).
80. J. Schmalian, Phys. Rev. Lett. **81**, 4232 (1998).
81. C.H. Pao and N.E. Bickers, Phys. Rev. Lett. **72**, 1870 (1994).
82. P. Monthoux and D.J. Scalapino, Phys. Rev. Lett. **72**, 1874 (1994).
83. S. Grabowski, M. Langer, J. Schmalian, and K.H. Bennemann, Europhys. Lett. **34**, 219, (1996).
84. J. Singleton et al., unpublished.

# COMPLEX COOPERATIVE BEHAVIOUR IN FRUSTRATED SYSTEMS

DAVID SHERRINGTON
*Department of Physics, University of Oxford, 1 Keble Road, Oxford, OX1 3NP, U.K.*

## 1. Introduction

In this lecture I shall briefly

(i) introduce and remark on the energy landscape paradigm
(ii) discuss recent progress on non-equilibrium dynamics and thermodynamics of some spin glass models, including extension to include a large ferromagnetic attractor, and applications to glasses without quenched disorder
(iii) describe recent progress on a model system involving competitive cooperation via collectively determined macro-data.

## 2. Energy landscapes

The landscape paradigm describes the structure of control functions relevant to the cooperative behaviour of an assembly of many interacting units [1]. In its simplest form the 'height' corresponds to the energy E in a high-dimensional 'position' or 'phase' space $\{\phi_i\}$; $i = 1...N$, where the $i$ label the units/particles and the $\phi$ their coordinates. Motion in the phase space is determined by changes $\delta E$ under moves $\phi \rightarrow \phi + \delta\phi$.

Examples of the $\phi_i$ are atomic positions, spin orientations and neural firing rates. Moves can be local (eg. spin-flips), or non-local (eg. genetic algorithm); sequential or parallel; deterministic (accepting $\delta E \leq 0$) or stochastic (with probability of acceptance dependent upon $\exp(-\beta\delta E)$. Our interest is in rugged $E(\phi)$, which can lead to complex cooperative behaviour. Ruggedness is normally the consequence of frustration or competition and can arise from quenched random interactions (as in a spin glass), or from an attempt to store many competing attractors (as in a

71

*A.T. Skjeltorp and S.F. Edwards (eds.), Soft Condensed Matter: Configurations, Dynamics and Functionality, 71-82.*
© 2000 *Kluwer Academic Publishers. Printed in the Netherlands.*

neural network), or simply through the existence of many non-equivalent quasi-stable amorphous states (as in a glass).

Rugged landscapes have many hills and valleys. It is thus perhaps not surprising that ruggedness can lead to many different long-lived macrostates. More surprising is the fact that it can be shown that these may not equilibrate, a consequence of the presence also of many saddles in a high-dimensional space [3].

## 3. Methodology

Discussion here is restricted to random sequential dynamics. A standard method to treat equilibrium is via the partition function

$$Z = Tr_\phi \exp(-\beta E); \beta = T^{-1}. \tag{1}$$

An analagous generating functional can be used to consider macrodynamics

$$Z_d = \int \mathcal{D}\phi(t)\Pi\delta \text{ (eqn. of motion)}, \tag{2}$$

where the integral and the product are over all phase space and time, and 'eqn. of motion' is short-hand for the quantity to put equal to zero in describing the microscopic dynamics. Stochastic disorder is averaged. In problems involving quenched disorder it is also convenient to average observables formally over the disorder before performing the microscopic traces, at a cost of introducing replicas or multiple times [2]. Interactions can be replaced by auxiliary fields corresponding to macroscopic order parameters, exactly in the case of infinite-range problems. If furthermore the interaction connectivities are extensive, two replica or two-time order functions suffice and extremal domination leads to closed coupled equations for them.

## 4. The p-spin spherical model

Of particular recent interest have been studies of an apparently highly artificial model system, the $p(> 2)$-spin spherical spin glass [4, 5, 6, 7]. This model is exactly soluble in the large N limit, for both thermodynamics and long-time dynamics, and has highlighted the relevance of a feature known as 1 step replica-symmetry breaking (1RSB). In its dynamics it demonstrates that a one-time macroscopic measure can apparently equilibrate while a two-time quantity shows that in fact equilibration never occurs in the low temperature state; this feature is known as aging and is now taken as a signature of a true thermodynamic glass [8]. For example, the averaged energy E(t) appears to equilibrate after a time of order $\tau_{eq}$, yet the two-time autocorrelation $C(t_w + \tau, t_w) \equiv N^{-1} \sum_i < \phi_i(t_w + \tau)\phi_i(t_w) >$ for $t_w \gg \tau_{eq}$ shows interesting counter-evidence. More specifically, $C(t_w + \tau, t_w)$ first decays to a plateau value after a time $\tau \sim \tau_{eq}$ and apparently behaves in an expected stationary fashion, until $\tau$ becomes of order $t_w$ when it decays further (to zero in the case of a pure spin glass, but to a larger value in a system with an external field [5] or with a self-consistent ferromagnetic component [7]). This second decay, being dependent on the initial waiting time, shows that equilibration has not occurred but rather that the initial waiting time $t_w$ also characterises a 'cage time' during which the system stays near one macroscopic 'state', but after which it moves to another (essentially orthogonal to the first in the absence of a global orienting field). Above a characteristic temperature $T_g$ the aging feature disappears, either by the height of the intermediate plateau reducing to join the final one, or by the time-spread of the plateau becoming $\tau$-dependent and reducing as $T$ is increased until it eventually disappears at a higher temperature $T^*$.

The low temperature aging is reminiscent of that found in many real glassy systems. The feature of a shrinking stationary plateau between $T_g$ and $T^*$ with different decays on either end is reminiscent of the predictions of mode-coupling theory and experimental observations for glasses [9]. This has suggested new ways to model glasses, also below $T_g$.

The dynamical behaviours are correlated with replica-symmetry breaking in the thermodynamics. The model shows 1RSB in its thermodynamic overlap function

$$q^{SS'} = N^{-1} \sum_i < \phi_i >^S < \phi_i >^{S'} \tag{3}$$

where the $< \quad >^S$ denotes a thermodynamic average over macrostate S, which occurs with probability $W_S$. A relevant observable is the overlap distribution

$$P(q) = \sum_{SS'} W_S W_{S'} \delta(q - q^{SS'}). \tag{4}$$

For a simple system $P(q)$ has a single delta function indicating a single macrostate (either with $q = 0$, as in a paramagnet, or $q \neq 0$, as in a ferromagnet). For RSB systems $P(q)$ has more structure indicating non-equivalent macrostates. 1RSB systems have

$$P(q) = x\delta(q - q_0) + (1 - x)\delta(q - q_1) \quad ; q_1 \geq q_0. \tag{5}$$

Here $q_1$ denotes the self-overlap and $q_0$ the mutual overlap of two different macrostates. $q_1$ and $q_0$ are also the values of the plateaux of $C(t_w + \tau, t_w)$. When marginal stability against fluctuations in replica-space is employed

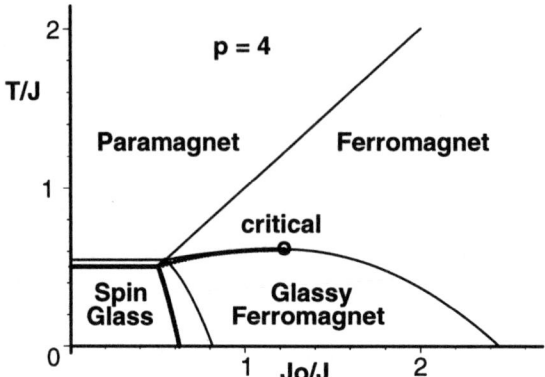

*Figure 1.* Phase diagram of a spherical model with $p = 4$ spin glass interactions of variance $J^2 p!/2N^{p-1}$ and $r = 2$ ferromagnetic interactions $J_0/N$ [7]. The thin phase lines are based on the use of marginal stability/dynamics, the bold lines show a static calculation extremised with respect to $x$. Glassy phases are characterized by RSB and aging. The critical point on the phase line between glassy and non-glassy ferromagnetism separates onsets via discontinuous (smaller $J_0$) and continuous (larger $J_0$) 1 RSB.

as a satisfiability criterion this two-peak feature occurs only beneath $T_g$. Above $T_g$ there is a single delta function at $q_0$. However, there are two scenarios for the onset of the two peaks as $T$ is lowered. In continuous 1RSB the single peak splits continuously with $q_1 - q_0$ growing gradually as $T$ is lowered beneath $T_g$, with $(1 - x)$ jumping discontinuously at the transition. On the other hand, in discontinuous 1RSB $(q_1 - q_0)$ appears discontinuously at a finite value as $T_g$ is crossed, while $(1 - x)$ grows continuously. These two scenarios correlate with those for the plateaux in the auto-correlation function. $x$ also has its counterpart in the dynamics in that one finds a modified fluctuation-dissipation relation

$$\partial_t C(t, t') = T_{eff} G(t, t') \qquad (6)$$

where $G(t, t')$ measures the response of $\phi$ at $t$ to a conjugate field applied at $t'$, with $T_{eff} = T$ in the stationary regime but $T_{eff} = T/x$ in the aging regime. Experiments are underway to test this prediction on real glasses.

A summary of the situation allowing for a 2-spin ferromagnetic interaction as well as a p-spin spin glass interaction is given in [7] and illustrated in Figs 1, 2; the second of these figures also illustrates the long-time aging solution $C(t, t') \sim C(\lambda)$, where $\lambda = t'/t$, with $C(\lambda) = \lambda^\gamma (0 < \gamma < 1)$ [6] in the spin glass phase and for $\lambda \to 1$ the non-analyticity $C(\lambda) = 1 - B(1 - \lambda)^b + O[(1 - \lambda)^{2b}]$ [7] in the glassy ferromagnet phase.

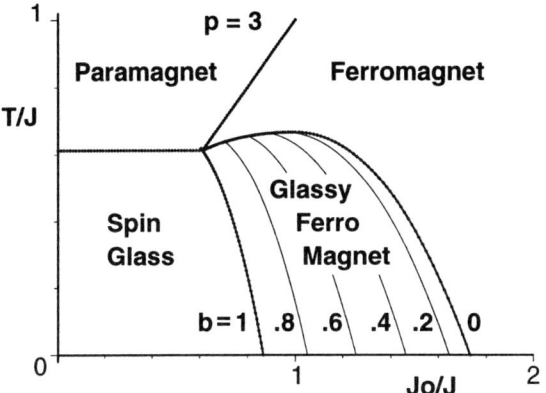

*Figure 2.* Dynamical phase transition lines (solid) for $p = 3$ and lines of constant $b$ in the glassy ferromagnet.[7]

Discontinuous 1RSB also suggests the presence of many mutually orthogonal macroscopic attractor basins in the energy landscape even above $T_g$ [10]. This can be further probed by an examination of the configurational entropy, or complexity,

$$\Sigma(f) = N^{-1} \ln \mathcal{N}(f) \tag{7}$$

where $\mathcal{N}(f)$ is the number of metastable states of free energy density $f$ [11, 12, 4, 13, 14]. Thermodynamic equilibrium is given by minimizing $\Phi = f - T\Sigma$. In the case of discontinuous 1RSB and at low enough temperatures $\Sigma(f)$ is non-zero over some range of $f$, growing monotonically from zero at $f_o$ to a maximum at $f_{th}$, beyond which it is zero again. The temperature at which this behaviour onsets is now usually referred to as $T_{TAP}$ [15]. $T_{TAP}$ is identified with $T^*$ introduced earlier. Above $T_g$ the paramagnet has the lowest free energy. At $T_g$ the states with $f = f_{th}$ become the relevant ones with progressively lower $f$ becoming relevant until a lower temperature $T_S$ at which an entropy crisis occurs ($\Sigma \to 0$) and the global free energy density sticks at $f_0$ [16]. This temperature can also be obtained from the replica thermodynamics by extremizing the free energy with respect to $x$, rather than using marginal stability. Again a comparison can be made with conventional glasses, with $T_K$ the Kauzman temperature (at which the extrapolation of the entropy of a supercooooled liquid would continue below that of the crystal). Additionally, in the region between $T_K$ and $T_g$ the global free energy of the glass is identical with that of the paramagnet.

## 5. Glasses without Hamiltonian disorder

The above picture of discontinuous 1RSB and its implications extends to several other spin glass systems. In particular it extends to many other systems lacking a Hamiltonian symmetry between the consequences of ferromagnetic and antiferromagnetic interactions, first considered for Potts spin glasses [17, 18]. It can also apply to systems without quenched disorder, where $P(q)$ can have similar properties without the need for replicas [19]. This observation, together with another trick [13], has opened the way to application to the analytic study of the low temperature properties of simple Lennard-Jones glasses [20].

One key aspect in the analytical modelling of a glass without Hamiltonian disorder is to recognize that it will have many non-equivalent metastable amorphous states. In an approximation in which the barriers between these states can be considered high they may be specified by their internal free energy densities $f$ and their configurational entropies $\Sigma(f)$, with the global free energy $\Phi = f - T\Sigma(f)$. It seems reasonable to assume that again $\Sigma(f)$ rises monotonically from zero at some $f_0$ to some $f_{th}$. The Kauzman temperature is then given by the temperature at which the thermodynamically relevant $f$, $f^*$, experiences an entropy crisis $\Sigma(f^*) \to 0$ and $f^*$ sticks at $f_0$. Another characteristic temperature is $T_D > T_K$ at which an extensive set of metastable states (labelled by $f_{th}$) becomes thermodynamically relevant; this is identifiable with $T_g$ and marks the onset of long correlation times.

Remarkably, one can obtain not only an approximate value for $T_K$ but also for the free energy in the glassy region $T < T_K$ in terms of an analysis of a model liquid [20]. The procedure employs a consideration of $m$ real replicas restricted to be in the same metastable states [13]. For such a system

$$\Phi(m, T) = \min_f \{mf - T\Sigma(f)\}. \qquad (8)$$

For $T < T_K$ this replicated (or cloned) system has an entropy crisis-driven phase transition at $m = m_c(T)$ and for $m > m_c$ the free energy density is constant at $\Phi = mf_0$. However, for $m < m_c(T)$ the configurational entropy can be calculated from $\Phi$;

$$\Sigma = (m^2/T)\partial(\Phi/m)/\partial m, \qquad (9)$$

$$f^* = \partial\Phi/\partial m. \qquad (10)$$

Assuming furthermore (by analogy with the p-spin glass) that in this region $(m < m_c(T))$ the free energy is that of the corresponding cloned liquid, one can calculate $\Phi$ and $\Sigma$ using liquid-state approximations. Even more, knowing that for $m > m_c$ the system free energy sticks at its freezing value,

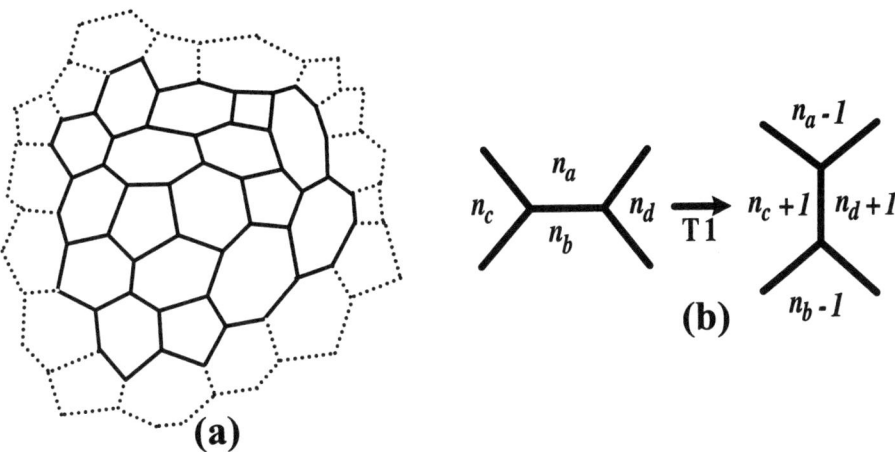

*Figure 3.* (a) A topologically stable cellular pattern (b) A T1 move

one can calculate the free energy of the glass at $T < T_K$ from $\Phi(m_c(T), T)$ determined from the limit of the cloned liquid calculation; recall that the physical glass has $m = 1 > m_c(T < T_K)$. For further details the reader is referred to [20].

## 6. Glass transition in self-organizing cellular structures

The above considerations also prompt a re-assessment of glassy behaviour in continuous networks such as epitomize foams or covalently-bonded amorphous solids. In this brief section I report some recent work on a simple model with purely topological energetic determination [25], devised as the potentially simplest non-trivial model. It is made up of 3-fold vertices with 'arms' paired together to make cells, and hence with the appearance of a two-dimensional foam (Fig 3(a)) and satisfying Euler's theorem that the average number of sides per cell is 6. However, unlike conventional foams or covalent glasses, the energy is taken to be purely topological

$$E \equiv \mu_2 N = \sum_{i=1}^{N}(6 - n_i)^2 \qquad (11)$$

so that lengths and angles are unconstrained. The ground state is thus topologically hexagonal. The system is taken to be subject to T1 dynamics which exchange neighbours between four cells (Fig. 3(b)) but conserve total cell and vertex numbers. These T1 moves are implemented randomly, sequentially and stochastically with probability $(1+\exp(\beta\Delta E))^{-1}$, where $\Delta E$

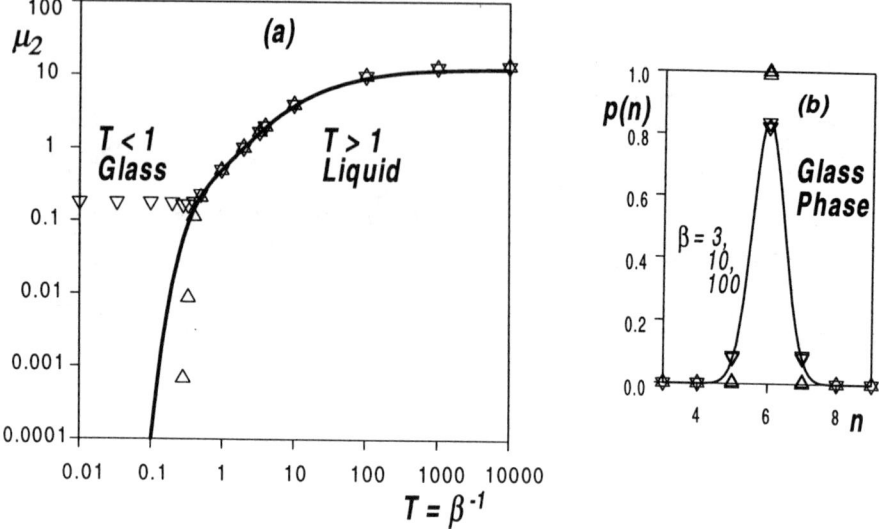

Figure 4. (a) Values of $\mu_2$ versus temperature for patterns with starting configurations fully ordered ($\triangle$) or fully random ($\nabla$), as 'equilibrated' after 1100N steps; N=100172. The curve is a theoretical equilibrium prediction. (b) $p(n)$ at low temperatures in the glassy phase. The curve is the equilibrium value for $\beta = 2.4$ [25].

is the energy change, and moves leading to two-sided or self-neighbouring cells are forbidden.

Two principal issues have been probed. One concerns glassy freezing as demonstrated by different dynamical consequences for one-time global measures of runs at the same temperature but starting respectively from the ordered ground state ($\mu_2 = 0$) or a high temperature random network ($\mu_2 \sim 13$); the second looks for evidence of aging in a two-time quantity. The one-time quantity studied is $p_t(n)$, the distribution of n-sided cells, and particularly the moment $\mu_2$. This shows that for finite but long runs the system equilibrates from either start above some characteristic temperature, but beneath that temperature the examples starting from the disordered state effectively freeze at the $p(n)$ characteristic of equilibrium at that temperature (see Fig. 4). Hence we identify this temperature as the glass temperature $T_g$ associated with the time-scale employed. The two-time quantity studied is a persistence function $C(t_w + \tau, t_w)$ which measures the fraction of cells which have not been involved in an accepted T1 move between $t_w$ and $t_w + \tau$. For $T > T_g$ $C$ is found to be stationary and to decay exponentially with $\tau$. However, for runs performed at $T \ll T_g$ $C(t_w + \tau, t_w)$ depends on $t_w$ and shows a slow decay (see Fig. 5). Thus this simple model exhibits glass-like features, although longer runs will be needed to probe properly regimes with $t_w \gg \tau_{eq}$ and test for issues such as fragility or

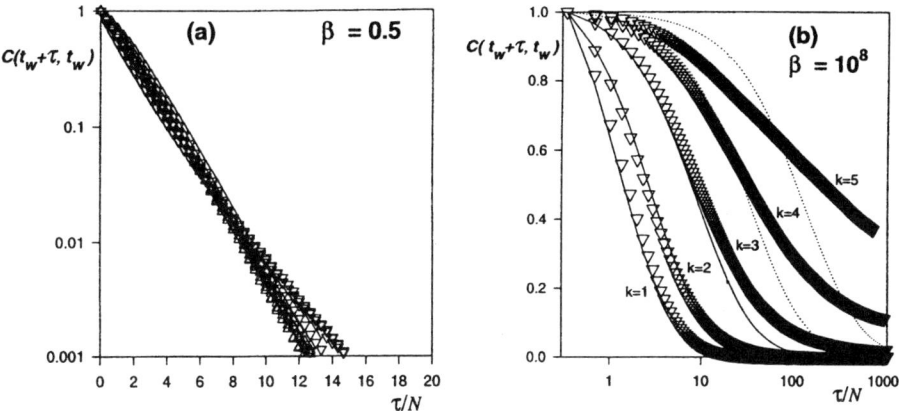

*Figure 5.* Persistence function for $t_w = (4^k)N/3; k = 1...5$, (a) at high temperature, (b) at low temperature [25].

strong-glass character [26].

## 7. Adaptive cooperation in a market

Cooperative behaviour can occur even without direct interaction but via a common macroscopic observable. Frustration enters when the individual objectives are to be in a minority or restricted group, requiring radical diversity in individual responses to the common data. Such problems occur naturally in models of competitive markets. Here we describe briefly recent study of one such model which leads to an unconventional and novel master equation.

This model is an extension of the so-called minority game [21]. In its simplest formulation $N$ agents, each with $s$ randomly chosen but fixed strategies, decide at each time-step to be a buyer or a seller. Overall this results in an excess of buyers over sellers or vice-versa (or occasionally, for $N$ even, a balance). The direction of this excess over the last $m$ steps is the input data on which their strategies act and the choice of which strategy to use is given by awarding 'points' to each strategy which would have been successful given the actual global outcome each time. At each time-step an agent uses his strategy which has the largest number of points. The relevant macroscopic observable to assess the global efficiency of the market is the volatility, the standard deviation of the number of buyers minus sellers. An intriguing observation in simulations is that in the steady-state this quantity becomes significantly less than that given by each agent making a random choice, $(\sqrt{N})$ over a range of $m$ which is neither too low nor too high; in particular for $m$ too low the performance is worse than random, while for $m$ too high it approaches random performance [22].

The problem poses serious challenges for a first-principles analytic solution because it is non-trivially non-local in time. In an effort to move towards such a solution we have extended the original model [23]. First we note a simplification; the global behaviour is identical if the time history on which the agents' strategies act is replaced by a random 'history', provided each agent receives the same information [24]. Second, we employ a continuous formulation in which (i) the space $\Gamma$ of macroscopic data is of dimension $D$, (ii) the possible strategies are vectors $\boldsymbol{R}$ of length $\sqrt{D}$ in the space $\Gamma$, (iii) each agent has $s$ strategies $R_i^r; r = 1...s$, chosen randomly (iv) the strategies act on a unit-length vector $\boldsymbol{\eta}(t)$ in $\Gamma$ chosen randomly each time-step, (v) the response of strategy $\boldsymbol{R}$ is a bid $b(\boldsymbol{R}, t)$ given by $\boldsymbol{R}.\boldsymbol{\eta}(t)$, (vi) the actual strategy used by agent $i$ at time $t$ is $\boldsymbol{R}_i^*(t)$, chosen from his personal suite of $s$ strategies (vii) the total bid at each time-step is $A(t) = \sum_i b_i(t) = \sum_i \boldsymbol{R}_i^*(t).\boldsymbol{\eta}(t)$, (viii) the strategies are weighted with points $P(\boldsymbol{R})$ which evolve according to $\partial_t P(\boldsymbol{R}, t) = -A(t)b(\boldsymbol{R}, t)$. Simulations (Fig. 6) show that this extension maintains the characteristic features discussed earlier. Third, we replace the use of the most successful strategy at each step by a thermal analogue in which the probability that agent $i$ uses his strategy $\boldsymbol{R}_i^a$ is $\Pi_i^a(t) \propto \exp(\beta P(\boldsymbol{R}_i^*, t))$ where the normalization is for each agent independently. Simulations (Fig 6 insert) show that such thermal noise can reduce the market volatility for all $D$ less than that having the minimum volatility in the deterministic case. Indeed for small $D$ it can replace worse than average behaviour by better than average behaviour.

This thermal continuous formulation now provides microscopic analytic evolutionary equations for the $\Pi_i^a(t)$. These have a form

$$\partial_t \Pi_i^a(t) = \beta \Pi_i^a(t)[W_i^a(t) - < W(t) >_i] \tag{12}$$

where

$$W_i^a(t) = -A(t)b_i^a(t) \tag{13}$$

$$b_i^a(t) = \boldsymbol{R}_i^a(t). \boldsymbol{\eta}(t) \tag{14}$$

$$A(t) = \sum_i \boldsymbol{R}_i^*(t).\boldsymbol{\eta}(t) \tag{15}$$

$$< W(t) >_i = \sum_{b=1}^s \Pi_i^b(t)W_i^b(t) \tag{16}$$

with the $\boldsymbol{\eta}(t)$ randomly chosen with $< \boldsymbol{\eta}(t).\boldsymbol{\eta}(t') >= \delta(t - t')$ and the $\boldsymbol{R}_i^*(t)$ also randomly chosen at each time $t$ with probability $\Pi_i^a(t)$. These equations offer the hope of a macroscopic solution via a generating functional as

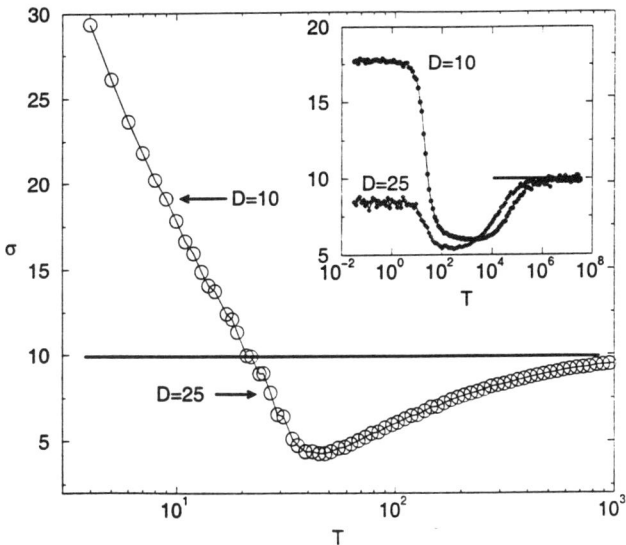

*Figure 6.* Steady state volatility as a function of the dimension $D$ in the continuous minority game for $s = 2$, $N = 100$. The horizontal line shows the result of uncorrelated random choices. Inset: the consequences of temperature $T$ for cases which for zero-temperature are worse than average ($D = 10$) and better than average ($D = 25$) [23].

in Eq.(2), but are significantly non-trivial (and novel) since they involve multiplicative stochastic noise and a second stochastic dependence on the microscopic dynamical parameters themselves, as well as non-linearity and quenched disorder; thus, in spite of its reminiscence of [13] to a conventional replicator equation it is in fact much more complicated. The macroscopic dynamical solution has yet to be performed. Since there is no detailed balance, even a steady state solution presents a challenge.

## 8. Conclusion

I hope that in this brief report I have demonstrated that even simple models can show novel, challenging and potentially relevant complex cooperative behaviour. Their solution can guide extension to more realistic models.

## 9. Acknowledgement

I would particularly like to thank my colleagues Tomaso Aste, Andrea Cavagna, Juan Pedro Garrahan, Irene Giardina, John Hertz and Theo Nieuwenhuizen with whom my recent work reported here was performed, also for valuable and educative discussions and advice. Andrea Cavagna specially prepared Fig. 6 for this paper.

82

## References

1. Sherrington, D. (1997), Physica **D107**, 117.
2. See, e.g. Sherrington,D. (1998) in M.P. Das (ed) "Physics of Novel Materials" World Scientific, 156 (cond-mat 9806289).
3. Cavagna, A., Giardina, I. and Parisi, G. (1998), Phys. Rev. **B53**, 11251.
4. Crisanti, A. and Sommers, H-J. (1992), Z. Phys. **B87**, 341
5. Crisanti, A., Sommers, H-J. and Horner, H. (1993), Z. Phys. **B92**, 257.
6. Cugliandolo, L.F. and Kurchan, J. (1993), Phys. Rev. Lett. **71**, 173.
7. Hertz, J., Sherrington, D. and Nieuwenhuizen, Th.M. (1999), to be published in Phys. Rev. E. Rapid Comm.
8. See, e.g. Bouchaud, J-P., Cugliandolo, L., Kurchan, J. and Mézard, M. (1998), in A.P. Young (ed.) "Spin Glasses and Random Fields, World Scientific (cond-mat 9702070).
9. Goetze, W. (1993) in T. Riste and D. Sherrington (eds) "Phase Transitions and Relaxation in Systems with Competing Energy Scales", Kluwer-Academic, Dordrecht, 191.
10. Kirkpatrick, T.R., Thirumalai, D. and Wolynes, P.G. (1987), Phys. Rev. **A40**, 1045.
11. Tanaka, F. and Edwards, S.F. (1980), J. Phys. **F10**, 1769.
12. Bray, A.J. and Moore, M.A. (1980), J. Phys **C13**, L469.
13. Monasson, R. (1995), Phys. Rev. Lett. **75**, 2847.
14. Cavagna, A., Garrahan, J-P. Giardina, I. (1999), J. Phys. **A32**, 711.
15. Thouless, D.J., Anderson, P.W. and Palmer, R.G. (1977), Phil. Mag. **35**, 593.
16. Both $f_0$ and $f_{th}$ are temperature-dependent but there is no chaotic evolution of states with temperature in the p-spin model. States of $f_{th}(T = 0) \geq f(T = 0) \geq f(T = 0)$ become locally stable at reducing temperatures between $T_M$ and $T_{TAP}$ but $T_g < T_M$.
17. Elderfield, D. and Sherrington, D. (1983), J. Phys. **C16**, L497.
18. Gross, D.J., Kanter, I. and Sompolinsky, H. (1985), Phys. Rev. Lett. **55**, 304.
19. e.g. Chandra, P., Ioffe, L.B. and Sherrington, D. (1995), Phys. Rev. Lett. **75**, 713.
20. Mézard, M. and Parisi, G. (1998), cond-mat 9812180.
21. Challet, D. and Zhang, Y-C. (1997), Physica **A246**, 407.
22. Savit, R., Manuca, R. and Riolo, R. (1999), Phys. Rev. Lett. **82**, 2203.
23. Cavagna, A., Garrahan, J-P., Giardina, I. and Sherrington, D. (1999), cond-mat 9903415.
24. Cavagna, A. (1999), Phys. Rev. **E59**, R3783.
25. Aste, T. and Sherrington, D. (1998), cond-mat 980721.
26. Angell, A. (1995), Science **267**, 1924.

# LINKING THE MESSENGER TO THE PROTEIN, A KEY TO *IN VITRO* EVOLUTION

ALBERT LIBCHABER & SHUMO LIU
*NEC Research Institute, Princeton, New Jersey*
*Rockefeller University, New York*

**Abstract.** Molecular evolution in the laboratory is becoming a feasible process. We show that by physically linking messenger and protein, genotype to phenotype, a protein can be part of an evolution cycle, the time scale being of the order of a day per cycle. The method used is to link the ribosome, the messenger RNA and the nascent protein together and is called ribosome display. We discuss the historical development of the concept, its feasibility and time scale.

## 1. Introduction

Biological evolution has three basic elements: selection, mutation and replication of an individual member of a species. Through the interplay of these elements, the distribution of species evolves. The same ideas can be applied to evolve molecules in vitro and in particular proteins.

We shall simply describe in this article a novel approach to laboratory protein evolution and its possible applications. The basic idea is simple and is based on creating a bound pair mRNA-protein. Mutation is done on the messenger RNA. The messenger is then translated into a protein using the ribosome machinery in vitro but the messenger and the protein are kept bound to the ribosome. The protein then goes through an affinity selection process and the bound mRNA is used for the next selection cycle. This complex cycle looks complicated but is in fact much simpler than one could imagine. The bound pair mRNA-protein is called ribosome display. The display is a clever alteration of the translation machinery from the messenger RNA to the protein. In some eukaryotes, if the stop codon is taken out from the mRNA reading frame, the RNA and the newly synthesized

83

*A.T. Skjeltorp and S.F. Edwards (eds.), Soft Condensed Matter: Configurations, Dynamics and Functionality,* 83-87.

protein are not released from the ribosome at the end of the translation process. The mRNA stays physically coupled to its encoded protein via the ribosome. It forms a complex protein-ribosome-mRNA. This is the beauty of ribosome display, where the "product" is attached to its "blueprint". The protein product is used for selection, and the mRNA blueprint for mutation and amplification. This defines the evolution cycle.

This laboratory approach to evolution of molecules is fairly recent. It was realized, first in 1979, that the nascent polypeptide chain might fold on its active conformation state while still bound to the ribosome [1, 2]. A protein ribosome bound state can thus lead to an effective protein. In other word a protein attached at its end to a substrate can still fold. The next step was to show that the ribosome could be stopped during translation, giving a stable mRNA-protein-ribosome object [3]. Then very large libraries of peptides linked to plasmid complexes were produced, and were called polysome display [4]. Finally display of proteins on prokariotic ribosomes [5] and eukariotic ribosomes [6] was recently demonstrated. Now the technique is used in many applications and review papers are even written [7, 8, 9] showing the time scale of evolution of a concept in molecular biology. The cell translation machinery has been diverted and is used for fast evolution of proteins! Let us now describe in detail the method. One starts with a DNA sequence. As the evolution cycles starts, mutations are introduced in the DNA starting sequence using mutating PCR (a noisy amplifier introducing mutations), in order to broaden the initial distribution of coding sequences. In effect the broadening used is about one mutation for ten bases. In our case the starting sequence codes for a piece of a protein, the NH2- terminal fragment of lacI protein (N-lacI); lacI is the repressor protein in E. coli lactose operon. The NH2-terminal of the protein lacI contains the DNA binding domain. It binds specifically to the DNA sequence lacO [10]. This starting DNA represents a very narrow distribution of coding sequences, a very homogenous gene pool. We will evolve this fragment to increase its binding to the specific DNA sequence it binds to.

## 2. Selection

We then subject the protein to affinity selection. The ligand in the experiment is lacO a sequence of double stranded DNA (dsDNA). Normally, in E coli a lacO site binds to a lacI dimer. Recently the lacI-lacO complex structure has been studied in detail [11]. In this ribosome display experiment, we use only the 65 amino acids of the NH2-terminal, which form a monomer but not a dimer nor a tetramer. This display of lacI NH2-terminal fragment and affinity binding has a counter part in vivo as was experimentally tested [12].

There are several lacI mutant that have higher affinity to lacO [10, 13, 14, 15]. These mutants present point mutations to the starting wild type lacI. They are expected to appear through selection and mutation.

For affinity selection, the ligand is immobilized on the surface of a solid media, such as polystyrene beads or the surface of glass tubing. By mixing the beads and the in vitro translation solution or drawing in and out the solution from the tubing, the protein- ribosome-mRNA complex is brought in contact with the ligand. The proteins of high affinity to the ligand will likely adhere while the low affinity proteins will remain in solution. After a brief wash of the solid media, the low affinity protein and its mRNA are washed away, and the high affinity proteins (in the form of protein-ribosome-mRNA complex) are retained on the solid surface. The mRNA is then eluted from the ribosome complex. This supposes that the low affinity proteins are not present in high concentration. This selection step is delicate to quantify and we have not yet found a precise determination of the selection level. Also the construct is such that mRNA is attached to the protein. This mRNA can hybridized to another mRNA who will then be transported without being selected.

## 3.  Mutation and amplification

After affinity selection mRNA is amplified. The standard way is to use three reaction steps. First the messenger is reversed transcribed into DNA using an enzyme called reverse transcriptase. DNA is then amplified by PCR and finally transcribed back into mRNA using T7 RNA polymerase. Mutations arise in reverse transcription and if more severe mutations are needed mutagenic PCR is used.

An alternative and faster way is to directly amplify mRNA, using a reaction which combines reverse transcription and transcription; this method is often called NASBA, nucleic acid sequence-based amplification [16]. In this reaction, the mRNA (the 5' untranslated region and the coding region) is transcribed into its complementary DNA. This complementary DNA is in turn transcribed into mRNA. A primer with the promoter and the 5' untranslated region is necessary for the transcription. In principle, each mRNA molecule generates one complementary DNA molecule, and each DNA can generate about 50 mRNA molecules. The overall mRNA amplification is about 106. Mutation can be introduced in the amplification.

## 4.  Discussion

Besides the known mutants, other may turn out in the evolution. In fact a large space search may develop. In cells these other mutants may be either prevented or very rare and therefore unnoticed. Evolution in vitro may be an efficient tool to find such mutants. In principle, given enough cycles of evolution, a more adapted solution (higher affinity proteins) will replace the starting one. In an experiment, can such in vitro evolution generate such new distribution? What are the individual elements of the new solution, are they known or are they new proteins?

Mutation rate is an important parameter in evolution. In a given selection, there should be an optimal mutation rate to evolve to the desired mutants.

For small changes, such as from wild type lacI to a higher affinity mutant with only a few amino acids replaced, a low rate of point mutation is more efficient. This situation is like tuning up existing solutions. Very high mutation rate will prevent from converging to the solution.

High mutation rate may be needed for large changes to generate the mutants quickly. The mutation in species evolution is like the energy in a mechanical dynamic system. High mutation rate can disperse the individuals in species space. Perhaps, high mutation in early evolution cycles followed by lower mutation to tune up is more efficient. There is a general problem with PCR under high mutation rates. The primers may hybridize to regions where reasonably complementary sequences are created by the mutations. This, of course, leads to shortened solutions, a cascade of shorter solutions. In this study, the effect of the mutation rates is somewhat qualitative. The rates are set either moderately low ( 1/1000 nucleotides) or high ( 1/10 nucleotides). Deletion, insertion and recombination are also present. These mutations may occur in nucleic acid replication reactions and they present very drastic changes in the protein sequence. They are unintended. Recombination is a very useful process in evolution but we are not addressing it in this study.

The solutions, after several cycles of evolution, are analyzed in two ways: competition with the starting lacI sequence in a new round of evolution and affinity to lacO measured by fluorescence polarization measurement.

In this study, the experimental procedures are adaptation and modification of existing techniques: PCR, in vitro transcription and translation, solid phase affinity selection. As the evolution cycles proceed, some of the techniques are slightly modified. The experimental procedure itself is evolving slowly, in order to be faster and cleaner.

## References

1.  Bergman, L.W. and Kuehl,W.M. (1979) Formation of an intrachain disulfide bond on nascent immunoglobulin light chains, *J. Biol. Chem.* **254**, 8869-8876
2.  Kolb, V.A., Makeyev E.V. and Spirin, A.S. (1995) Cotranslational folding of proteins, *Biochem. Cell Biol.* **73**, 1217-1220
3.  Haeuptle, M.T. Frank, R. and Dobberstein, B. (1986) *Nucleic Acids Res.* **14**, 1427-1448

4.  Mattheakis, L.C., Bhatt, R.R. and Dower, W.J. (1994) An in vitro Polysome display system for identifying ligands from very large peptide libraries, *Proc. Natl. Acad Sci. USA* **91**, 9022-9026

5.  Hanes, J. and Pluckthun, A. (1997) In vitro selection and evolution of functional proteins by using ribosome display *Proc. Natl. Acad Sci. USA* **94**, 4937-4942

6.  He, M. and Taussig. M. J. (1997) Antibody-ribosome-mRNA (ARM) complexes as efficient selection particles for in vitro display and evolution of antibody combining sites *Nucleic Acids Res.* **25**, 5132-5134.

7.  DallAcqua, W. and Carter, P. (1998) Antibody engineering *Curr. Opin. Struct. Biol.* **8**, 443-450

8.  Jermutus, L., Lyubov, A.R. and Pluckthun, A. (1998) Recent advances in producing and selecting functional proteins by using cell-free translation *Curr. Opin. in Biotechn.* **9**, 534-548

9.  Makeyev E.V., Kolb, V.A. and Spirin, A.S. (1999) Cell-free immunology: construction and in vitro expression of a PCR-based library encoding a single-chain antibody repertoire *FEBS Letters* **444**, 177-180

10. Miller, J. H. and Reznikoff W. S. (1990) *The Operon*, 2nd ed., Cold Spring Harbor, NY: Cold Spring Harbor Laboratory

11. Lewis, M., Chang G., Horton N. C., Kercher M. A., Pace H. C., Schumacher M. A., Brennan R. G., and Lu P. (1996) Crystal structure of the lactose operon repressor and its complexes with DNA and inducer *Science* **271**, 1247-1254.

12. Khoury, A. M., Nick H. S. and Lu P. (1991) In vivo interaction of Escherichia coli lac repressor N-terminal fragments with the lac operator *Journal of Molecular Biology* **219**, 623-634

13. Pfahl, M. (1979) Tight-binding repressors of the lac operon: selection system and in vitro analysis *Journal of Bacteriology* **137**, 137-145, J. (1991)

14. Pfahl, M. (1981) Mapping of I Gene mutations which lead to repressors with increased affinity for lac operator *Journal of Molecular Biology* **147**, 175-178

15. Kolkhof, P. (1992) Specificities of three tight-binding Lac repressors *Nucleic Acids Research* **20** 5035-5039.

16. Compton, J. (1991) Nucleic acid sequence-based amplification *Nature* **350**, 91-92.

# A MODEL FOR THE THERMODYNAMICS OF PROTEINS

ALEX HANSEN
*Department of Physics, Norwegian University of Science and Technology*
*NTNU, N–7034 Trondheim, Norway*

AND

MOGENS H. JENSEN, KIM SNEPPEN AND GIOVANNI ZOCCHI
*Niels Bohr Institute and NORDITA*
*Blegdamsvej 17, DK-2100 Ø, Denmark*

## 1. Abstract

We review a statistical mechanics treatment of the stability of globular proteins based on a model Hamiltonian taking into account protein self interactions and protein-water interactions. The model contains both hot and cold folding transitions. In addition it predicts a critical point at a given temperature and chemical potential of the surrounding water. The universality class of this critical point is new.
Key words: Protein folding, protein thermodynamics, cold denaturation, folding pathways

Biologically relevant proteins are macromolecules [1] whose structures are determined by the evolutionary process [2, 3]. The folded conformation of a globular protein is compact, where each amino have highly specific positions. In addition, unlike any other known solids, globular proteins are not really rigid, but are able to perform large conformational motions while retaining locally the same folded structure. Finally, proteins are mesoscopic systems, consisting of a few thousand atoms.

Quantitatively, the peculiarities of this state of matter are perhaps best appreciated from thermodynamics. Delicate calorimetric measurements [4, 5, 6] on the folding transition of globular proteins reveals the following picture. The transition is first order, at least in the case of single domain proteins. The stability of the folded state, i.e., the difference in Gibbs po-

*A.T. Skjeltorp and S.F. Edwards (eds.), Soft Condensed Matter: Configurations,*
*Dynamics and Functionality, 89-99.*

tential $\Delta G$ between the unfolded and the folded state is at most a fraction of $kT_{room}$ per amino acid. This is referred to as "cooperativity". The Gibbs potential difference $\Delta G$, as a function of temperature, is non monotonic: it has a maximum around room temperature (where $\Delta G > 0$ and consequently the folded state is stable), then crosses zero and becomes negative both for higher *and* lower temperatures. Correspondingly, the protein unfolds not only at high, but also at low temperatures. The melting transition under cooling is referred to as "cold unfolding" or "cold denaturation." For temperatures around the cold unfolding transition and below, the enthalpy difference $\Delta H$ between the unfolded and the folded state is negative; this means that cold unfolding proceeds with a release of heat (a negative latent heat), as is also observed experimentally; at the higher unfolding transition, on the contrary, $\Delta H > 0$ which corresponds to the usual situation of a positive latent heat. There are two peaks in the specific heat, corresponding to the two unfolding transitions, and a large gap $\Delta C$ in the specific heat between the unfolded and the folded state (Figure 1). This gap is again peculiar to proteins: usually, for a melting transition $\Delta C$ is small.

It is (however not universally) believed that from the microscopic point of view, the main driving force for folding is the hydrophobic effect; in the native state of globular proteins hydrophobic residues are generally found in the inside of the molecule, where they are shielded from the water, while hydrophobic residues are typically on the surface.

As in other branches of physics, once the thermodynamics of a system is known it is desirable to develop a corresponding statistical mechanics model. In the following, we describe a recently proposed model of this kind that accounts for the strange thermodynamical behavior described above [7, 8, 9, 10]. Its starting point is the simple but appealing "zipper model" [11], which was introduced to describe the helix – coil transition. In this model, the relevant degrees of freedom (conformational angles) are modeled through binary variables. Each variable is either matching the ordered structure (helix), or in a "coiled" state. A related parametrization for the 3-d folding transition has been proposed by Zwanzig [12], describing it in terms of variables $\psi_i$, each of which is "true" (1) when there is local match with the correct ground state, or "false" (0) if there is no match. The term "local" is here defined through the parametrization index $i$. A zipper scenario that deals with the initial pathway of protein folding has been proposed by Dill et al. [13]. We can parametrize this model in the same way as done by Zwanzig by assigning the value one to each of the binary variables $\psi_i$ describing closed contacts in the zipper. Build into the model is that opening and closing of contacts occur in a particular order: They behave as the individual locks in a zipper. This ordering is characterized through imposing the constraints

$$\psi_i \geq \psi_{i+1} \, . \tag{1}$$

The variables $\psi_i$ alone cannot describe the degrees of freedom that become liberated when a portion of the zipper is open. The open part of the zipper may move freely ($\psi_i = 0$) whereas they cannot move in the part of the zipper where the contacts are closed ($\psi_i = 1$). In order to take into account this effect, we introduce a second, independent set of variables $\xi_i$. For simplicity, we also make these variables binary, taking the values 1 or $-B$. We are now in the position to propose a Hamiltonian for this zipper model,

$$H = -\sum_{i=1}^{N} \psi_i \xi_i \, , \tag{2}$$

subjected to the constraints (1). In this formulation, the assumption that all ground state coefficients in $\xi_i$ takes the same value 1, corresponds to a maximal guiding principle, where the protein folds fastest. In reality, one would expect some variation in $\xi_i$, in particular one would expect that good folders assign a somewhat larger binding to the last contact $i = N$. This we explore further in [14] using the van't Hoff quantification of measurement of protein folding cooperativity by Privalov [6].

We note that for any finite value of $B$, parts of the protein may unfold inside the already folded region i.e. in the parts of the zipper where $\psi_i = 1$. In order to prevent this, we assume $B$ to be sufficiently large compared to any other energy scale in the system — in particular $kT$, where $T$ is the temperature — so that the $\xi_i$ variables never assume the value $-B$ as long as $\psi_i = 1$.

We will in the following use this Hamiltonian as a starting point for analyzing the hot and cold denaturation transitions of proteins when dissolved in water [8]. It is awkward to work with the Hamiltonian (2) directly because of the constraints (1). We therefore make a transformation to a different set of variables where the constraints (1) are implicitly taken into account. We define a set of binary, unconstrained variables $\varphi_i$, by the following relation:

$$\psi_i = \varphi_1 \cdots \varphi_i \, . \tag{3}$$

In particular, $\psi_1 = \varphi_1$. In the limit when $B \to \infty$, the Hamiltonian (2) becomes

$$H = -\varphi_1 - \varphi_1 \varphi_2 - \varphi_1 \varphi_2 \varphi_3 - \cdots - \varphi_1 \varphi_2 \cdots \varphi_N \, , \tag{4}$$

where there are no additional constraints [9]. The role of the variables $\xi_i$ — which is to provide entropy to the unfolded part ($\psi_i = 0$) of the zipper

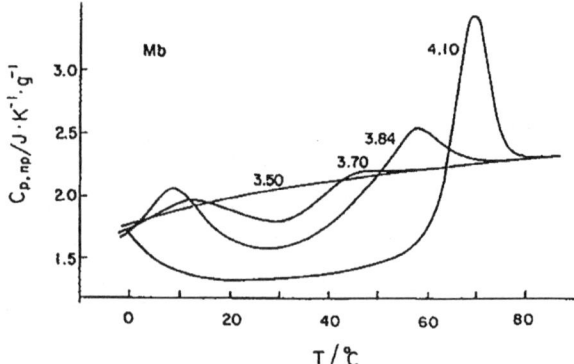

*Figure 1.* Calorimetric measurements of the specific heat of Myoglobin at four different values of pH, as presented by Privalov in ref. [6]. At sufficiently low pH the native structure of the protein never becomes stable, thus the protein remains in its unfolded structure with approximately constant heat capacity over the measured temperature range. By increasing pH the native structure becomes stabilized for intermediate temperatures, defining a transition to an unfolded state at both low and high temperatures, denoted respectively cold and warm denaturation. There is a gap in the specific heat between the folded and the unfolded states.

— is now played by the degeneracy introduced into the Hamiltonian in the following way: When a particular $\varphi_j = 0$, the Hamiltonian (4) will be degenerate with respect to the variables $\varphi_i$ where $i > j$.

The interactions between protein and water may be taken into account by adding to (4) a coupling parametrized through water variables $w_1, w_2, ..., w_N$ [8]. Returning for a moment to the original variables $\psi_i$, we propose an interaction $(1 - \psi_i \xi_i) w_i$. The rationale behind this form is that when a contact is open ($\psi_i = 0$), the part of the protein parametrized by $i$ is exposed to water and interact, while if the contact is closed ($\psi_i = 1$), there is no access to the water and the interaction is zero. Returning to the new variables $\varphi_i$, the resulting Hamiltonian is

$$H = -\mathcal{E}_0 \left( \varphi_1 + \varphi_1\varphi_2 + \varphi_1\varphi_2\varphi_3 + \cdots + \varphi_1\varphi_2 \cdots \varphi_N \right)$$

$$+\left[(1 - \varphi_1)w_1 + (1 - \varphi_1\varphi_2)w_2 + ... + (1 - \varphi_1\varphi_2 \cdots \varphi_N)w_N\right], \qquad (5)$$

where we have introduced a scale parameter $\mathcal{E}_0$ in order to vary the relative strength of the protein self interactions and the protein-water interactions. In order to model hydrophobicity, we assume the $w_i$ variables take values $\mathcal{E}_{min} + s\Delta$, $s = 0, 1, ..., g - 1$. Here, $\Delta$ is the spacing of the energy levels of the water-protein interactions. The equidistant energy levels reflect the experimentally observed approximate constant heat capacity at intermediate temperatures, whereas the finite number of levels $g$ takes into account that protein-water interactions vanish at high temperatures, in practice above

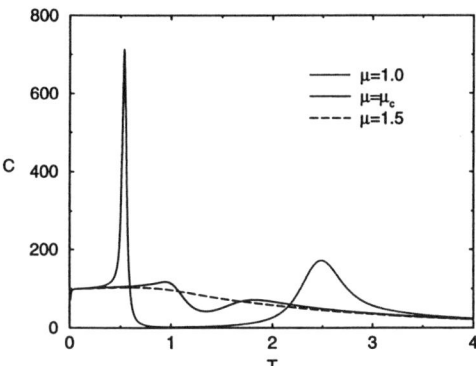

*Figure 2.* Heat capacity, $C$, as a function of $T$ for three values of the chemical potential $\mu$. Here $g = 350$, $\Delta = 0.02$ and $N = 100$. The value $N = 100$ has been chosen as to be close to realistic values for this parameter.

120 degree celsius. eps $N$, each unit will be larger and energies and entropies appropriately increased (inversely proportional to $N$).

The calculation of the partition function is straightforward. We parametrize the states of the system by the number $n$ of consecutive matches $\varphi_1 = 1, \varphi_2 = 1, ..., \varphi_n = 1$ and ending with $\varphi_{n+1} = 0$ and the values $\{s_{n+1}, ..., s_N\}$ where each $s_i \in \{0, 1, 2, ..., g-1\}$ for the $(N-n)$ $\mu$ variables coupled to the unfolded portion of the protein. The energy of this state is

$$\varepsilon(n, s_{n+1}, ..., s_N) = -n\,\mathcal{E}_0 + \sum_{i=n+1}^{N} (\mathcal{E}_{min} + \Delta\mathcal{E}\,s_i) \qquad (6)$$

where we have introduced the energy scale $\mathcal{E}_0$ for the protein variable in order to make the formulas dimensionally more transparent (up to now we used $\mathcal{E}_0 = 1$). Denoting $\beta = 1/T$ as the reciprocal temperature, the partition function is

$$Z = \sum_{n=0}^{N-1} 2^{N-n-1} g^n \sum_{s_{n+1}=0}^{g-1} \sum_{s_{n+2}=0}^{g-1} \cdots \sum_{s_N=0}^{g-1} \exp(-\beta\varepsilon(n, s_1, \cdots, s_N))$$

$$+ g^N \exp(\beta\mathcal{E}_0 N) \qquad (7)$$

In the above equation the factor $2^{N-n-1}$ is the degeneracy of the unfolded protein degrees of freedom and the factor $g^n$ is the degeneracy of water which is not exposed to the inside of the protein. Factorizing the sums over $s_i$ into partition functions $Z_w$

$$Z = \frac{1}{2}(2Z_w)^N \sum_{n=0}^{N-1} \left(\frac{g\exp(\beta\mathcal{E}_0)}{2Z_w}\right)^n + (g\exp(\beta\mathcal{E}_0))^N \qquad (8)$$

94

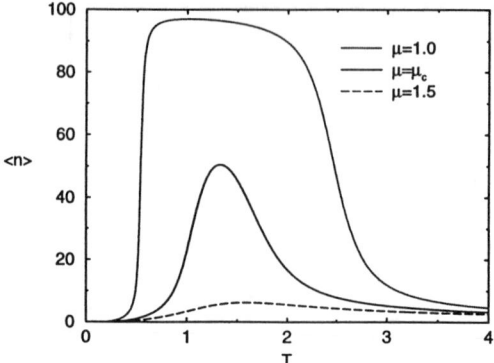

*Figure 3.* Degree of folding, $\langle n \rangle$, as a function of $T$ for three values of the chemical potential $\mu$. The parameters are chosen as in Fig. 2.

where the phase space for a water degree of freedom exposed to a unfolded protein degree of freedom is

$$Z_w = \sum_{s=0}^{g-1} \exp(-\beta(\mathcal{E}_{min} + s\Delta\mathcal{E})) = \frac{(\exp(-\beta\mathcal{E}_{min}) - \exp(-\beta\mathcal{E}_{max}))}{(1 - \exp(-\beta\Delta\mathcal{E}))} \quad (9)$$

where $\mathcal{E}_{max} = \mathcal{E}_{min} + g\Delta\mathcal{E}$. ¿From Eq. 8 one sees directly that the state of the system is determined by the size of the quantity

$$r = \frac{g\exp(\beta\mathcal{E}_0)}{2Z_w} = \exp(\beta\Delta f) \quad (10)$$

If $\Delta f > 0$ then the system will be in the folded state because the sum in Eq. (8) is dominated by the last term, whereas for $\Delta f < 0$ the system will be unfolded.

The sum in Eq. (8) can be readily performed and the total partition function is

$$Z = \frac{1}{2}(2Z_w)^N \frac{1 - (g\exp(\mathcal{E}_0\beta) / (2Z_w))^N}{1 - (g\exp(\mathcal{E}_0\beta) / (2Z_w))} + (g\exp(\beta\mathcal{E}_0))^N \quad (11)$$

The free energy is $F = -T\ln(Z)$, the energy $E = -d\ln(Z)/d\beta$ and the heat capacity $C = dE/dT$. Because there is no pressure in the model, the energy $E$ takes the place of the enthalpy $H = E + pV$ and the free energy $F = E - TS$ takes the place of the Gibbs potential $G = H - TS$.

In Fig. 2 we show the heat capacity the three different choices of $\mathcal{E}_{min}$, representing three different values of the chemical potential as we discuss later. The characteristic feature is that there are two peaks corresponding to warm and cold unfolding, and a gap $\Delta C$ in the heat capacity between the unfolded and the folded form. At higher temperatures, i.e., $T > g\Delta\mathcal{E}$, the gap goes to zero because the water becomes effectively degenerate again. In Fig. 3 we show the order parameter $\langle n \rangle$ as function of temperature for three values of the chemical potential. The figure indeed confirms that the protein is folded between the two transitions.

We now calculate explicitly the difference in the thermodynamic functions between the unfolded and the folded state. We consider these quantities per degree of freedom. The thermodynamic functions associated to a folded (f) protein variable is the energy $e_f = -\mathcal{E}_0$, the entropy $s_f = \ln(g)$ and the free energy $f_f = -\mathcal{E}_0 - T\ln(g)$. The free energy associated to an unfolded (u) protein variable is given by the corresponding partition function of water multiplied by the degeneracy factor of an unfolded part of the protein: $f_u = -T\ln(Z_w\,2)$. The difference in free energy between folded and unfolded state is accordingly

$$\Delta f = f_u - f_f = T\,\ln(\frac{g\,\exp(\beta\mathcal{E}_0)}{2\,Z_w}) \qquad (12)$$

which is the quantity we earlier identified as the one which decides whether the system cooperatively selects the folded or the unfolded state. To clarify the physical contents of this formula we rewrite it for small energy level spacings $\Delta\mathcal{E} << T$:

$$\Delta f = \mathcal{E}_0 + \mathcal{E}_{min} + T\ln(\frac{g\Delta\mathcal{E}}{2\,T}) - T\ln\left(1 - \exp(-(\mathcal{E}_{max} - \mathcal{E}_{min})/T)\right) \quad (13)$$

¿From this expression for the difference in free energy one easily obtains the corresponding differences in energy, entropy and specific heat. In particular, we obtain a gap in the specific heat between the folded and unfolded state of a protein degree of freedom $\Delta c = (\Delta\mathcal{E}/T)^2/(e^{\Delta\mathcal{E}/T} - 1)^2\,e^{\Delta\mathcal{E}/T} \sim \exp(\Delta\mathcal{E}/T) \sim 1$ for temperatures $T \in [\Delta\mathcal{E}, \mathcal{E}_{min} + \mathcal{E}_{max}]$, see Fig. 2.

To simplify the discussion let us consider the limit of large $\mathcal{E}_{max}$ in (13). It is easily seen that $\Delta f$ has a maximum at the temperature $T_m \approx g\Delta\mathcal{E}/2e$. The corresponding value of $\Delta f$ is $\Delta F(T_m) \approx (\mathcal{E}_{min} + \mathcal{E}_0) + g\Delta\mathcal{E}/2e$, so the condition for the existence of a region of stability of the ordered structure $(\Delta f > 0)$ is:

$$\frac{g\Delta\mathcal{E}}{2e} > -(\mathcal{E}_{min} + \mathcal{E}_0). \qquad (14)$$

This is of course always satisfied if $(\mathcal{E}_{min} + \mathcal{E}_0) > 0$, however the more interesting situation is $(\mathcal{E}_{min} + \mathcal{E}_0) < 0$, since then $\Delta F < 0$ at sufficiently low temperature, i.e. the phenomenon of cold unfolding appears. Under these conditions $\Delta E$ is also negative at sufficiently low temperature which means that we have a negative latent heat for cold unfolding.

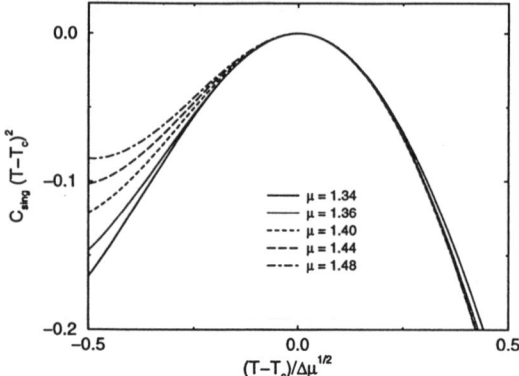

Figure 4. $C_{sing}(T - T_c)^2$ vs. $(T - T_c)/\Delta\mu^{1/2}$. We have chosen $N = 100$, $g = 350$ and $\Delta = 0.02$. Note the good quality of the data collapse in spite of smallness of the system.

Coming back to the partition function (7) and (8), we may write:

$$\mathcal{E} = -N\mathcal{E}_0 + (N - n)(\mathcal{E}_0 + \mathcal{E}_{min}) + \sum_{i=n+1}^{N} \Delta\mathcal{E}\ s_i$$

$$= -N\mathcal{E}_0 + \sum_{i=n+1}^{N} [\Delta\mathcal{E}\ s_i + \mathcal{E}_0 + \mathcal{E}_{min}] \tag{15}$$

and

$$Z = e^{\beta N\mathcal{E}_0} \sum_{n=0}^{N-1} 2^{N-n-1} g^n \sum_{\{s_i\}} e^{-\beta \sum_{i=n+1}^{N} (\mathcal{E}_i - \mu)} + g^N \exp(\beta\mathcal{E}_0 N) \tag{16}$$

where we have set $\mathcal{E}_i = \Delta\mathcal{E}\ s_i$ , $\mu = -(\mathcal{E}_0 + \mathcal{E}_{min})$. ¿From this expression for $Z$ we can identify $\mu$ with the chemical potential *of the water* , or, to be more precise, the difference in chemical potential of the water when it is in contact with the hydrophobic interior of the protein and when it is not. Therefore, $\mu > 0$ is the physically relevant situation. Experimentally, $\mu$ can be changed by adding denaturants, changing pH, etc., which indeed alters the stability of the ordered structure.

For an intermediate value of the chemical potential, $r$ — defined in Eq. (10) — just touches the line $r = 1$, that is $dr/dT = 0$ when $r = 1$, corresponding to a merging of two first order transitions. This defines a critical point. Around this point, $r$ varies quadratically in $T - T_c$ and linearly in $\mu - \mu_c$, as seen from expanding Eq. (10). In experiments of protein folding this point is accessible by changing the pH value of the solution. In fact, Privalov's data on low pH values indeed indicate that such a critical point exists. The scaling properties around this point thus opens for a possibility to gain insight into the nature of the folding process, in particular whether the pathway scheme we suggest can be falsified.

In Fig. 2 we show heat capacity as a function of temperature for chemical potential at the critical value $\mu = \mu_c$. For the chosen values of $\mathcal{E}_0 = 1$ and level density $\Delta = 0.02$ and $g = 350$ the critical point is situated at $T_c = 1.33303\ldots, \mu_c = 1.2838\ldots$. That is, it is situated at a *minimum* of the heat capacity curve. This is at first sight surprising, usually heat capacity has a pronounced increase at the critical point. The minimum reflects a partial ordering, as envisioned in Fig. 3 where we show the degree of folding, counted by the average number of folded variables $\varphi_i = 1, i = 1, ..., n$ from $i = 1$ until the first variable $i = n + 1$ which takes value $\varphi_{n+1} = 0$. The average value of this $\langle n \rangle$ is $N/2$ at the critical point, reflecting that the system is on average half ordered at this point. Correspondingly the heat capacity dips to a value in between the value of an unfolded and a completely folded state.

To characterize the functional form of the dip in the heat capacity, we investigate analytically $C_{sing}(T) = C(T, \mu) - C(T, \mu_c)$ with $\mu >> \mu_c$ for different values of the size $N$. For finite $N$ we may express the singular part of the heat capacity in the form:

$$C_{sing} = |T_c - T|^{-\alpha} \, g\left((T_c - T)N^{1/\nu}\right) \qquad (17)$$

where $g(x) \to const$ when $x \to \infty$ and $g(x) \propto x^\alpha$ when $x \to 0$. We find analytically $\alpha = \nu = 2$ from differentiating the partition function (8).

In terms of experiments on proteins, the relevant scaling behaviour is the how the degree of folding (order parameter) and the heat capacity behaves as function of temperature, when one changes chemical potential away from its critical value. The qualitative prediction is that the width of the singular part of the heat capacity has a minimum at the critical value $\mu = \mu_c$. The broadening of the heat capacity is

$$C_{sing}(T - T_c)^2 = h\left(\frac{T - T_c}{\Delta\mu^{1/2}}\right) \quad for \ \mu > \mu_c \qquad (18)$$

where $h(x) \propto x^{-2}$ for $x \to \infty$ and $h(x) = const$ for $x \to 0$ and where $\Delta\mu = \max(\mu - \mu_c, \Delta\mu_{min})$ with $\Delta\mu_{min} \propto 1/N$ takes into account the finite size sensitivity of the scaling. We show in Fig. 4, an example of such a data collapse. These predictions are experimentally accessible through the use of standard calorimetric techniques, where one should seek to obtain a data collapse above the critical point, i.e. the point of minimal width. The heat capacity below the critical $\mu$ is complicated by the merging of two first order transitions. However, the distance between these moves away from each other in $T$ as $\Delta\mu^{1/2}$.

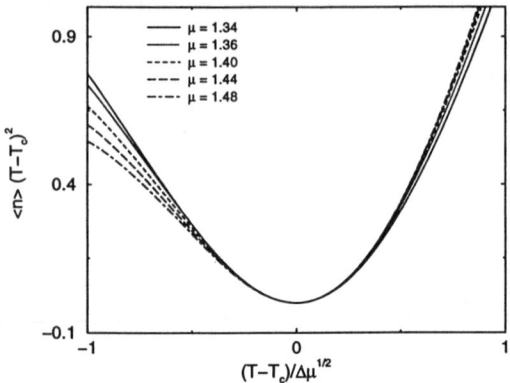

*Figure 5.* $\langle n \rangle (T - T_c)^2$ *vs.* $(T - T_c)/\Delta\mu^{1/2}$. We have chosen $N = 100$, $g = 350$ and $\Delta = 0.02$. Note the good quality of the data collapse in spite of smallness of the system.

Likewise, we expect the degree of folding $\langle n \rangle$ to show data collapse of the form

$$\langle n \rangle (T - T_c)^2 \mu > \mu_c \qquad (19)$$

where $k(x)$ behaves asymptotically as $h$. We show this in Fig. 5. This quantity can be observed experimentally through fluorescence measurements.

In summary we have proposed an effective model that optimizes the rate of self assembly for a protein. The model works with variables that are associated to events along the folding pathways, and thereby differs from models where single amino acids or atoms are the variables. Thus a key problem will be to identify a mapping between the atomic description and the proposed coarse graining. However even without this mapping the model have specific predictions, in particular in connection to the combined description of both hot and cold unfolding of proteins. Furthermore the model opens for a new characterization of good and bad folders in terms of fluctuations in contact energies [14].

# References

1. B. Alberts, D. Bray, J. Lewis, M. Raff, K. Roberts and J. D. Watson, *Molecular biology of the cell* (Garland Publ., New York, 1994).
2. H. Neurath, Science, **224**, 350 (1984).
3. W. Gilbert, Science, **228**, 823 (1985).
4. P. L. Privalov and N. N. Khechinashvili, J. Mol. Biol. **86**, 665 (1974).
5. P. L. Privalov, E. I. Tiktopulo, S. Yu. Venyaminov, Yu. V. Griko, G. I. Makhatadze and N. N. Khechinashvili, (1989) J. Mol. Biol. **205** 737 (1989).
6. P. L. Privalov in *Protein Folding*, ed. by T. E. Creighton (W. H. Freeman, New York, 1992).
7. A. Hansen, M. H. Jensen, K. Sneppen and G. Zocchi, Physica A, **250**, 355 (1998).
8. A. Hansen, M. H. Jensen, K. Sneppen and G. Zocchi, Euro. Phys. Jour. B, **6**, 157 (1998).

9. A. Hansen, M. H. Jensen, K. Sneppen and G. Zocchi, Euro. Phys. Jour. B, **10**, 193 (1999).

10. A. Hansen, M. H. Jensen, K. Sneppen and G. Zocchi, submitted to Europhys. Lett. (1999).

11. J. A. Schellman J. Phys. Chem. **62** 1485 (1958); B. H. Zimm and J. K. Bragg, J. Chem. Phys. **31** 526 (1959); C. R. Cantor and P. R. Schimmel *Biophysical Chemistry*, Chapter 20 (W. H. Freeman, San Fransisco, 1980).

12. R. Zwanzig, Proc. Natl. Acad. Sci. **92**, 9801 (1995).

13. K. A. Dill, K. M. Fiebig and H. S. Chan, Proc. Natl. Acad. Sci. **90** 1942 (1993).

14. A. Bakk, J.S. Høye, A. Hansen, K. Sneppen and M. H. Jensen, preprint (1999).

15. P. A. Lindgård and H. Bohr, Phys. Rev. E **56**, 4497 (1997).

16. H. E. Stanley, *Phase Transitions and Critical Phenomena* (Cambridge Univ. Press, Cambridge, 1971).

# ATTACHMENT AND SPREADING OF MAMMALIAN CELLS IN VITRO

IVAR GIAEVER and CHARLES R. KEESE
*School of Science, RPI, and Applied Biophysics, Inc.*
*Troy NY, 12180-3590 USA*

## 1. Abstract

When mammalian cells are seeded out in tissue culture they will drift down to the surface and firmly attach. The cells attach to a preadsorbed protein layers and not to the naked surface. Normally the protein layer will be a mixture of protein obtained from serum added to the tissue culture medium, but it is also possible to preadsorb pure protein layers onto the surface. This talk will discuss the many factors that affect the interaction of cells with surfaces including the time course of attachment and spreading and the forces that bind the cells to the protein layers.

## 2. Introduction

Biology is a very complex experimental science because it is difficult to control many of the factors influencing the experiment at hand. The field of tissue culture represents a major advancement in the study of animals, because it is much cheaper and simpler to deal with cells than studying the whole animal. Since the 1950's it has been possible to routinely grow mammalian cells isolated from a variety of different tissues and organisms in the laboratory [1]

In tissue culture, cells divide and carry on a variety of biological activities while feeding on a rich nutrient medium that supplies all of the necessary molecules for their survival and growth. Unlike bacterial cultures, most commonly cultured normal mammalian cells, such as fibroblasts, are anchorage dependent, which means they will not grow in suspension but require attachment to a rigid surface. Traditionally the substrate has been glass or polystyrene that has been treated to render it hydrophilic. In most tissue culture work, the medium contains in addition to defined components (salts, sugars, vitamins etc.) a large amounts (often 10%) of serum. Before the cells attach to the substratum, the protein in the serum has already rapidly adsorbed to the surface in a monolayer. It is this protein layer that forms the attachment sites for the cells, and not the bare glass or plastic surface.

*A.T. Skjeltorp and S.F. Edwards (eds.), Soft Condensed Matter: Configurations,*
*Dynamics and Functionality, 101-109.*
© 2000 *Kluwer Academic Publishers. Printed in the Netherlands.*

In order to achieve a spread morphology and to move about on a surface, cells in culture exert forces at their points of attachment. These forces are generated by an intracellular system of muscle-like fibers referred to as microfilaments and composed mainly of the protein actin. The attachment of the cells will be sensitive to the protein adsorbed on the surface, and for the cells to spread successfully the protein layer must be able to withstand the forces exerted by the cells. Several experiments that elucidate cell attachment and spreading on protein layers are discussed in this paper.

## 3. Protein films on liquid surfaces

That cells indeed attach to a protein film and not directly to a solid surface can be demonstrated by culturing cells at a liquid-liquid interface. Under proper conditions cells will attach and spread on such surfaces, and since a liquid can not support a shear force the cells much attach to a protein film at the surface.

In 1964 Rosenberg introduced the use of a fluid substrate for the growth of both transformed and anchorage-dependent cells [2]. In this method, a cell suspension was introduced over fluorocarbon oil having a density greater that that of the aqueous medium, and the cells, being of intermediate density, settled to the interface. Rosenberg found that such interfaces could serve as supports for attachment, spreading, and growth of a variety of cell lines

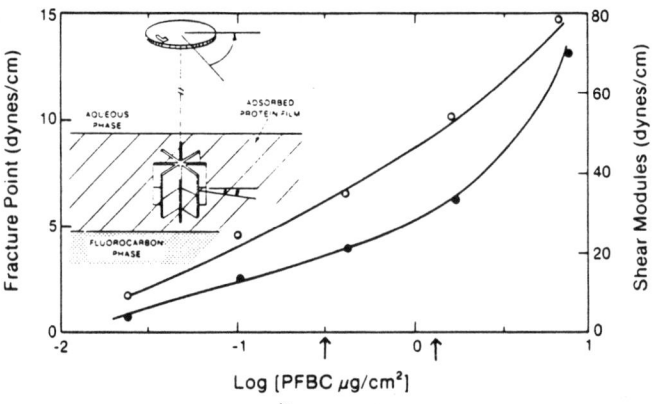

Figure 1. Mechanical measurement of BSA layers reacted with PFBC using a modified viscometer shown in insert. Following 20 hours incubation, torque was applied to the paddle wheel and the angular displacement was measured until the wheel broke free. From the data the fracture point (o) and the shear modulus (•) were calculated. The arrows encompass surface concentration where 100% spreading was observed.

Since this initial observation, we have demonstrated that the adsorbed proteins on highly purified fluorocarbon fluids do not form adequate interfacial substrates unless the oil contains small amounts of specific, surface active compounds. The compound we have found to be most effective in this capacity is pentafluorobenzoyl chloride (PFBC). To produce an interfacial substrate that is adequate for the growth of most fibroblastic mammalian cells, this compound is added to the oil phase to yield a final surface concentration of at least 0.25 micrograms per square centimeter. The necessity for this compound (or similar compounds) has been thoroughly investigated in our laboratory. It is believed that PFBC cross links the adsorbed proteins on the surface and thereby increases the mechanical strength of the layer. The cells will only spread properly when the adsorbed protein layer satisfies the minimal mechanical strength required to sustain the forces involved in cell spreading [3].

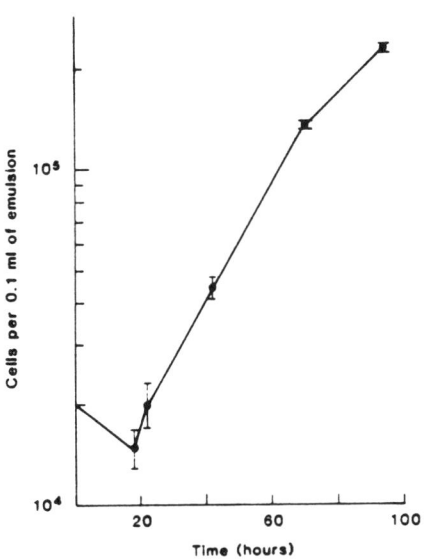

We have investigated the alteration in mechanical properties of the protein layer caused by the PFBC using a modified surface viscometer. The protein film was placed under a shearing stress by the application of a small torque through a torsion wire to "a Teflon paddle wheel" inserted into the interfacial boundary, and the angular deformation of the film was measured. From this data it was possible to obtain stress-strain curves and to determine the surface shear modulus and surface fracture point for the protein layer [4].

Figure 2. Cell growths on liquid microcarriers are exponential in time. Cell were counted at the indicated times by using a citric acid-crystal violet nuclear staining solution for 1 hour. Stained nuclei released from the cells were counted in a hemacytometer.

In most studies, protein was adsorbed to the interface of perfluorotributylamine from either a buffered bovine serum albumin (BSA) solution or culture medium containing 10% serum. Following a 24-hour period, to allow diffusion of PFBC to the interface, measurements were carried out. The results are summarized in figure 1 where the conversion of an essentially fluid layer of adsorbed protein to an elastic film by the presence of the acid chloride is shown.

## 4. Liquid Microcarriers

By utilizing fluorocarbon fluids containing pentafluorobenzoyl chloride, a liquid microcarrier system has been developed capable of use with a variety of cell types including normal human fibroblast. In this configuration, cells on the surface of a coarse oil dispersion (approx. 150-micrometer diameter) exhibit exponential growth as shown in figure 2. In addition, a microcarrier based on silicone oil has been formed and used to culture mouse fibroblasts [5].

These novel interfacial substrates allow manipulation of cells in culture in ways previously not possible using solid-substrate culturing. The liquid microcarrier system permits mass culturing of cells in arrangements commonly used with solid microcarriers, but also allows one to mechanically harvest the cells by breaking the dispersion into its two component phases for example by centrifugation. Such an arrangement may prove to be of particular value in studies involving purification of large quantities of receptors and other surface molecules, where chemical methods of cell harvesting could damage the components of interest.

## 5. ECIS: electric cell-surface impedance sensing

It is possible to detect small differences in cell-substrate interactions using weak electric fields and in this manner to quantitatively measure the dynamics of cell attachment and spreading to defined protein monolayers. The details of the system have been previously described [6,7,8].

In brief, cells are cultured on gold electrodes under standard tissue culture conditions. To minimize the effect of solution resistance, the system is designed with

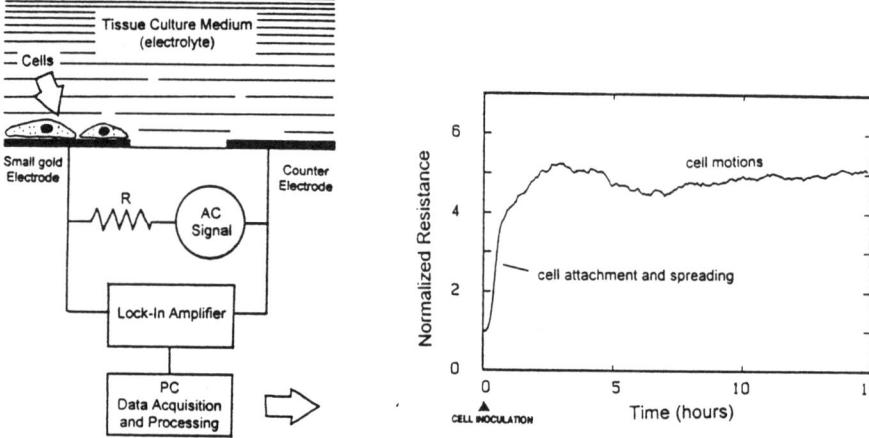

Figure 3. Schematic diagram of the ECIS principle. The heart of the system is the small electrode that acts as a bottleneck in the circuit, and the cells on this electrode causes the increased impedance.

one small electrode (area $0.0005cm^2$) and one large electrode. The small electrode acts as a bottleneck and dominates the measured impedance. Applying a 1.0 µA AC current (normally at 4000 Hz) produces a voltage drop at the boundary of the solution and the small electrode of a few mV. Under these conditions, there are no detectable effects of the electric fields on cell morphology, length of generation time, etc. As fibroblasts attached and spread on this surface, the impedance increases as the cells block the current. Because of the insulating cell membranes most of the small current is now forced to flow underneath and between the cells. Figure 3 shows a schematic drawing of the ECIS principle.

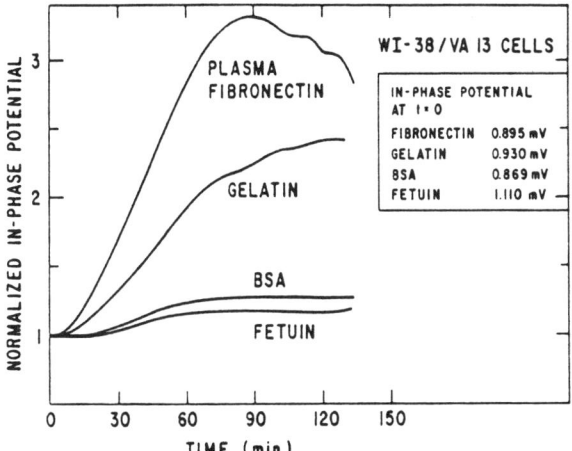

Figure 4. A normalized in-phase potential measured during attachment of cells to electrode surfaces precoated with different proteins. As seen the cells prefer fibronectin to the other proteins. However, after about 20-30 hours the in-phase potential will be similar on all the four electrodes.

We have applied this new means of monitoring cell behavior to study the interaction of cells in culture medium with defined layers of adsorbed protein. Before the addition of the tissue culture medium containing serum, the small electrode was exposed for 15 minutes to a 100 µg/ml solution of a selected protein. Following protein adsorption, the electrode was thoroughly rinsed and inoculated with a fibroblast suspension. Figure 4 presents data obtained when electrodes coated with adsorbed

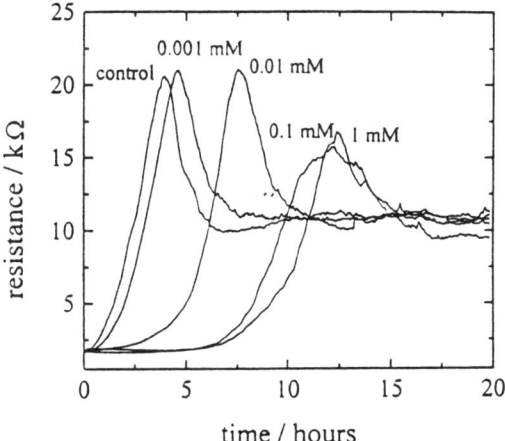

Figure 5. MDCK cells seeded out in a serum free medium onto a fibronectin covered surface. Increasing concentration of RGDS peptides inhibits cell attachment and spreading because the peptide competes with similar sites on the fibronectin.

layers of plasma fibronectin, gelatin, BSA and fetuin were inoculated with WI-38/VA 13 cells; a transformed (cancerous like) cell line derived from WI-38, a normal human lung fibroblast. As can be seen, there was a pronounced difference in the response of the cells to each specific protein layer. Although the initial rate of change in the resistive component of the impedance was greatly reduced for BSA and fetuin, eventually the final impedance level, and hence cell-substrate interaction, appeared to be equivalent [9].

The rate of impedance increase will be different for different cell lines, but common to all fibroblastic cell lines tested so far is that the rate of attachment is most rapid on fibronectin covered surfaces. This is not surprising, as fibronectin has long been the leading prototype of "a glue" that connects cells to a surface. That cells attach using specific recognition sites on the fibronectin molecule can be shown by inhibiting cell attachment using small peptides that binds to the same integrins (molecules on the cell surface involved in cell attachment) as fibronectin. As seen from figure 5 the concentrations of the peptide Arg-Gly-Asp-Ser (known also as RGDS) have al large

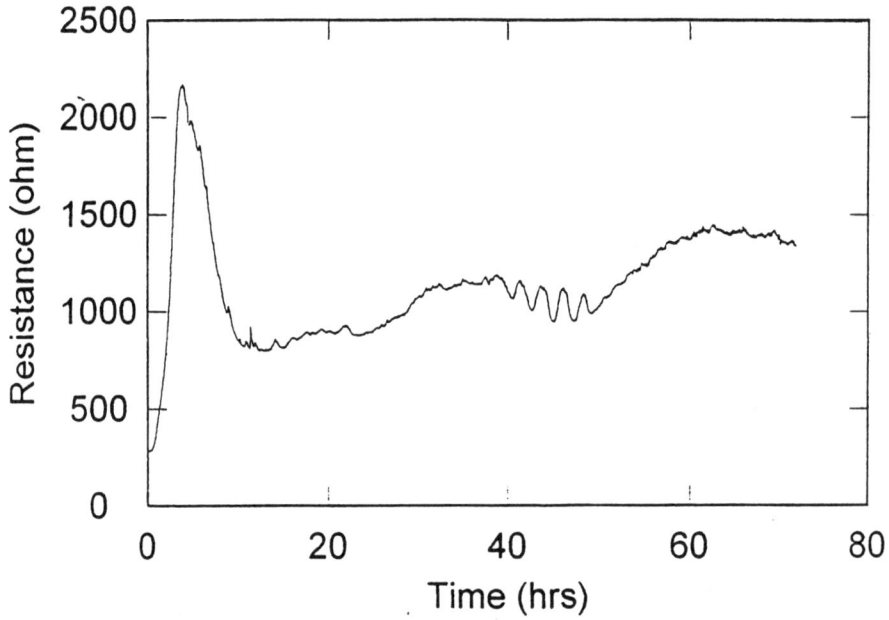

Figure 6 MDCK cells seeded out in serum. After about 40 hours in culture the cells will oscillate for about 10 hours, for again to revert to random micromotion.

influence on the time it takes cell to attach and spread.

In addition to studies involving the dynamics of cell attachment and spreading on protein-coated substrates, the ECIS system has also been employed to study cell locomotion and changes in cell morphology. Figure 6 is an example of both fluctuation and regular oscillation caused by MDCK cells in an ECIS assay. That these fluctuations are related to motion has been confirmed with model calculations and also by the use of drugs [10].

## 6. Forces binding the cells to the protein layers

It is of great interest to measure the forces that bind spread cells to the adsorbed protein layers. This was accomplished by using a modification of the atomic force microscope referred to as the manipulation force microscope. This instrument was first used to measure adhesion forces of protein-covered silica spheres adsorbed to polystyrene surfaces [11]. It was later used to measure the force necessary to displace cells attached to various substrates [12]. In figure 7 is shown the effect that various adsorbed proteins have on the attachment of cervical carcinoma cells. The cells were seeded either in a medium containing serum on clean surfaces (giving a surface with adsorbed serum proteins) or on surfaces where BSA, fibronectin, or laminin had been preadsorbed. The median adhesion forces are shown as a function of time. As can be seen the cells attach faster and stronger on fibronectin than on any of the other protein layers. Also both the force and the

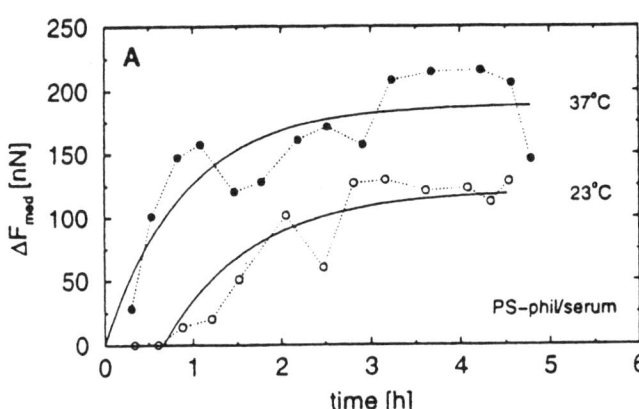

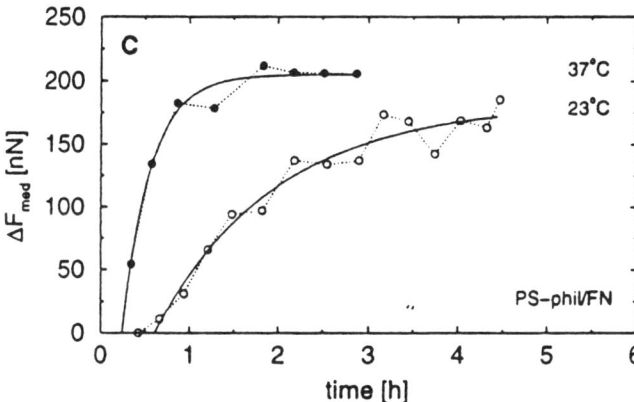

Figure 7. Attachment on a fibronectin-covered surface is shown at right and to serum protein covered surface on the left for two different temperatures. As can be seen the attachment and spreading is faster and somewhat stronger on fibronectin.

speed of attachment is much slower at 23°C than at 37°C as would be expected.

## 7. Discussion

Most mammalian cells are anchorage dependent in tissue culture; this means that they need to attach and spread on a surface to undergo cell division. Exactly how cells attach is not known in detail, some of the experiments described here try to elucidate this problem. From the liquid-liquid interface data, it is clear that cells attaches to a protein film formed at the interface. From the ECIS data it is clear that some protein are more desirable than other proteins. It is also shown that cells attaches to specific sites on fibronectin by inhibit cell attachment with small peptides. The force data also clearly show that fibronectin is the "best" substrate. When serum is present, cells will attach on practically any surface; the only exception known to us is the Fc fragment of specifically bound antibody. The most probable explanation is that fibronectin or other attachment molecules either replace or nonspecifically attach to the absorbed protein layer on the surface, forming specific attachment sites for cells. It is of great importance to understand these problems in detail, as they have bearing on wound healing, metastatic behavior of cancer cells, and embryology.

## 8. Acknowledgment

This work was performed pursuant to a contract with the National Foundation for Cancer Research. We are grateful to Dr. Joachim Wegener for the data in figure 5 and for useful discussions of some of the content in this paper.

## 9. References

1.   Freshney,R.I. (1983) Culture of Animal Cells: A .Manual of Basic Techniques, Alan R. Liss, Inc. New York
2.   Rosenberg, M. D.(1964), in Emmelott, P. and Muhlbock, O. (eds.) Cellular Control Mechanisms and Cancer Elsevier, Amsterdam, pp 146-164.
3.   Keese, C. R. and Giaever, I. (1983) Cell growth on liquid interfaces: Role of surface active compounds, Proc. Natl. Acad. Sci. USA 80 pp5622-5626
4.   Keese, C. R. and Giaever, I (1991) Substraté Mechanics and Cell Spreading, Exp. Cell Research·195, pp 528-532
5.   Keese, C. R. and Giaever, I, (1983) Cell Growth on Liquid Microcarriers Science 219 pp1448-1449
6.   Giaever, I. and Keese, C. R. (1984) Monitoring fibroblast in tissue culture with an applied electrical field Proc. Natl. Acad. Sci. USA 81 pp3761-3764
7.   Giaever, I. and Keese, C. R. (1984) Use of electrical fields to monitor dynamical aspects of cell behavior in tissue culture. IEEE Trans. Biomed. Eng. 33 pp 242-246
8.   Giaever, I. and Keese, C. R. (1991) Micromotion of mammalian cells measured electrically Proc. Natl. Acad. Sci. USA 88 pp7896-7900
9.   Giaever, I. and Keese, C. R (1989) Fractal motion of mammalian cells, in Aharony, A. and Feder, J. Fractals in Physics North-Holland, Amsterdam pp128-133
10.  Mitra, P. Keese, C. R. and Giaever, I. (1991) Electrical Measurements Can Be Used to Monitor the Attachment and Spreading of Cells in Tissue Culture BioTechniques 11 pp504 -510

11.  Sagvolden, G. Giaever, I. And Feder, J.(1998) Characteristic Protein Adhesion Forces on Glass and Poluystyrene Substrates by Atomic Force Microscopy Langmuir 14 pp5984-5987
12.  Sagvolden, G. Giaever, I. Pettersen, E. O. and Feder, J. (1999) Cell adhesion force microscopy Proc. Natl. Acad. Sci. USA 96 88 pp471-476

# 'SAUSAGE-STRING' PATTERNS IN BLOOD VESSELS AT HIGH BLOOD PRESSURES

P. ALSTRØM[1], R. MIKKELSEN[1],
F. GUSTAFSSON[2] AND N.-H. HOLSTEIN-RATHLOU[2]
[1] *CATS, The Niels Bohr Institute,*
*DK-2100 Copenhagen, Denmark*
[2] *Department of Medical Physiology, The Panum Institute,*
*DK-2200 Copenhagen, Denmark*

ABSTRACT. At high blood pressure conditions, 'sausage-string' patterns of alternating constrictions and dilatations may form in blood vessels. We propose that a new Rayleigh-type instability explains the pattern formation. Our theory provides predictions for the conditions under which the normal cylindrical geometry of a blood vessel becomes unstable. The theory is related to experimental observations in rats, where high blood pressure is induced by intravenous infusion of angiotensin II.

## 1. Introduction

10% or more of the population in Western societies has high blood pressure and roughly half recieves pharmacological treatment. Despite the development of potent blood pressure lowering drugs, elevated blood pressure remains a major risk factor for development of stroke and heart disease. High blood pressure is primarily caused by an elevation of resistance to blood flow in the circulation. This elevation in resistance is in turn induced by an increase in the tone of the small arteries and arterioles (diameter $30\text{-}300\mu$) by contraction of the smooth muscle cells surrounding the vessels. The contraction may be induced by several vasoconstricting agents, one of the most potent being the octapeptide angiotensin II. Standard medication for high blood pressure includes drugs that lower the production or effects of vasoconstricting agents on the smooth muscle cells.

Patients with a kidney disease or pregnant women suffering from preeclampsia are in danger of suffering from a sudden and extreme increase in blood pressure. The underlying large increase in arterial resistance may be accompanied by severe damage to the circulatory system and may result in irreversible and even fatal organ dysfunction, particularly of the kidneys and the brain. In almost every case the increase in blood pressure is accompanied by considerable eleva-

*A.T. Skjeltorp and S.F. Edwards (eds.), Soft Condensed Matter: Configurations,*
*Dynamics and Functionality,* 111-126.
© 2000 *Kluwer Academic Publishers. Printed in the Netherlands.*

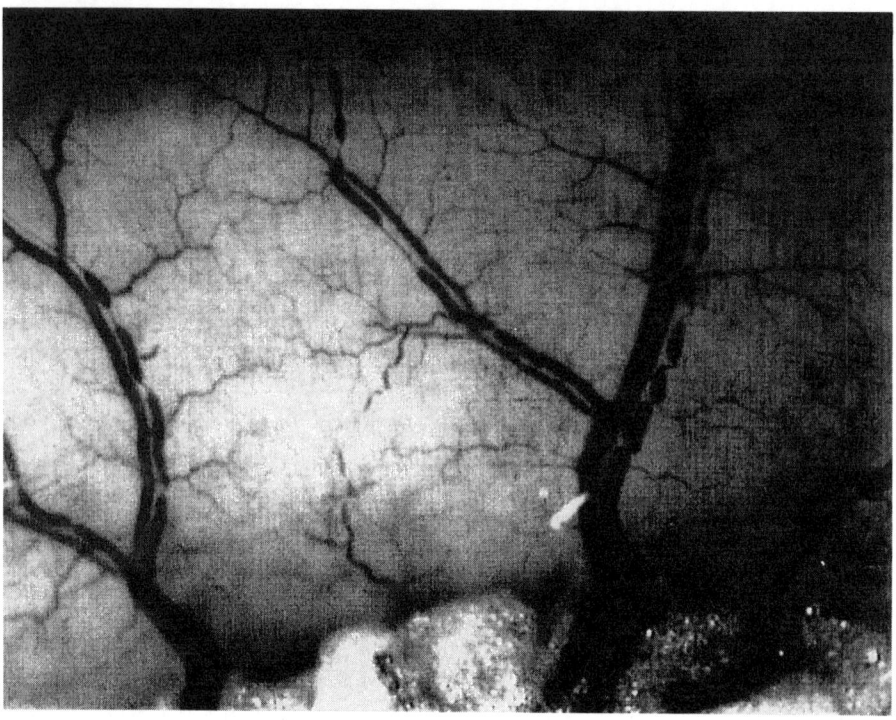

Figure 1: *In vivo* micrograph of rat intestinal arterioles showing a typical 'sausage-string' pattern following an acute increase in blood pressure induced by intravenous infusion of angiotensin II. The neighboring vessels not showing constrictions and dilatations are the corresponding venules. From [2].

tions of the blood levels of angiotensin II [1]. The vascular damage associated with substantial increase in blood pressure is confined to small arteries and arterioles, and it is preceded by a peculiar vascular reaction pattern, where the blood vessels develop alternating constriction and dilatation, giving the vessel a "sausage-string appearance" (Fig. 1). This pattern has been demonstrated in many organs, including the brain, the gut, and the kidney [2], and it is essential for the subsequent development of vascular damage, since damage to the vascular wall occurs exclusively in the dilated regions of the vessel [3, 4, 5].

In the experimental studies presented here of extreme high blood pressures in rats, high blood pressure is induced by intravenous infusion of angiotensin II [3, 4, 5]. When the infusion is at a sufficient level, the vessels may develop the sausage-string pattern (Fig. 1). Despite several decades of experimental research of the phenomenon, the mechanism causing the 'sausage-string' pattern has remained unknown [2]. It has been speculated that the pattern is due to a 'blow out' of the vessel wall caused by the high blood pressure [6]. However, the sausage-string pattern is observed only in the small arteries and large arterioles (Fig. 2), and here the pressure elevation is relatively small compared to that in the larger arteries, cf. Fig. 3. Also, we find that the phenomenon is highly reproducible, see also [4]. If the infusion of angiotensin II is stopped, the normal, uniform cylindrical geometry is restored without remaining deformations, as generally would be expected if the phenomenon was a breakdown due to mechanical failure of the elastic tissue. Restoring the infusion, the sausage-string pattern reappears. Another intriguing feature of the pattern is its overall periodicity with constrictions and dilatations occurring in an almost regular and repetitive pattern. We argue that the sausage-string pattern is caused by an instability, and not by a mechanical breakdown.

Here, we use a simple anisotropic, elastic model of the vessel wall. We show that under certain hypertensive conditions a new type of Rayleigh instability occurs which leads to a periodic sausage-string pattern of constrictions and dilatations along the vessel. Our theory provides predictions for the conditions under which the cylindrical form of a blood vessel becomes unstable. In accordance with experimental observations, we show that the appearance of the sausage-string pattern is limited to smaller blood vessels, because the pressure elevation there is relatively small (Fig. 3). In addition, we find that the instability only occurs for vessels where the wall-to-lumen ratio (vessel-wall thickness divided by the inner radius) is sufficiently small. In agreement herewith, the sausage-string pattern has not been observed in the small arterioles where the wall-to-lumen ratio is 0.3 - 0.5, compared to a value of 0.1 - 0.2 for larger blood vessels. The borders of the instability window, where the sausage-string pattern appears, are determined by two different conditions, one related to the pressure, the other related to the geometry.

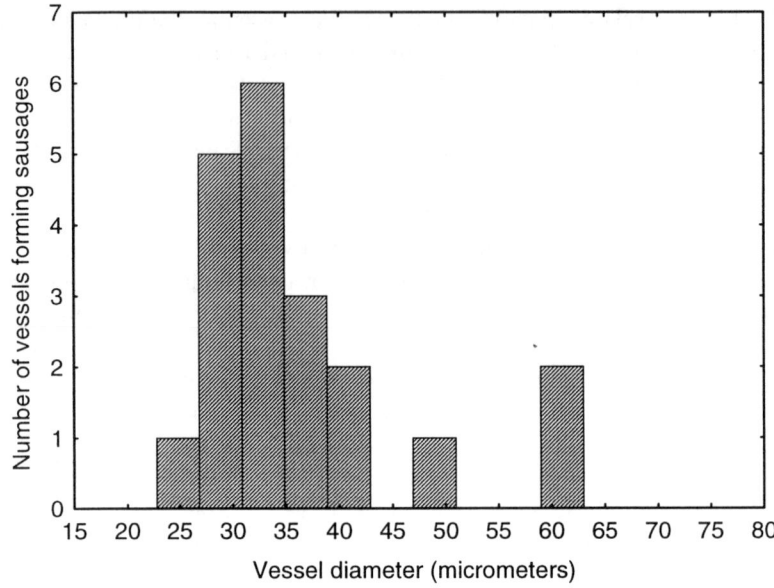

Figure 2: Size distribution of blood vessels undergoing the sausage-string formation, from observations of gut arterioles (15-80 $\mu$m) in rats with high blood pressure.

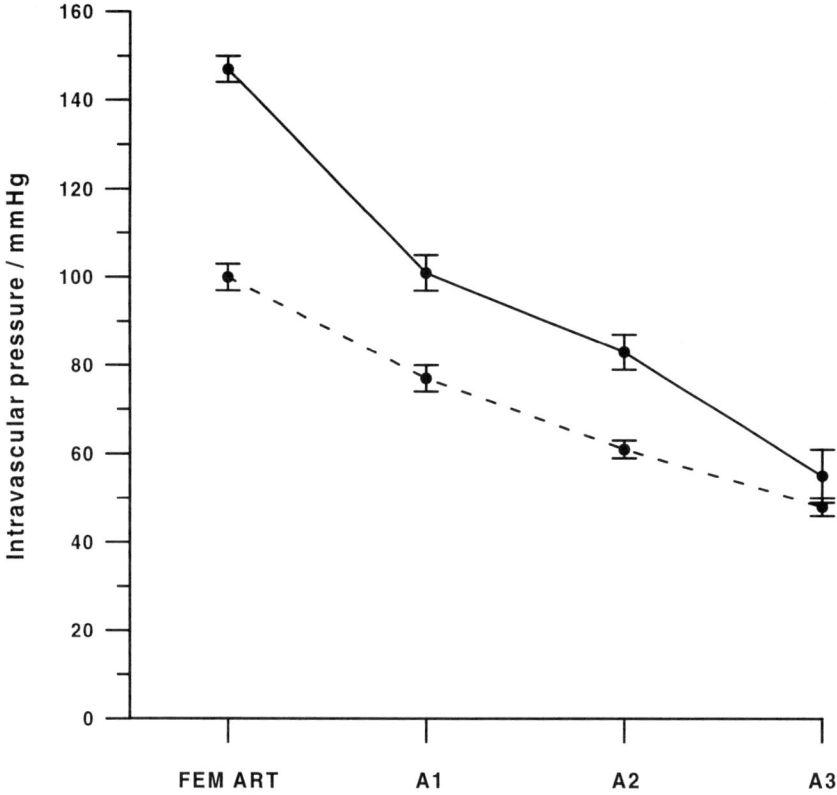

Figure 3: Intravascular pressure in a large (femoral) artery (FEM ART) of diameter $\sim 1$ mm, and in three categories of gut arterioles (A1-A3) in normal rats (dashed line) and rats with high blood pressure (solid line). Mean resting vessel diameters were: A1: 78 $\mu$m; A2: 30 $\mu$m; A3: 15 $\mu$m. From [7].

## 2. Instability Theory

To investigate the stability of a cylindrically shaped blood vessel, we consider an initial small axial symmetric perturbation of the inner radius, $r \to r + u(z)$ (Fig. 4a). If the perturbation grows in time, the cylindrical form is unstable, if it decreases toward zero, the cylindrical form is stable. To determine the stability, we must know the dynamic equation for the perturbation $u(z,t)$. First, we invoke the continuity equation,

$$\partial_t(\pi r^2) = -\partial_z J . \tag{1}$$

A change of cross-sectional area at a downstream site $z$ gives rise to a fluid flux $J(z)$. The flux is related to the transmural pressure $P$, by

$$J = -c(r)\partial_z P . \tag{2}$$

Here $c(r)$ is the fluid conductance, which generally depends strongly on $r$ (in the Hagen-Poiseuille approximation, the fluid conductance is $c(r) = \pi r^4/(8\eta)$, where $\eta$ is the dynamic viscosity of the blood). The specific form of $c(r)$ is not important for our purpose.

From the continuity equation and the flux-pressure relation, the equation of motion to lowest order in the perturbation follows,

$$\partial_t u = \frac{c(r)}{2\pi r}\partial_z^2 P . \tag{3}$$

We have neglected the pressure drop along the vessel because this is much smaller than the transmural pressure.

Consider first a very thin vessel wall, for which the pressure is given by the Laplace form [8, 9],

$$P = (T/R) + (T_z/R_z) . \tag{4}$$

$T$ and $T_z$ are the tensions in the principalangular and vessel directions, and $1/R$ and $1/R_z$ are the corresponding curvatures,

$$\frac{1}{R} = \frac{1}{r[1+(\partial_z r)^2]^{1/2}} , \quad \frac{1}{R_z} = \frac{-\partial_z^2 r}{[1+(\partial_z r)^2]^{3/2}} . \tag{5}$$

Next, assume that the tensions are constant and identical, $T_z = T$. From the above expression, Eq. (4), for the pressure, we get from Eq. (3), keeping only linear terms in $u$,

$$\partial_t u = -\frac{Tc(r)}{2\pi r^3}[\partial_z^2 u + r^2\partial_z^4 u] . \tag{6}$$

For a perturbation of the form

$$u = \sum_k u_k(t)\cos(kz) , \tag{7}$$

(a)

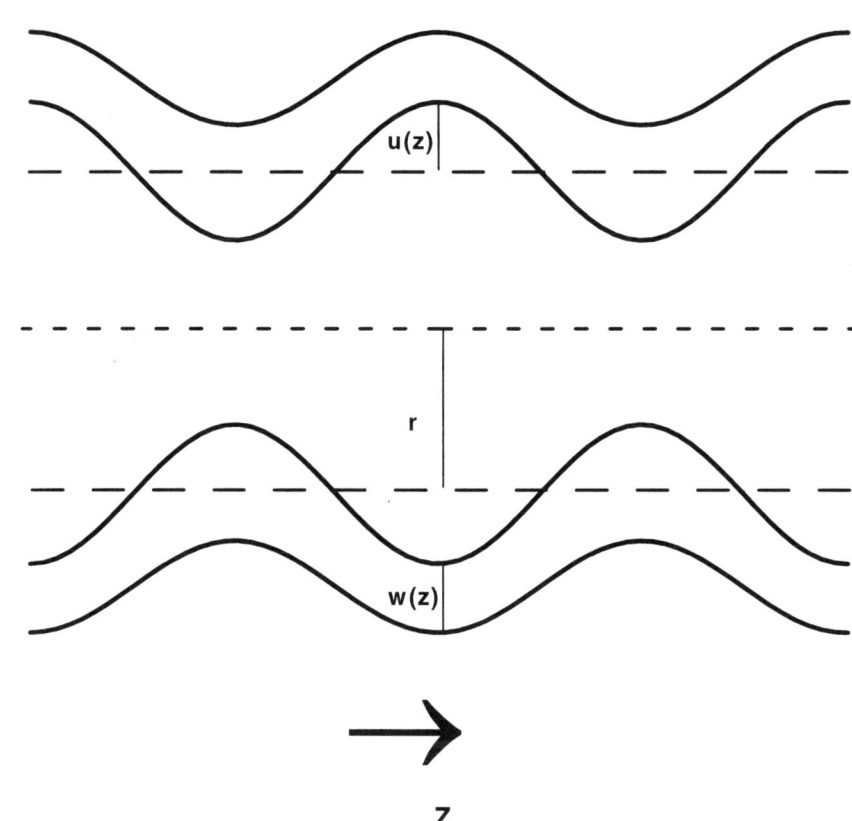

(b)                                                                                    (c)

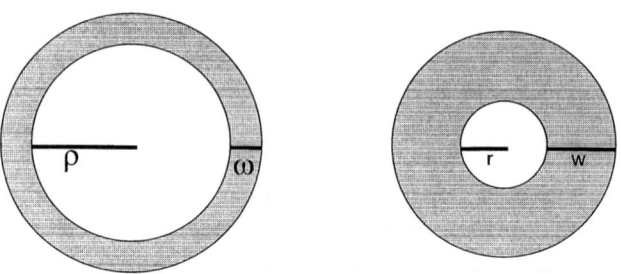

Figure 4: (a) A schematic picture of a blood vessel of inner radius $r$ undergoing a perturbation $u(z)$. The wall thickness $w(z)$ is larger at smaller radii since the circumference is smaller. (b) Schematic cross-section of a blood vessel in relaxed state. (c) Schematic cross-section of a blood vessel in activated state. The cross-sectional area is assumed fixed, so the wall thickness is larger than in (b).

it follows that $u_k(t) \sim u_k(0)e^{\lambda_k t}$, where

$$\lambda_k = \frac{Tc(r)}{2\pi r^3} k^2[1 - r^2 k^2] .$$                                   (8)

Hence, the vessel wall is unstable to modes with $rk < 1$. The fastest growing mode is at $k = 1/(\sqrt{2}r)$, where $\lambda_k$ is maximal.

The above instability is known as the Rayleigh instability [10, 11], and explains why a cylindrical column of water with surface tension $T$ is unstable. Another way to see this is by varying the surface area

$$A = \int da = \int 2\pi r[1 + (\partial_z r)^2]^{1/2} dz$$                             (9)

for fixed volume

$$V = \int \pi r^2 dz ,$$                                                              (10)

in order to find the minimal surface energy

$$F = \int T da .$$                                                                   (11)

Using $u = u_0 + u_k \cos(kz)$, where $u_0 = -u_k^2/(4r)$ [to lowest order in $u_k$] is determined by volume conservation, one finds an energy change

$$\delta F = -F_0[1 - r^2 k^2](u_k/2r)^2$$                                             (12)

[to lowest order in $u_k$], which is negative when $rk < 1$.

The cylindrical geometry may be stable if stabilizing terms are added to the energy functional $F$. For example, in the recent studies of the so-called pearling instability [12, 13], reluctance against bending in tubular lipid membranes stabilizes the cylindrical geometry. In the simplest form, an additional term [12]

$$\Delta F = \frac{1}{2}\kappa \int [\frac{1}{R} + \frac{1}{R_z}]^2 da \ , \tag{13}$$

$\kappa$ being the bending modulus, is added to the surface energy. For $Tr^2/\kappa$ sufficiently small, the cylindrical geometry remains stable (for all $k$). The instability occurs when $Tr^2/\kappa$ exceeds a certain critical value.

In small blood vessels it is not the reluctance against bending but rather a nonlinear stress-strain relation (Fig. 5) that is responsible for the stability of the cylindrical form. Under normal conditions, the stress is exponentially increasing with circumferential strain, and an increase in area by an amount $\delta a$ is energetically much more expensive than gained by a similar decrease in area. This is why the cylindrical form remains stable. However, when the smooth muscle cells contracts, the exponential behavior is replaced by a more slow variation, with a smaller area dependence in energy cost. At sufficiently strong contractions, energy is gained by reducing area, and the cylindrical form becomes unstable.

For blood vessels the width $w$ of the vessel wall must be taken into account. Assuming that the only stresses are in the principal angular and vessel directions [9], the Laplacian form for the pressure is replaced by an integral,

$$P = \int_r^{r+w} [\ S\frac{1}{\tilde{r}[1 + (\partial_z\tilde{r})^2]^{1/2}} - S_z\frac{\partial_z^2\tilde{r}}{[1 + (\partial_z\tilde{r})^2]^{3/2}}\ ]\ d\tilde{r} \ , \tag{14}$$

where $S$ and $S_z$ are the stresses in the angular direction and in the vessel direction. The stresses, $S$ and $S_z$ (forces per actual cross-sectional area), are related to experimentally measured idealized stresses, $\sigma$ and $\sigma_z$, defined as the forces per relaxed cross-sectional area [15],

$$S = \gamma\gamma_z\sigma \ , \quad S_z = \gamma\gamma_z\sigma_z \ . \tag{15}$$

Here, $\gamma$ and $\gamma_z$ are the normalized lengths in the angular and vessel directions (e.g. $\gamma_z$ is equal to $L/L_0$ where $L$ and $L_0$ is the actual and resting length of a tissue strip in the angular direction). We assume that $\gamma_z$ is constant, $\gamma_z = \gamma_0$, because the length of a vessel remains almost constant during a contraction. Correspondingly, the stress $\sigma_z$ is replaced by a constant $\sigma_0$.

In order to relate the width $w$ of the vessel wall to the inner radius $r$ we assume that the cross-sectional area of the vessel wall is constant. In this case the radius dependence of $w$ is given, when the inner radius $\rho$ and wall thickness $\omega$ are known for the angularly relaxed state ($\gamma = 1$). We have (Fig. 4b,c)

$$(r + w)^2 - r^2 = (\rho + \omega)^2 - \rho^2 \ . \tag{16}$$

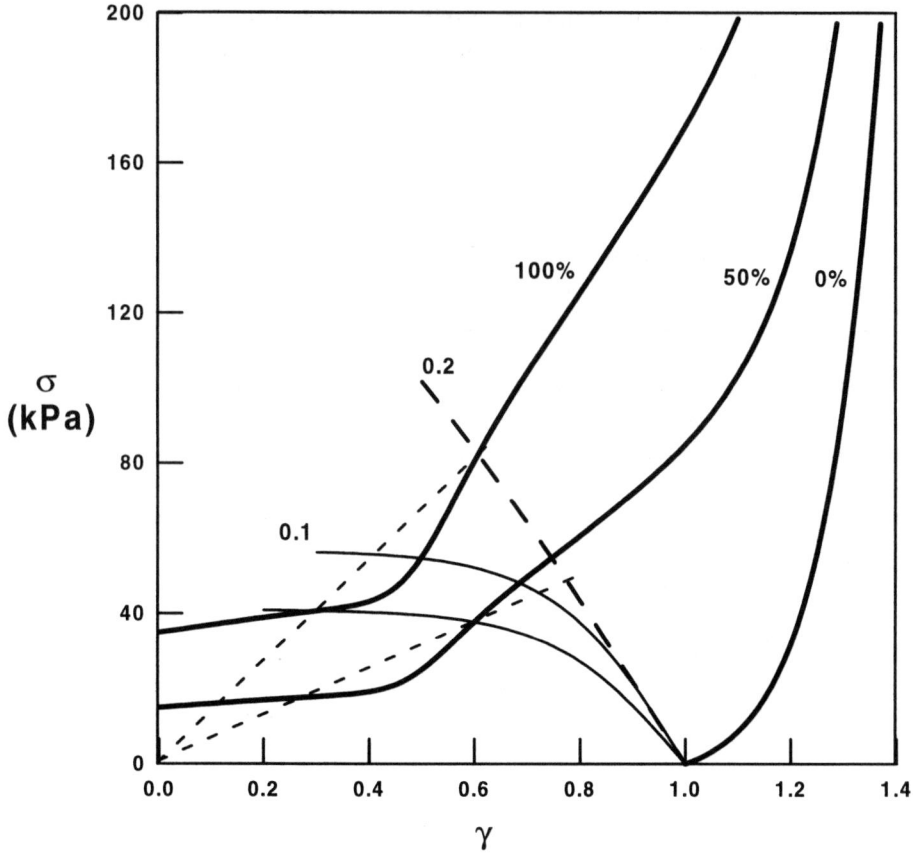

Figure 5: A schematic plot of typical stress-strain relations for arterioles (adapted from [14]). The three heavy solid curves correspond to a completely relaxed vessel (0%), a vessel where the smooth muscle cells are half maximally activated (50%), and a vessel where the smooth muscle cells are maximally activated (100%). The thin solid lines indicate how the points $(\gamma_r, \sigma(\gamma_r))$ and $(\gamma_w, \sigma(\gamma_w))$ [marked 0.1] move with muscle cell activation for an arteriole with wall-to-lumen ratio $\omega/\rho$ $= 0.1$. The point of instability for the cylindrical form of the blood vessel can be illustrated geometrically by thin dashed lines from $(0,0)$ through $(\gamma_r, \sigma(\gamma_r))$ [see text]. The instability point is where $\sigma(\gamma_w)/\gamma_w$ equals $\sigma(\gamma_r)/\gamma_r$. The thick dashed line [marked 0.2] shows how the point $(\gamma_w, \sigma(\gamma_w))$ move with muscle cell activation for an arteriole with $\omega/\rho = 0.2$, keeping the same curve for $(\gamma_r, \sigma(\gamma_r))$.

Equation (16) suggests a useful change of variable, from $\tilde{r}$ to $\tilde{\rho}$, where

$$\tilde{r}^2 - r^2 = \tilde{\rho}^2 - \rho^2 . \tag{17}$$

When $\tilde{r}$ varies between values $r$ and $r + w$, which under perturbations changes along the $z$ axis, $\tilde{\rho}$ varies between the fixed values $\rho$ and $\rho + w$. The normalized length $\gamma$ at a radius $\tilde{r}$ is simply the ratio between $\tilde{r}$ and its relaxed value $\tilde{\rho}$: $\gamma = \tilde{r}/\tilde{\rho}$.

The expression for the pressure reduces for small perturbations to

$$P = \gamma_0 \int_\rho^{\rho+w} [\sigma - \sigma_0 r \partial_z^2 r][\tilde{\rho}^2 - \rho^2 + r^2]^{-1/2} \, d\tilde{\rho} . \tag{18}$$

The stress $\sigma$ (angular direction) depends on the normalized length

$$\gamma = [\tilde{\rho}^2 - \rho^2 + r^2]^{1/2}/\tilde{\rho} . \tag{19}$$

One finds to lowest order in the perturbation $u(z,t)$,

$$P = P_0(r) + I(r)u - I_0(r)\partial_z^2 u , \tag{20}$$

where

$$P_0(r) = \gamma_0 \int_\rho^{\rho+w} \sigma[\tilde{\rho}^2 - \rho^2 + r^2]^{-1/2} \, d\tilde{\rho} , \tag{21}$$

$$\begin{aligned} I_0(r) &= \gamma_0 \sigma_0 r \int_\rho^{\rho+w} [\tilde{\rho}^2 - \rho^2 + r^2]^{-1/2} \, d\tilde{\rho} \\ &= \gamma_0 \sigma_0 r \log[1 + (\omega + w)/(\rho + r)] , \end{aligned} \tag{22}$$

and

$$I(r) = \frac{d}{dr} P_0(r) = \gamma_0 \int_\rho^{\rho+w} \tilde{\rho}^{-1} \frac{d}{d\gamma}\left[\frac{\sigma}{\gamma}\right] \frac{\partial\gamma}{\partial r} \, d\tilde{\rho} . \tag{23}$$

The partial derivatives of $\gamma$ with respect to $r$ are related with $\tilde{\rho}$,

$$\tilde{\rho}^{-1}(\partial\gamma/\partial r) = r(\rho^2 - r^2)^{-1}(\partial\gamma/\partial\tilde{\rho}) . \tag{24}$$

Thus, $I(r)$ can be expressed in terms of the normalized length $\gamma$,

$$I(r) = \frac{\gamma_0 \gamma_r}{\rho(1 - \gamma_r^2)}\left[\frac{\sigma(\gamma_w)}{\gamma_w} - \frac{\sigma(\gamma_r)}{\gamma_r}\right] , \tag{25}$$

where

$$\gamma_r = r/\rho , \quad \gamma_w = (r + w)/(\rho + \omega) , \tag{26}$$

are the normalized inner and outer radius ($I(r)$ is not singular at $\gamma_r = 1$, where also $\gamma_w = 1$).

According to Eq. (20), the dynamic equation, Eq. (3), for $u(z,t)$ takes the form

$$\partial_t u = -\frac{c(r)}{2\pi r}[-I(r)\partial_z^2 u + I_0(r)\partial_z^4 u] \ . \tag{27}$$

For a perturbation of the form (7), we have $u_k(t) \sim u_k(0)e^{\lambda_k t}$ with

$$\lambda_k = \frac{c(r)}{2\pi r} \ k^2[-I(r) - I_0(r)k^2] \ . \tag{28}$$

Since the value of $I_0(r)$ is always positive, it is the sign of $I(r)$ that determines the stability of the vessel wall. When $I(r)$ becomes negative, the cylindrical geometry becomes unstable to modes with $k^2 < |I|/I_0$.

## 3. Experimental Results

The fastest growing mode, where $\lambda_k$ is maximal, is at the value $k = [|I|/(2I_0)]^{1/2}$. This corresponds to 'sausages' of length

$$\ell = 2\pi[2I_0/|I|]^{1/2} \ . \tag{29}$$

For $\omega/\rho \sim 0.1$, and $\sigma_0 \sim 100$ kPa, we find $|I|\rho \sim I_0/\rho \sim 10$ kPa. Hence, $\ell \sim 10\rho$, i.e. the length of the 'sausages' will be about 10 times the radius of the relaxed vessel. This is in good agreement with our experimental observations (Fig. 6) (see also [4]).

The cylindrical form becomes unstable when $\sigma/\gamma$ calculated at the inner radius equals the value of $\sigma/\gamma$ at the outer radius [Eq. (25)]. This is illustrated geometrically (Fig. 5), by drawing a line (thin dashed) in the plot of $\sigma$ versus $\gamma$ from $(0,0)$ through $(\gamma_r, \sigma(\gamma_r))$. If the point $(\gamma_w, \sigma(\gamma_w))$ lies above this line, $\sigma(\gamma_w)/\gamma_w > \sigma(\gamma_r)/\gamma_r$, and the cylindrical form is stable. If on the other hand the point $(\gamma_w, \sigma(\gamma_w))$ lies below the line, $\sigma(\gamma_w)/\gamma_w < \sigma(\gamma_r)/\gamma_r$, and the cylindrical form is unstable.

The angular stress $\sigma$ in blood vessels increases exponentially with the normalized length at normal physiological conditions [9, 14] (Fig. 5). The value of $I(r)$ is positive (Fig. 7) and the blood vessel keeps its cylindrical form. However, when extreme hypertension is induced by infusion of angiotensin II, the operating point for the vessel will move to the less steep part of the $\sigma - \gamma$ curve (Fig. 5), and the value of $I(r)$ becomes negative (Fig. 7). The cylindrical form becomes unstable, and the 'sausage-string' pattern appears.

In our experimental studies, we find that the 'sausage-string' pattern following infusion of angiotensin II occurs predominantly in arterioles of diameter 30-40 $\mu$m (Fig. 2). In accordance herewith, our analysis predicts that large vessels will be stable, because the operating point $(\gamma, \sigma(\gamma))$ lies on the steep part of the $\sigma - \gamma$ curve, i.e. at high pressure elevation and rather small contraction (Figs. 3,5). For small arterial vessels, the wall-to-lumen ratio $\omega/\rho$ is generally large [9]. From the

123

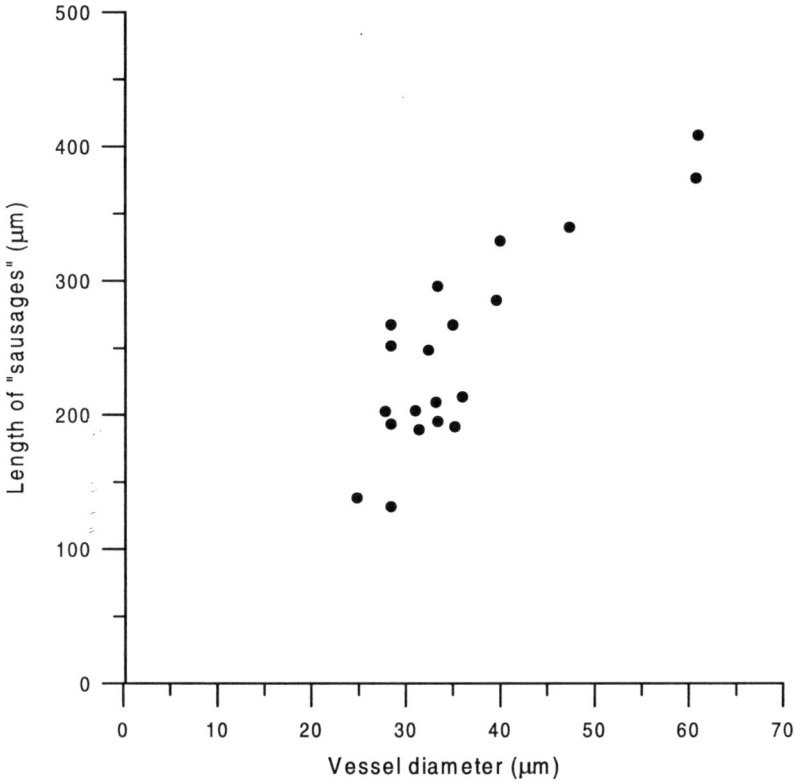

Figure 6: Length of the 'sausages' observed in rats with high blood pressure, plotted versus vessel diameter before infusion of angiotensin II.

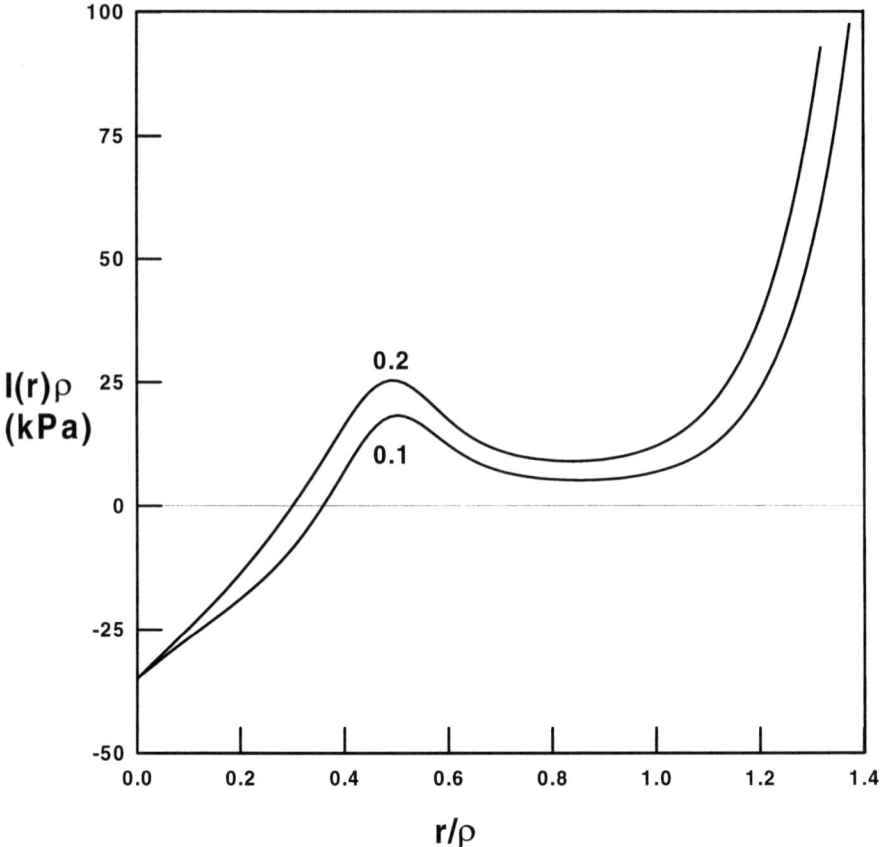

Figure 7: A plot of the stability measure $I(r)$ at large muscle cell activation for two different wall-to-lumen ratios, $\omega/\rho = 0.1$ and $\omega/\rho = 0.2$. The cylindrical form of a blood vessel becomes unstable when $I$ becomes negative. Note that the part of the stress-strain relation which is close to linear gives rise to a decay of $I$. Above $\gamma = 1$, where the stress increases exponentially, also $I(r)$ increases exponentially.

expression for $I(r)$, Eq. (25), we find that the instability point decreases with increasing wall-to-lumen ratio $\omega/\rho$ (Fig. 7). In Fig. 5, the thin solid lines are examples of how the points $(\gamma_r, \sigma(\gamma_r))$ and $(\gamma_w, \sigma(\gamma_w))$ [marked 0.1] may move with muscle cell activation for an arteriole with wall-to-lumen ratio $\omega/\rho = 0.1$. For a given activation of the smooth muscle cells (50%, 100%), a thin dashed line is drawn from $(0,0)$ through $(\gamma_r, \sigma(\gamma_r))$. When the point $(\gamma_w, \sigma(\gamma_w))$ lies above the thin dashed line, the cylindrical form is stable; when it lies below, the cylindrical form is unstable. For $\omega/\rho = 0.1$, we see from Fig. 5 that the cylindrical form is stable at 50% activation, but unstable at 100% activation of the smooth muscle cells surrounding the blood vessel. For $\omega/\rho = 0.2$ $((\gamma_w, \sigma(\gamma_w))$ illustrated by thick dashed line marked 0.2), the cylindrical form is only barely unstable at 100% activation. For larger wall-to-lumen ratios, the cylindrical form remains stable. The 'sausage-string' instability does not appear in small blood vessels with large wall-to-lumen ratios. We see that the transmural pressure and the contractile potential sets an upper limit and the wall-to-lumen ratio sets a lower limit for vessels that undergo the 'sausage-string' instability in response to an acute increase in blood pressure.

When $I(r)$ becomes negative, the pressure at slightly larger radii is smaller than at slightly smaller radii. Accordingly, the resulting flow $J$ will be directed from low-radii regions to high-radii regions, causing the small radii to become even smaller, and the large radii to become larger. This continues until the pressure stabilizes at a value which is the same for both the large radius $r_{max}$ and the small radius $r_{min}$. The stabilization is possible because the pressure for radii above the instability again increases with $r$. Since the almost linear part of the stress function (Fig. 5) gives rise to a decay of $I(r)$ in the same region (Fig. 7), $r_{max}/r_{min}$ can become quite large. However, close to $\gamma = 1$, the stress increases exponentially due to the elastic properties of the vessel wall [9], and the value of $I(r)$ will also increase rapidly. This prevents $r_{max}$ from becoming much larger than the relaxed radius, $\rho$, of the vessel. Our experimental data confirm this.

## 4. Conclusion

In conclusion, we have demonstrated that at a large increase in the smooth muscular tone of blood vessels, the normal cylindrical vessel geometry may become unstable, giving rise to the appearance of a sausage-string pattern of alternating constrictions and dilatations. The sausage-string pattern is the expression of a novel Rayleigh-type instability, involving the nonlinear stress-strain characteristics of the vessel wall, and not caused by a mechanical failure of the vessel wall due to a high blood pressure. The developed theory explains some key features which we observe experimentally, particularly why the instability is only observed in small arteries and large arterioles. Our theory shows that the instability predominantly arises in blood vessels with high contractile potential, limited pressure elevation, and small wall-to-lumen ratios. The observed corespondence between the 'sausage' length and the vessel diameter is also consistent with the theory.

The present study was supported by grants from the Danish Natural Science Research Council, the Danish Medical Research Council, and the Danish Heart Association.

# References

[1] C. Kitiyakara and N.J. Guzman, J. Am. Soc. Nephrol. **9**, 133 (1998).

[2] F. Gustafsson, Blood Pressure **6**, 71 (1997).

[3] F. B. Byrom, Lancet **2**, 201 (1954); F.B. Byrom, Prog. Cardiovasc. Dis. **1**, 31 (1974).

[4] J. Giese, Acta Pathol. Microbiol. Scand. **62**, 497 (1964).

[5] J. Giese, *The Pathogenesis of Hypertensive Vascular Disease* (Munksgaard, Copenhagen, 1966).

[6] L.J. Beilin and F.S. Goldby, Clin. Sci. Mol. Med. **52**, 111 (1977).

[7] G. A. Meininger, K. L. Fehr, M. B. Yates, J. L. Borders, and H. J. Granger, Hypertension **8**, 66 (1986).

[8] A. E. Green and J. E. Adkins, *Large Elastic Deformations* (Clarendon Press, Oxford, 1960).

[9] Y.C. Fung, *Biomechanics. Mechanical Properties of Living Tissues, 2nd Ed.* (Springer-Verlag, New York, 1990); Y.C. Fung, *Biomechanics. Motion, Flow, Stress, and Growth* (Springer-Verlag, New York, 1990).

[10] J. Plateau, *Statique Experimentale et Theorique des Liquides Soumis aux Seules Forces Moleculaires* (Gautier-Villars, Paris, 1873).

[11] Lord Rayleigh, Philos. Mag. **34**, 145 (1892).

[12] R. Bar-Ziv and E. Moses, Phys. Rev. Lett. **73**, 1392 (1994).

[13] P. Nelson, and T. Powers, Phys. Rev. Lett. **74**, 3384 (1995); R.E. Goldstein, P. Nelson, T. Powers, and U. Seifert, J. Phys. II France **6**, 767 (1996).

[14] R.W. Gore, Circ. Res. **34**, 581 (1974); M.J. Davis, and R.W. Gore, Am. J. Physiol. **256**, H630 (1989).

[15] R. Feldberg, M. Colding-Jørgensen, and N.-H. Holstein-Rathlou, Am. J. Physiol. **269**, F581 (1995).

# EQUATIONS OF GRANULAR MATERIALS: DEPOSITION THEORY

S. F. EDWARDS
*Cavendish Laboratory*
*University of Cambridge, Madingley Road,*
*Cambridge CB3 OHE, United Kingdom*

## 1. Introduction

Suppose that we take a bin and gently and uniformly pour in a granular material. As the material in the bin builds up we can identify a surface and ask the question what is the magnitude of the fluctuation in the height of surface? Also of interst is the length scale of the surface fluctuations and how they behave dynamically as more material is added. We assume that particles are deposited uniformly. There is a weak flux i.e. there are no correlations between incoming particles. Particles settle gently so that there is no cooperative reorganization. These assumptions can be considered as part of the definition of the problem.

## 2. Edwards-Wilkinson Model

In this, particles fall at random in space and time. They can roll over. Since there is no overhang the height $z$ can be written $z(x, y, t)$. If the average rate of landings per unit time per unit area is $r$, then the average increase in height is $rh$, and $z$ satisfies the simple differential equation

$$\frac{\partial z}{\partial t} = r\,h + r\,a^4\,\nabla^2 z + \eta \tag{1}$$

where $a$ is the particle radius and $\eta$ a random variable whose correlation function is approximately

$$\langle \eta(\vec{r}, t)\,\eta(\vec{r}_0, t_0) \rangle \sim h^2\, e^{-\frac{|\vec{r} - \vec{r}_0|^2}{4a^2}}\, \delta(t - t_0) \tag{2}$$

127

*A.T. Skjeltorp and S.F. Edwards (eds.), Soft Condensed Matter: Configurations, Dynamics and Functionality, 127-134.*

All is very simple and reasonably adequate. We can invoke the mathematical theorem: if $\frac{\partial Z}{\partial t} = \eta$ and $\eta$ is random i.e. $\langle \eta(t)\,\eta(t_0)\rangle = h\,\delta(t - t_0)$ then probability of finding $Z = z$ at $t$ is

$$\frac{\partial P}{\partial t} - h\frac{\partial^2 P}{\partial z^2} = 0 \tag{3}$$

More generally if

$$\frac{\partial Z}{\partial t} + AZ = \eta \tag{4}$$

then

$$\frac{\partial P}{\partial t} - \frac{\partial}{\partial z}\left(h\frac{\partial}{\partial z} + Az\right)P = 0 \tag{5}$$

If $Z = Z(\vec{r}, t)$ and conditions are homogeneous we can use Fourier transform

$$Z_k(t) = \frac{1}{2\pi}\int e^{i\vec{k}\cdot\vec{r}}\,Z(\vec{r}, t)\mathrm{d}^3\vec{r} \tag{6}$$

and

$$\frac{\partial P}{\partial t} - \sum_k \frac{\partial}{\partial z_k}\left(h_k\frac{\partial}{\partial z_{-k}} + A_k\,z_k\right)P = 0 \tag{7}$$

In our problem $h_k$ is independent of $k$, but in general it need not be. The Fokker-Planck equation () is well studied in the literature and it is easy to obtain the dynamic fluctuations:

$$(z(t + t') - z(t))^2 = \frac{h^2}{4\pi a^4}\,\log(1 + r\,t'\,a^2). \tag{8}$$

The time to add one layer is about $(r\,a^2)^{-1}$, $t' \ll (r\,a^2)^{-1}$ and $(z' - z)^2 = \frac{h^2\,r\,t'}{4\pi\,a^2}$, i.e. $z$ moves in a random walk. it is interesting to note that

$$\langle z(x, y, t)\,z(x + x', y + y', t + t')\rangle =$$
$$\frac{h^2}{8\pi^2\,a^4}\int \frac{\mathrm{d}k\,\mathrm{d}j}{k^2 + j^2}\,e^{ik(x-x')+ij(y-y')}\,e^{-a^2(k^2+j^2)(1+rt'a^2)}. \tag{9}$$

This diverges unless cut-off at small $(k, j)$, i.e. surface fluctuations depend on size of the box, $L$:

$$\langle z^2 \rangle = \frac{h^2}{4\pi\, a^4}\, \log\frac{L}{\pi a} \tag{10}$$

No escape from this, but it is difficult to even measure it.

## 3. A Class of Nonlinear Stochastic Equations

Use a variable $h(\vec{r}, t)$ which can be vector e.g. velocity $u_i(\vec{r}, t)$ or height of powder $h(x, y; t)$ etc. The basic equation is

$$\frac{\partial h_k}{\partial t} + \nu k^2 h_k + M_{kjl}\, h_j h_l = \eta_k \tag{11}$$

For nonlinear corrections of the Edwards-Wilkinson model a much studied equation is Kardar-Parisi-Zhang equation [4] which has the form

$$\frac{\partial h}{\partial t} = \nu \nabla^2 h - \frac{\lambda}{2}(\nabla h)^2 + \eta(\vec{r}, t) \tag{12}$$

$$\langle \eta(\vec{r}, t)\, \eta(\vec{r}_0, t_0) \rangle = 2D\, \delta(\vec{r} - \vec{r}_0)\, \delta(t - t_0) \tag{13}$$

or

$$\frac{\partial h_k}{\partial t} + \nu(k)\, h_k + \sum_{l,j} l\, j\, h_l\, h_j\, \delta(k + l + j) = \eta_k. \tag{14}$$

For hydrodynamics we have Navier-Stokes equation

$$\frac{\partial \vec{u}}{\partial t} - \nu \nabla^2\, \vec{u} + (\vec{u} \cdot \nabla)\vec{u} - \nabla p = \eta \tag{15}$$

where $div\, \vec{u} = 0$ fixes $p$.

$$\frac{\partial u_k^i}{\partial t} + \nu k^2 u_k^i + \sum_{l,m,j,l} M_{kjl}^{ilm}\, u_k^l\, u_j^m = \eta_k^i \tag{16}$$

when $M = k^i O_k^{lm} \delta(k + l + j)$, where $O_k^{lm}$ is the Oseen tensor.

In 1-D NS equation is called Burgers equation. It can be obtained from KPZ equation by the transformation $\vec{u} = -\nabla h$.

Expect

$$\langle u_k(t)\, u_{-k}(0)\rangle \;=\; \phi_k\, F(kt^{\beta}) \tag{17}$$

and

$$\langle h_k(t)\, h_{-k}(0)\rangle \;=\; \phi_k\, F(kt^{\gamma}) \,. \tag{18}$$

Dimensional analysis gives the Kolmogoroff solution to NS equation $\phi_k = k^{-11/3}$ and $\beta = \frac{3}{2}$. From eqn.(11) we can derive

− an equation for $P(h; t)$

$$\frac{\partial P}{\partial t} + \sum_k \frac{\partial}{\partial h_{-k}} \left( -\nu_k\, h_k - \sum_{l,m} M_{klm}\, h_l\, h_m + \eta_k \right) P = 0 \tag{19}$$

$\langle P\rangle_\eta$ satisfies (write $P$ for $\langle P\rangle_\eta$ from now on)

$$\frac{\partial P}{\partial t} - \sum_k \left( \frac{\partial^2}{\partial h_k h_{-k}} D_0 + \frac{\partial}{\partial h_{-k}} \left( \nu_k\, h_k + \sum_{l,m} M_{klm}\, h_l\, h_m \right) \right) P = 0 \,. \tag{20}$$

or

− an equation for $P([h])$ the probability of whole history

$$\left( \frac{\partial h}{\partial t} + \nu\, h + M\, h\, h - \eta \right) P = 0 \,. \tag{21}$$

Consider the first option. If terms $M\, h\, h$ were a random variable, it would simply enhance $D_0$ i.e. $\frac{\partial^2 D_0}{\partial h^2} \rightarrow \frac{\partial^2}{\partial h_k h_{-k}} (D_0 + O((Mhh)^2))$.

Peierls [5] derived a Boltzmann equation for $\phi = \langle h_k\, h_{-k}\rangle$,

$$\frac{\partial n_k}{\partial t} + \int \mathrm{d}j^3\, K_1(k,j,l)\, n_k\, n_j - \int \mathrm{d}j^3\, K_2(k,j,l)\, n_l\, n_j = \text{s.t.} \tag{22}$$

where $h$ is the displacement of a lattice and $n_k$ is the number of phonons For NS $\phi = \langle u_k\, u_{-k}\rangle$ is the energy in mode $k$. If $\omega_k$ is the (approximate) lifetime of mode $k$, the Boltzmann equation is, for steady state

$$\nu_q \, \phi_q \, - \, D_{0q} \, = \, \int \mathrm{d}^d j \mathrm{d}^d l \, \frac{\tilde{M}_{qjl} \, \tilde{M}_{qjl} \, \phi_j \, \phi_l}{\omega_q + \omega_j + \omega_l} \, \delta(q + j + l)$$

$$+ \int \mathrm{d}^d j \mathrm{d}^d l \, \frac{\tilde{M}_{qjl} \, \tilde{M}_{jql} \, \phi_q \, \phi_l}{\omega_q + \omega_j + \omega_l} \, \delta(q + j + l)$$

$$+ \int \mathrm{d}^d j \mathrm{d}^d l \, \frac{\tilde{M}_{qjl} \, \tilde{M}_{ljq} \, \phi_q \, \phi_j}{\omega_q + \omega_j + \omega_l} \, \delta(q + j + l) \tag{23}$$

and for time dependent equation

$$\frac{\partial \Phi_k}{\partial t} + \nu \Phi_k \nu_q = \int \mathrm{d}^d j \mathrm{d}^d l \, \frac{\tilde{M}_{kjl} \, \tilde{M}_{kjl} \, \Phi_j \, \Phi_l}{\omega_k + \omega_j + \omega_l} \, \delta(k + j + l)$$

$$+ \int \mathrm{d}^d j \mathrm{d}^d l \, \frac{\tilde{M}_{kjl} \, \tilde{M}_{jkl} \, \Phi_k \, \phi_l}{\omega_k + \omega_j + \omega_l} \, \delta(k + j + l)$$

$$+ \int \mathrm{d}^d j \mathrm{d}^d l \, \frac{\tilde{M}_{kjl} \, \tilde{M}_{ljk} \, \Phi_k \, \phi_j}{\omega_k + \omega_j + \omega_l} \, \delta(k + j + l) \tag{24}$$

with boundary condition

$$\frac{\partial \Phi_k(0)}{\partial t} = D_{0k} \, . \tag{25}$$

Thus for steady state we expect

$$\nu \, k^2 \, \phi_k + \int \mathrm{d}j^3 \, K_1(k, j, l) \, \phi(t) \, \phi(t) - \int \mathrm{d}j^3 \, K_2(k, j, l) \, \phi(t) \, \phi(t) \, = \, D_k^0 \tag{26}$$

$$\frac{\partial \Phi_k(t)}{\partial t} + \nu \, k^2 \, \Phi_k(t) + \int \mathrm{d}j^3 \, K_1 \, \Phi_k(t) \, \phi_l - \int \mathrm{d}j^3 \, K_2 \, \Phi_j(t) \, \Phi_l(t) \, = \, 0 \tag{27}$$

with boundary condition

$$\frac{\partial \Phi_k(0)}{\partial t} = D_k^0 \, . \tag{28}$$

Note that $\Phi = \langle h(t) \, h(0) \rangle$ and not $\langle h(t) \, h(t) \rangle$. The derivation of eqn.(24) is via a self-consistent expansion based on the model

$$\frac{\partial P}{\partial t} + \sum_q \frac{\partial}{\partial h_{-q}} \left( D_q \frac{\partial}{\partial h_q} + \omega_q \, h_{-q} \right) P \, = \, L_0 \, P \, = \, 0 \tag{29}$$

which leads to eqn.(23) and $\omega_q$ is either derived from the time dependent equation (27) which gives

$$\omega_k = \int d^d j \, d^d l \, \frac{M_{kjl} \, M_{ljk} \, \phi_j \, \phi_l}{\phi_k(\omega_k + \omega_j + \omega_l)} \left( \frac{\omega_l + \omega_j - \omega_k}{omega_l + \omega_j} \right) \tag{30}$$

or by the scaling relation which ensures the consistency of the self consistent expansion

$$\omega_q \sim \sum_{l,m} M_{qlm} \frac{M_{lmq} \, \phi_m + M_{mlq} \, \phi_l}{\omega_l + \omega_m} \tag{31}$$

i.e. $\omega_k \sim M^2(k)k^d k^\alpha$. If $\omega_k \sim k^{\frac{1}{\beta}}$ and $\phi_k \sim k^{-\alpha}$, $M(k) \sim k^m$ and $2 = \beta(m + d - \alpha)$.

For eqn.(30) we define a mean $\omega_k$ by

$$\omega_k^{-1} = \frac{\int_0^\infty \Phi_k(t) \, dt}{\Phi_k(0)} \tag{32}$$

## 4. Asymptotic Power-Law solution of the Steady-State Equation

We can write eqn.(24) as

$$D_0 - \nu q^2 \, \phi_q - I_1(q) + I_2(q) = 0 \tag{33}$$

and eqn.(30) as

$$\omega_q - J(q) = 0. \tag{34}$$

In eqn.(24) $D_0$ drives the equation and $\nu$ makes $\phi \to 0$ as $q \to \infty$. The friction $\nu$ simply cuts $\phi$ off at large $q$, but for intermediate $q$ either $I_1 - I_2 = D_0$ or $I_1 = I_2$ and $D_0$ simply fixes the behaviour at small $q$. If $\phi_q = A q^{-\Gamma}$ and $I_1(\Gamma) = I_2(\Gamma)$ which is the case for the KPZ equation then $\Gamma = 2.6$. For Navier-Stokes equation $\Gamma = 11/3$ (Kolmogoroff) and $\omega_q = B q^\mu$ where $\mu = 2/3$. The solution of eqn.(27) can now be shown to be $e^{-kt^{\frac{1}{\mu}}}$. For KPZ it is a stretched exponential for $\mu > 0$, but for Navier-Stokes equation it is a compressed exponential $e^{-kt^{\frac{3}{2}}}$.

## 5. Model of the Boltzmann Equation

Since the behaviour of the transport equations is novel we offer a model:

$$\nu(x)\Phi(x) + \lambda \int_0^\infty \frac{\Phi(y)}{x+y}\,dy - \Phi(x) = S(x)\,. \tag{35}$$

There is the relation

$$\int_0^\infty \frac{y^{-\alpha}}{x+y}\,dy = \frac{\pi}{\sin\pi\alpha}\, x^{-\alpha}\,, \tag{36}$$

which suggests a Stieltjes transform

$$\Phi(y) = \int_0^1 F(\beta)\, x^{-\beta}\,d\beta\,. \tag{37}$$

If $\nu(x) = x^\gamma$

$$F(\beta - \gamma) + \frac{\lambda\,\pi}{\sin\pi\beta}\, F(\beta) - F(\beta) = S(\beta)\,. \tag{38}$$

Handle the term $F(\beta - \gamma)$ by boundary condition, but rest is nontrivial. We expect the solution to decrease monotonically, the starting point being fixed by $S(\beta)$ and the tail (uninterestingly) by $\nu$.

Crudely speaking

$$\nu\,\Phi + \int K\,\Phi = S\,. \tag{39}$$

Either $\lambda < \frac{2}{\pi}$ when there is a solution to

$$\int K\,\Phi = 0 \tag{40}$$

and that $\Phi$ is modified for $\nu$ and $S$, or $\lambda > \frac{2}{\pi}$ and there is no solution to the eqn.(39) but $K$ has an inverse, so

$$\Phi = K^{-1}S \tag{41}$$

and is modified at large $k$ to cover $\nu$. For Navier-Stokes equation this is the case, but input near $k = 0$ (Kolmogoroff) is pathological. Another interesting problem is point probaility $P(h(r) = H)$ is not gaussian

$$P = e^{-H^\gamma c} \tag{42}$$

For fluids $\langle (u(r) - u(0))^2 \rangle \sim r^{\frac{2}{3}}$ but $P(u(r) - u(0) = U) \neq e^{\frac{-U^2}{r^{2/3}}}$.

## References

1. Edwards, S.F., Wilkinson, D.R. (1982) The surface statistics of a granualr aggregate, *Proc. R. Soc. Lond.*, **A 381**, pp. 17–31.

2. Schwartz, M., Edwards, S.F. (1992) Nonlinear deposition: a new approach, *Europhys. Lett.*, **20**, pp. 301–305.
3. Schwartz, M., Edwards, S.F. (1998) Peierls-Boltzmann equation for ballistic deposition, *Phys. Rev. E*, **57**, pp. 5730–5739.
4. Kardar, M., Parisi, G., Zhang, Y-C. (1986) Dynamic scaling of growing interfaces, *Phys. Rev. Lett.*, **56**, pp. 889–892.
5. Peierls, R. (1955) *Quantum Theory of Solids*, Oxford University Press, London.

# EQUATIONS OF GRANULAR MATERIALS: TRANSMISSION OF STRESS

S. F. EDWARDS AND D. V. GRINEV

*Cavendish Laboratory*
*University of Cambridge, Madingley Road,*
*Cambridge CB3 OHE, United Kingdom*

## 1. Introduction

Transmission of stress and statistics of force fluctuations in static granular arrays are fundamental, but unresolved problems in physics. Despite several theoretical attempts [1, 2, 3] and a vast engineering literature [4, 5] the connectivity of granular media is still poorly understood at a fundamental level. We propose a theory of stress transmission in disordered arrays of rigid cohesionless grains with perfect friction. A real granular aggregate (e.g. sand or soil) is a very complex object [5]. However, simple models are easier to comprehend, and extra complexities can always be incorporated subsequently. In our case the rigid grain paradigm provides a crucial starting point from which to appreciate the theoretical physics of the problem. We model the granular material as an assembly of discrete rigid particles whose interactions with their neighbours are localized at pointlike contacts. Therefore the description of the network of intergranular contacts is essential for the understanding of force transmission in granular assemblies. Grain $\alpha$ exerts a force on grain $\beta$ at a point $\vec{\mathcal{R}}^{\alpha\beta} = \vec{R}^{\alpha} + \vec{r}^{\alpha\beta}$. The contact is a point in a plane whose normal is $\vec{n}^{\alpha\beta}$. The vector $\vec{R}^{\alpha}$ is defined by:

$$\vec{R}^{\alpha} = \frac{\sum_{\beta} \vec{\mathcal{R}}^{\alpha\beta}}{z}, \tag{1}$$

so that $\vec{R}^{\alpha}$ is the centroid of contacts, and hence

$$\sum_{\beta} \vec{r}^{\alpha\beta} = 0, \quad \vec{R}^{\alpha\beta} = \vec{r}^{\alpha\beta} - \vec{r}^{\beta\alpha}, \tag{2}$$

135

*A.T. Skjeltorp and S.F. Edwards (eds.), Soft Condensed Matter: Configurations, Dynamics and Functionality, 135-144.*
© 2000 *Kluwer Academic Publishers. Printed in the Netherlands.*

136

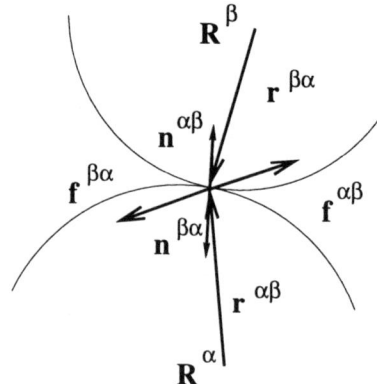

*Figure 1.* Detail of two grain contact

where $z$ is the number of contacts per grain and $\sum_\beta$ means summation over the nearest neighbours. Hence $\vec{R}^\alpha$, $\vec{r}^{\alpha\beta}$ and $\vec{n}^{\alpha\beta}$ are geometrical properties of the aggregate under consideration and the other shape specifications do not enter.

Friction is assumed to be infinite and the geometry is frozen after the deposition and can not be changed by applying or removing an external force on the boundaries. In a static array Newton's equations of integranular force and torque balance are satisfied. Balance of force around the grain $\alpha$ requires

$$\sum_\beta f_i^{\alpha\beta} = g_i^\alpha, \tag{3}$$

$$f_i^{\alpha\beta} + f_i^{\beta\alpha} = 0, \tag{4}$$

where $i = 1, 2, 3$ are cartesian indices and $\vec{g}^\alpha$ is the external force acting on grain $\alpha$. Further on $\vec{g}$ is used also for the external forces at the boundaries.

The equation of torque balance is

$$\sum_\beta \epsilon_{ikl} f_k^{\alpha\beta} r_l^{\alpha\beta} = C_i^\alpha. \tag{5}$$

The centroid of the contact points need not coincide with the centroid of the forces e.g. the centre of mass of a solid grain, but we will assume it is so in order to keep the analysis simple so that we ensure that the macroscopic stress tensor is symmetric, at least on average. It can be verified that, for the integranular forces in the static array to be determined by these equations, the coordination number $z = 3$ in 2-D and $z = 4$ in 3-D is required. In this paper we present the results for the 2-D case only. The microscopic version of stress analysis is to determine all of the intergranular forces, given the applied force, torque loadings on each grain and geometric specification of a granular array. The number of unknowns per grain is $zd/2$. Required force and torque equations give $d + \frac{d(d-1)}{2}$ constraints. The system of equations for the integranular forces is complete when the coordination number is $z_m = d+1$. Theory which confirms this observation has been proposed for periodic arrays of grains with perfect and zero friction [6]. It is clear that the coordination number $z$ controls the connectivity of granular media. We will assume that $z$ is indeed 3 in 2-D, for this is surely the simplest situation, and one which is physically possible. The ultimate goal, however, is to determine the macroscopic stress tensor at every point of a granular array, given external loadings and geometric specification. The macroscopic state of stress is a function of the distribution of contact forces. For any aggregate of discrete grains subjected to external loading, the transmission of stress from one point to another can only occur via the intergranular contacts. Therefore it is clear that the network of contacts determines the distribution of stresses within the granular array. The network of contacts is determined by the deposition history of the sample and the external loading on the boundaries. We define the tensorial force moment:

$$S_{ij}^\alpha = \sum_\beta f_i^{\alpha\beta}\, r_j^{\alpha\beta}, \tag{6}$$

which is the microscopic analogue of the stress tensor. With $C_i^\alpha = 0$, $S_{ij}^\alpha$ will be symmetric. Our goal is to find a complete system of equations for the macroscopic stress tensor $\sigma_{ij}$, which is supported by the given network of contacts in the state of mechanical equilibrium. Given an assembly of discrete grains which is represented by a very complex network of contacts, we associate a continuous medium to have continuously distributed properties. Such spatial smoothing or coarse-graining can be accomplished formally. To obtain the macroscopic stress tensor from the tensorial force moment to the macroscopic stress tensor we coarse-grain i.e. average it over an ensemble of configurations:

$$\sigma_{ij}(\vec{r}) = \langle \sum_{\alpha=1}^{N} S_{ij}^\alpha\, \delta(\vec{r} - \vec{R}^\alpha) \rangle. \tag{7}$$

In the simplest cases of isotropic and homogeneous arrays this is not a problem. The difficulties appear when the array under consideration is anisotropic or inhomogeneous. Within the confines of this paper we explore only the simplest cases. Let us give a simple illustration of our theory. Suppose that there exist a granular packing for which we can "invert" $\frac{Nd(d+1)}{2}$ eqns.(6) and write

$$f_i^{\alpha\beta} = A_{ikl}^{\beta\alpha} S_{kl}^{\alpha} \tag{8}$$

After invoking Newton's third law (4) and eqns.(3,5) one obtains a complete system of equations for tensorial force moments $S_{ij}^{\alpha}$'s

$$\sum_{\beta} A_{ikl}^{\alpha\beta} S_{kl}^{\beta} + g_i^{\alpha} = 0 \tag{9}$$

and

$$\sum_{\beta} \epsilon_{ijk} A_{jlm}^{\alpha\beta} S_{lm}^{\beta} r_k^{\alpha\beta} = 0 \tag{10}$$

There are two ways to deal with these discrete equations. The first one is to average them and obtain continuous equations for the macroscopic stress tensor. The second way is to solve them and then average solution. The number of macroscopic equations required equals the number of independent components of a symmetric stress tensor $\sigma_{ij} = \sigma_{ji}$ and is $\frac{d(d+1)}{2}$. At the same time, the number of equations available is $d$. These are vector equations of the stress equilibrium $\frac{\partial \sigma_{ij}}{\partial x_j} = g_i$ which have their origin in Newton's second law. Therefore we have to find $\frac{d(d-1)}{2}$ equations, which possess the information from Newton's third law, to complete and solve the system of equations which governs the transmission of stress in a granular array. Thus in 2-D there is one missing equation, and we derive it in terms of the geometry of the system.

Given the set of equations $(3-5)$ we can write the probability functional for the integranular force $f_i^{\alpha\beta}$ as

$$
\begin{aligned}
P\{f_i^{\alpha\beta}\} &= \mathcal{N}\delta(\sum_{\beta} f_i^{\alpha\beta} - g_i^{\alpha}) \\
&\times \delta(\sum_{\beta} \epsilon_{ikl} f_k^{\alpha\beta} r_l^{\alpha\beta}) \\
&\times \delta(f_i^{\alpha\beta} + f_i^{\beta\alpha}),
\end{aligned} \tag{11}
$$

where the normalization, $\mathcal{N}$, which is a function of a configuration, is

$$\mathcal{N}^{-1} = \int \prod_{\alpha,\beta} P\{f_i^{\alpha\beta}\} \mathcal{D}f^{\alpha\beta}. \tag{12}$$

The probability of finding the tensorial force moment $S_{ij}^{\alpha}$ on grain $\alpha$ is

$$P\left\{S_{ij}^{\alpha}\right\} = \int \prod_{\alpha,\beta} \delta\left(S_{ij}^{\alpha} - \sum_{\beta} f_i^{\alpha\beta} r_j^{\alpha\beta}\right) P\left\{f_i^{\alpha\beta}\right\} \mathcal{D}f^{\alpha\beta} \qquad (13)$$

where $\int \mathcal{D}f^{\alpha\beta}$ implies integration over all functions $f^{\alpha\beta}$, since all the constraints on $f^{\alpha\beta}$ have been experienced. We assume that the $z = d + 1$ condition means that the integral exists. The algebra that follows, aims to transform eqn.(13) into

$$P\left\{S_{ij}^{\alpha}\right\} = \prod_{\alpha,\beta} \delta\left(K_{ijk}^{\alpha\beta} S_{jk}^{\beta} + g_i^{\alpha}\right) \delta\left(P_{ijk}^{\alpha\beta} S_{jk}^{\beta}\right) \qquad (14)$$

To start with, we exponentiate the delta functions in (13) and thus introduce the set of conjugate fields $\zeta_{ij}^{\alpha}$, $\gamma_i^{\alpha}$, $\lambda_i^{\alpha}$ and $\eta_i^{\alpha\beta}$.

$$P\left\{S_{ij}^{\alpha}\right\} = \int \prod e^{iA} \mathcal{D}f^{\alpha\beta} \mathcal{D}\zeta^{\alpha} \mathcal{D}\gamma^{\alpha} \mathcal{D}\lambda^{\alpha} \mathcal{D}\eta^{\alpha\beta}, \qquad (15)$$

where $A$ is

$$\begin{aligned}
A = &\sum_{\alpha} \zeta_{ij}^{\alpha}\left(S_{ij}^{\alpha} - \sum_{\beta} f_i^{\alpha\beta} r_j^{\alpha\beta}\right) \\
&+ \gamma_i^{\alpha}\left(\sum_{\beta} f_i^{\alpha\beta} - g_i^{\alpha}\right) \\
&+ \lambda_i^{\alpha}\left(\sum_{\beta} \epsilon_{ikl} f_k^{\alpha\beta} r_l^{\alpha\beta}\right) \\
&+ \eta_i^{\alpha\beta}\left(f_i^{\alpha\beta} + f_i^{\beta\alpha}\right).
\end{aligned} \qquad (16)$$

The $\lambda^{\alpha}$ field term gives the symmetry of $S_{ij}^{\alpha}$. After integrating out the $f^{\alpha\beta}$ and $\eta^{\alpha\beta}$ fields we find the following linear equation for the conjugate fields:

$$\zeta_{ij}^{\alpha} r_j^{\alpha\beta} - \gamma_i^{\alpha} = \zeta_{ik}^{\beta} r_k^{\beta\alpha} - \gamma_i^{\beta}. \qquad (17)$$

The idea of the conjugate fields method is to use these equations for the $\zeta$ field in the stress probability functional, in order to derive the complete system of equations for the stress tensor. The general solution of the above equation is a sum of the $\zeta^0$ field which is the particular solution and depends on $\gamma$, and $\zeta^*$ which is the complimentary function

$$\zeta_{ij}^{\alpha} = \zeta_{ij}^{\alpha 0} + \zeta_{ij}^{\alpha *}. \qquad (18)$$

If we introduce the fabric tensor $F_{ij}^\alpha$ and its inverse $M_{ij}^\alpha$:

$$F_{ij}^\alpha = \sum_\beta R_i^{\alpha\beta} R_j^{\alpha\beta}, \quad M_{ij}^\alpha = \left(F^\alpha\right)_{ij}^{-1}, \tag{19}$$

we can rewrite the equation (17) in the following form:

$$\zeta_{ij}^\alpha = M_{jl}^\alpha \sum_\beta R_l^{\alpha\beta}(\gamma_i^\alpha - \gamma_i^\beta) + M_{jl}^\alpha \sum_\beta R_l^{\alpha\beta} r_k^{\beta\alpha}(\zeta_{ik}^\beta - \zeta_{ik}^\alpha). \tag{20}$$

which permits an expansion based on the first two terms, i.e:

$$\zeta_{ij}^\alpha \simeq M_{jl}^\alpha \sum_\beta R_l^{\alpha\beta}(\gamma_i^\alpha - \gamma_i^\beta) +$$

$$+ M_{jl}^\alpha \sum_\beta R_l^{\alpha\beta} r_k^{\beta\alpha} M_{km}^\beta \sum_\delta R_m^{\beta\delta}(\gamma_i^\beta - \gamma_i^\delta) + ... \tag{21}$$

## 2. The Stress-Force Equation

The next step is to integrate out the $\gamma$ field which gives us the stress-force equation:

$$\sum_\beta M_{jl}^\alpha R_l^{\alpha\beta} S_{ij}^\alpha - \sum_\beta M_{jl}^\beta R_l^{\beta\alpha} S_{ij}^\beta = g_i^\alpha. \tag{22}$$

So by expanding $\beta$ quantities about $\alpha$ quantities we reach:

$$\nabla_j \sigma_{ij} + \nabla_j \nabla_k \nabla_m K_{ijkl}\sigma_{lm} + ... = g_i. \tag{23}$$

where $K_{ijkl} = \langle R_i^{\alpha\beta} R_j^{\alpha\beta} R_k^{\alpha\beta} R_l^{\alpha\beta} \rangle$ and gives a correction to the standard equation of stress equilibrium $\nabla_j \sigma_{ij} = g_i$ at the length-scale which is small compared to the size of the system. These corrections correspond to the presence of the second, third etc. nearest neighbours and topological correlations and must vanish in the $k \to 0$ limit.

## 3. The Stress-Geometry Equation

So far the well-known equations have been derived by using the information from Newton's second law. But we still have unused information from Newton's third law. By integrating out the $\zeta^*$ field we obtain the missing equations we are looking for. Let us consider that part of the eqn.(15) which contains the $\zeta^*$ field:

$$\int e^{i \sum_{\alpha=1}^N \zeta_{ij}^{\alpha*} S_{ij}^\alpha} \delta(\zeta_{ij}^{\alpha*} r_j^{\alpha\beta} - \zeta_{ij}^{\beta*} r_j^{\beta\alpha}) \prod_{\alpha=1}^N \mathcal{D}\zeta_{ij}^{\alpha*}, \tag{24}$$

and

$$\zeta_{ij}^{\alpha}{}^{*}r_j^{\alpha\beta} - \zeta_{ij}^{\beta}{}^{*}r_j^{\beta\alpha} = 0 . \tag{25}$$

Counting the degrees of freedom in this equation we note that it can only give two (scalar) equations in 2-D and three in 3-D. Using $\vec{R}^{\alpha\beta}$ and $\vec{Q}^{\alpha\beta} = \vec{r}^{\alpha\beta} + \vec{r}^{\beta\alpha}$, we can get these equations by projecting the vector equation equation into:

$$\zeta_{ij}^{\alpha}{}^{*} \sum_{\beta} R_i^{\alpha\beta} R_j^{\alpha\beta} + \sum_{\beta} (\zeta_{ij}^{\alpha}{}^{*} - \zeta_{ij}^{\beta}{}^{*}) r_j^{\beta\alpha} R_i^{\alpha\beta} = 0 , \tag{26}$$

$$\zeta_{ij}^{\alpha}{}^{*} \sum_{\beta} Q_i^{\alpha\beta} R_j^{\alpha\beta} + \sum_{\beta} (\zeta_{ij}^{\alpha}{}^{*} - \zeta_{ij}^{\beta}{}^{*}) r_j^{\beta\alpha} Q_i^{\alpha\beta} = 0 , \tag{27}$$

It should be emphasised that the system under consideration is disordered and therefore $\vec{Q}^{\alpha\beta} \neq 0$ (whereas for a honeycomb periodic array $\vec{Q}^{\alpha\beta} = 0$). Assuming as before that $\zeta_{ij}^{\alpha}{}^{*} - \zeta_{ij}^{\beta}{}^{*}$ gives rise to gradient terms we can exponentiate (26) and (27) by parametric variables $\phi^{\alpha}$ and $\psi^{\alpha}$:

$$\int e^{i \sum_{\alpha=1}^{N} \zeta_{ij}^{\alpha}{}^{*} (S_{ij}^{\alpha} - \phi^{\alpha} F_{ij}^{\alpha} - \psi^{\alpha} G_{ij}^{\alpha})} \prod_{\alpha}^{N} \mathcal{D}\zeta_{ij}^{\alpha}{}^{*} \mathcal{D}\phi^{\alpha} \mathcal{D}\psi^{\alpha} . \tag{28}$$

where $F_{ij}^{\alpha}$ is given by (19) and

$$G_{ij}^{\alpha} = \frac{1}{2} \left( \sum_{\beta} Q_i^{\alpha\beta} R_j^{\alpha\beta} + Q_j^{\alpha\beta} R_i^{\alpha\beta} \right) . \tag{29}$$

After integrating out the $\zeta^{\alpha}{}^{*}$, $\phi^{\alpha}$ and $\psi^{\alpha}$ fields, we find the following equation for $S_{ij}^{\alpha}$:

$$\begin{vmatrix} S_{11}^{\alpha} & F_{11}^{\alpha} & G_{11}^{\alpha} \\ S_{22}^{\alpha} & F_{22}^{\alpha} & G_{22}^{\alpha} \\ S_{12}^{\alpha} & F_{12}^{\alpha} & G_{12}^{\alpha} \end{vmatrix} = 0 . \tag{30}$$

Note that although there are explicit forms generalising (30) in 3-D, these are more complex algebraically as a consequence of the higher coordination number. $F_{ij}^{\alpha}$ and $G_{ij}^{\alpha}$ will depend on configuration and averaging (30) is quite complex. The simplest array will have $\vec{Q}^{\alpha\beta}$ orthogonal to $\vec{R}^{\alpha}$, i.e. if $\vec{R}^{\alpha\beta} = (X^{\alpha\beta}, Y^{\alpha\beta})$, then $\vec{Q}^{\alpha\beta} = (Y^{\alpha\beta}, -X^{\alpha\beta})$. It follows, that $F_{ij}^{\alpha}$ and $G_{ij}^{\alpha}$ can be written as

$$F_{ij}^{\alpha} = \begin{pmatrix} 1 & 0 \\ 0 & 1 \end{pmatrix} , \qquad G_{ij}^{\alpha} = \begin{pmatrix} \sin\theta^{\alpha} & \cos\theta^{\alpha} \\ \cos\theta^{\alpha} & -\sin\theta^{\alpha} \end{pmatrix} . \tag{31}$$

Then eqn.(30) can be rewritten:

$$S_{22}^\alpha - S_{11}^\alpha = 2S_{12}^\alpha \tan \theta^\alpha \qquad (32)$$

Thus if we are given $S_{12}^\alpha$, the probability of finding $S_{11}^\alpha - S_{22}^\alpha$ is

$$P\{S_{11}^\alpha - S_{22}^\alpha \,|\, S_{12}^\alpha\} = \frac{2}{\pi} \frac{|S_{12}^\alpha|}{(S_{11}^\alpha - S_{22}^\alpha)^2 + (S_{12}^\alpha)^2} . \qquad (33)$$

Mathematically it is more convenient to introduce $(\xi^\alpha)^2 = (S_{11}^\alpha - S_{22}^\alpha)^2 + (S_{12}^\alpha)^2$ and determine the probability of finding $S_{11}^\alpha - S_{22}^\alpha$ given $\xi^\alpha$.

$$P\{S_{11}^\alpha - S_{22}^\alpha \,|\, \xi^\alpha\} = \frac{1}{2\pi} \frac{1}{\sqrt{(\xi^\alpha)^2 - (S_{11}^\alpha - S_{22}^\alpha)^2}} . \qquad (34)$$

The mean values of $S_{11}^\alpha - S_{22}^\alpha$ and $S_{12}^\alpha$ are zero, hence we predict, rather obviously, hydrostatic pressure. However, notice that we are able to predict the fluctuations away from hydrostatic pressure, and would do more on correlations if one could find a pathway to measure them.     Another approach to deal with the system of discrete equations (22, 30) is to solve it for $S_{ij}^\alpha$, and then average the solution. This way seems to be feasible at least for the simplest granular systems (e.g. isotropic or periodic arrays) and may provide deep insight into the origins of the non-gaussian statistics of stress fluctuations. In complex cases this can be accomplished in some approximation, or by using computer simulations. By applying Fourier transformation to (22, 30) one can obtain $S_{ij}(\vec{k})$. The macroscopic stress tensor is obtained by averaging over the distribution of angles $\theta^\alpha$

$$i\sigma_{11}(\vec{k}) = \langle S_{11}(\vec{k}) \rangle_\theta = \frac{g_1(k_1^3 + 3k_2^2 k_1) + g_2(k_2^3 - k_1^2 k_2)}{|\vec{k}|^4} \qquad (35)$$

$$i\sigma_{22}(\vec{k}) = \langle S_{22}(\vec{k}) \rangle_\theta = \frac{g_2(k_2^3 + 3k_1^2 k_2) + g_1(k_1 k_2^2 - k_1^3)}{|\vec{k}|^4} \qquad (36)$$

$$i\sigma_{12}(\vec{k}) = \langle S_{12}(\vec{k}) \rangle_\theta = \frac{(g_1 k_2 - g_2 k_1)(k_2^2 - k_1^2)}{|\vec{k}|^4} \qquad (37)$$

where $|\vec{k}|^2 = k_1^2 + k_2^2$ and $\sigma_{ij}(\vec{r}) = \int \sigma_{ij}(\vec{k}) \, e^{i\vec{k}\vec{r}} \, d^3\vec{k}$. By doing the inverse Fourier transformation one can see that the macroscopic stress tensor is diagonal. There must also be constraints on the permitted configurations (due to the absence of tensile forces) which are not so easily expressed, for they affect each grain in the form

$$S_{ik}^\alpha M_{kl}^\alpha R_l^{\alpha\beta} n_i^{\alpha\beta} > 0 \qquad (38)$$

which has not yet been put into continuum equations other than Det $\sigma > 0$ and Tr $\sigma > 0$.

## 4. Discussion

In this paper we have derived the fundamental equations of stress equilibrium:

$$\nabla_j\,\sigma_{ij} + \nabla_j\nabla_k\nabla_m\,K_{ijkl}\sigma_{lm} + ... = g_i \tag{39}$$

$$P_{ijk}\sigma_{jk} + \nabla_j T_{ijkl}\sigma_{kl} + \nabla_j\nabla_l U_{ijkl}\sigma_{km} + ... = 0\,. \tag{40}$$

In order to solve these equations one needs to know the geometric quantities $K_{ijkl}$, $P_{ijk}$, $T_{ikl}$ and $U_{ijkl}$. In practice details of the distribution of intergranular contacts are not known in advance, but should be obtained from the deposition history of the system or experimental measurements of two-body correlation functions.

If the system is strongly anisotropic (i.e. there exists a preferred direction characterised by some angle $\phi$) and $\tan\theta^\alpha$ has an average value $\tan\phi$, then eqn.(30) becomes in the mean-field approximation

$$\sigma_{11} - \sigma_{22} = 2\sigma_{12}\tan\phi. \tag{41}$$

where $\phi$ is the angle of repose. It is known as the Fixed Principal Axes equation [3], and has been used with notable effect to solve the problem of the stress distribution in sandpiles. Explicit mathematical expressions for the 3-D case are more complex, and will be reported elsewhere. The issue of whether the derived system of equations (39-40) is robust against the inclusion of real friciton, softness of grains etc. illuminates the existence of a whole array of fascinating theoretical and experimental problems. Other important issues which are not addressed in this paper are that of stress fluctuations and the response of a granular aggregate to external perturbations. In general, cohesionless granular materials are quasistatic or "fragile"[7], which means that they cannot support certain types of infinitesimal changes in stress without configurational rearrangements.

In conclusion, our theory in its present form gives a simplified, but physical, picture of stress behaviour in cohesionless granular media. Further development is needed to make it a predictive tool which could be able to match experimental findings.

## 5. Acknowledgements

We acknowledge financial support from Leverhulme Foundation (S. F. E.), Shell (Amsterdam) and Gonville & Caius College (Cambridge) (D. V. G.). We thank Prof. Robin Ball for discussions.

144

## References

1. Edwards, S.F., Mounfield, C.C. (1996) A theoretical model for the stress distribution in granular matter, *Physica A*, **226**, pp. 1–25.
2. Bouchaud, J-P., Cates, M. E., Claudin, P. (1995) Stress distribution in granular media and nonlinear wave equation, *J. de Phys. I (France)*, **5**, pp. 639–656.
3. Wittmer, J. P., Clauidn, P., Cates, M. E. (1997) Stress propagation and arching in static sandpiles, *J. de Phys. I (France)*, **7**, pp. 39–80.
4. Nedderman, R. M. (1992) *Statics and Kinematics of Granular Materials*, CUP, Cambridge.
5. Wood, D. M. (1990) *Soil Behaviour and Critical State Soil Mechanics*, CUP, Cambridge.
6. Ball, R. C., Grinev, D. V. (1998) The Stress transmission universality classes of rigid grain powders, *cond-mat/9810124*.
7. Cates, M. E., Wittmer, J. P., Claudin, P. and Bouchaud, J. P. (1998) Jamming, Force chains, and Fragile Matter, *Phys. Rev. Lett.*, **81**, pp. 1841–1844.

# EQUATIONS OF GRANULAR MATERIALS: COMPACTIVITY AND COMPACTION

S. F. EDWARDS AND D. V. GRINEV

*Cavendish Laboratory*
*University of Cambridge, Madingley Road,*
*Cambridge CB3 OHE, United Kingdom*

## 1. Introduction

Traditionally, settling of powders under mechanical vibration has been a subject of soil mechanics and powder technology [1]. In recent years a large body of literature has emerged on the physics of vibrated granular media. A wide range of fascinating phenomena have been studied; vibration-induced size-segregation [2, 3], slow relaxation to a steady state density [4, 16], convection rolls [5], vibration history dependence of density [6], large density fluctuations about the steady state [7] and crystallization under horizontal shaking [8], to name but a few. A typical granular material is a system with a large number of individual grains and therefore it has a huge number of degrees of freedom. Grains interact with each other via contact forces which are determined by friction, gravitational loading and amplitude of an exernal force if the system is perturbed. Therefore one needs to invent a formalism that would allow us to calculate macroscopic averages in terms of microscopic (i.e. of individual grains) properties of the system. If we assume that it may be characterised by a small number of parameters (e.g. analogous to temperature ) and that this system has properties which are reproducible given the same set of extensive operations (i.e. operations acting upon the system as a whole, rather than upon individual grains) then we may apply the ideas of statistical averaging over the ensemble of configurations to granular systems [9].

## 2. Compactivity

We propose the statistical-mechanical approach which gives a clear physical explanation of experimental data [6]. It has been shown that external vibra-

145

*A.T. Skjeltorp and S.F. Edwards (eds.), Soft Condensed Matter: Configurations,*
*Dynamics and Functionality, 145-156.*

tions lead to a slow approach of the packing density to a final steady-state value. Depending on the initial conditions and the magnitude of the vibration acceleration, the system can either reversibly move between steady-state densities or can become irreversibly trapped into metastable states that is the rate of compaction and the final density depend sensitively on the history of vibration intensities that the system experiences (see Figure 3.1).

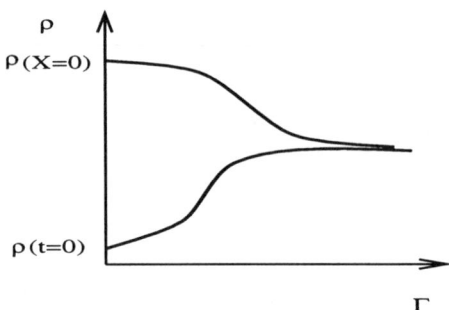

*Figure 1. Dependence of the steady-state packing density on the tapping history (Nowak et al.). Experimental values of density packing fraction are in the following correspondence with model parameters: $\rho(X = 0) = \frac{1}{v_0} \approx 0.64, \rho(t = 0) = \rho_0 = \frac{1}{v_1} \approx 0.58$ and $\rho(X = \infty) = \frac{2}{(v_0+v_1)} \approx 0.62$. The vibration intensity is parametrized by $\Gamma = \frac{a}{g}$*

In the present paper we consider the simplest model of a granular material by introducing the volume function $W$ and assume the simplest case that all configurations of a given volume are equally probable; in many cases the mechanism of deposition will leave a history in the configuration but this will not be considered here. $W$ will depend on the coordinates of the grains and their orientations and is the analogue of a Hamiltonian. Averaging over all the possible configurations of the grains in real space gives us a configurational statistical ensemble describing the random packing of grains. Since we are assuming that we are dealing with a system whose constituents are hard (i.e. impenetrable) we have to include some account of this in our formalism in order to reduce the number of possible configurations the system may occupy. Also for a packing which is stable under applied force we must consider the configurations restricting the number of possible volume states that the system may occupy to be only those configurations which are stable. Also grains cannot overlap and this condition produces very strong constraints (frustration) on their relative positions. This implies that all grains have to be in contact with their nearest neighbours. Of course in the real powder the topological defects can exist such as vacancies, voids or arches. But as these will be a subject of a future paper we do not consider them here. Thus we have a "microcanonical" probability distribution:

$$P = e^{-\frac{S}{\lambda}} \delta(V - W)\,\Theta(contacts)\,.\tag{1}$$

The normalization gives

$$e^{\frac{S}{\lambda}} = \int \delta(V - W)\,\Theta(contacts)\,\mathrm{d}(\text{all degrees of freedom})\tag{2}$$

where we define $\Theta$ as:

$$\Theta(contacts) = \begin{cases} 1 & \text{if } z \geq z_m \\ 0 & \text{if } z < z_m \end{cases}\tag{3}$$

where $z_m$ is the minimal coordination number of a grain. We have to introduce $\Theta$ because we consider the stable isotropic and homogeneous packings. Just as in conventional statistical mechanics with microcanonical distribution:

$$P = e^{-\frac{S}{k}} \delta(E - H)\,,\tag{4}$$

and temperature:

$$T = \frac{\partial E}{\partial S}\,.\tag{5}$$

We can define the analogue of temperature as:

$$X = \frac{\partial V}{\partial S}.\tag{6}$$

This fundamental parameter is called compactivity [9]. It characterises the packing of a granular material and may be interpreted as being characteristic of the number of ways it is possible to arrange the grains in the system into volume $\Delta V$ such that the disorder is $\Delta S$. Consequently the two limits of $X$ are 0 and $\infty$, corresponding to the most and least compact stable arrangements. This is clearly a valid parameter for sufficiently dense powders because one can in principle calculate the configurational entropy of an arrangement of grains and therefore derive the compactivity from the basic definition. One can expect despite the strong constraints resulting from the stability conditions, the number of packings to grow exponentially with the volume of a sample and the configurational entropy defined as a logarithm of this number is extensive.

As usual it is more convenient to introduce the canonical probability distribution:

$$P = e^{\frac{Y-W}{\lambda X}}, \tag{7}$$

where $\lambda$ is a constant which gives the entropy the dimension of volume, $Y$ we call the effective volume, it is the analogue of the free energy:

$$e^{-\frac{Y}{\lambda X}} = \int e^{-\frac{W(\mu)}{\lambda X}} \, d(\text{all}), \tag{8}$$

$$V = Y - X \frac{\partial Y}{\partial X}. \tag{9}$$

To illustrate this theory consider the simplest example of a $W$, the analogue of Bragg-Williams approximation: each grain has neighbours touching it with a certain coordination and angular direction. In order to set up an analogy with the statistical mechanics of alloys we assume that each grain has a certain property, which defines the "interaction" with its nearest neighbours. Taking the coordination number of a grain as such a property and assuming that there are just two types of coordination $z_0$ and $z_1$ we assign a volume $v_i$ to any grain with $z_i$ coordination number. Thus we write the volume function as:

$$W = n_0 v_0 + (N - n_0) v_1 \tag{10}$$

where N is the number of grains in the system, $n_i$ is the number of grains with the coordination number $z_i$ and $N = n_0 + n_1$. The simple calculation of $Y$ and $V$ gives us:

$$Y = N \frac{(v_0 + v_1)}{2} - N\lambda X \ln 2\cosh \frac{(v_0 - v_1)}{\lambda X} \tag{11}$$

$$V = N \frac{(v_0 + v_1)}{2} + N \frac{(v_0 - v_1)}{2} \tanh \frac{(v_0 - v_1)}{\lambda X}). \tag{12}$$

Thus we have two limits: $V = Nv_0$, when $X \to 0$ and $V = N(v_0 + v_1)/2$ when $X \to \infty$ ($N$ is a number of grains). Note that the maximum $V$ is not $Nv_1$ just as in the thermal system (say a spin in a magnetic field) with two energy levels $E_0$ and $E_1$ one has $E = E_0$ when $T \to 0$ and $E = (E_0 + E_1)/2$ when $T \to \infty$.

## 3. The Model

We consider the rigid grains powder dominated by friction deposited in a container which will be shaken or tapped(in order to consider the simplest case we ignore other possible interactions e.g. cohesion and do not distinguish between the grain-grain interactions in the bulk and those on the boundaries). We assume that most of the particles in the bulk do not acquire any non ephemeral kinetic energy i.e. the change of a certain configuration occurs due to continuous and cooperative rearrangement of a free volume between the neighbouring grains. Any such powder will have a remembered history of deposition and in particular can have non-trivial stress patterns, but we will confine the analysis of this paper to systems with homogeneous stress which will permit us to ignore it. The fundamental assumption is that under shaking a powder can return to a well defined state, independent of its starting condition. Thus in the simplest system, a homogeneous powder, the density characterises the state.

It is sensible to seek the simplest algebraic model for our calculation and to this end since the orientation of the grain must have at least two degrees of freedom, say $\mu_1$ and $\mu_2$, our volume function is:

$$W = v_0 + (v_1 - v_0)(\mu_1^2 + \mu_2^2) \tag{13}$$

implying a picture in two-dimensions (see Figure 3.2).

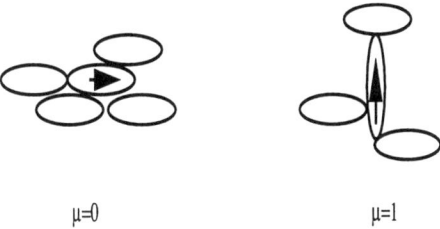

$$\mu=0 \qquad\qquad\qquad \mu=1$$

*Figure 2. Graphical representation for the limit values of the degree of freedom $\mu$ in 2-D*

When $\mu = 0$ we have $W = v_0$ then the grain is "well oriented" which means that a free volume is minimal and when $\mu = 1$ and $W = v_1$ then the grain is "not well oriented" (free volume is maximal). It is a self-consistent approximation since the parameters $v_0$ and $v_1$ are the average volumes of the grain in the presence of other grains. For the sake of simplicity we do not take into account the shape of particles. However, We note that this can be done explicitly in the cases of elongated grains [10], mixtures of spheres and elongated grains [11], and irregularly shaped grains [12]. In general we can write:

$$e^{-\frac{Y}{\lambda X}} = \int w(\mu)\, e^{-\frac{W(\mu)}{\lambda X}}\, d\mu \tag{14}$$

where $w(\mu)$ is the weight factor attached to $\mu$. From (8) we derive $Y$ and $V$:

$$Y = Nv_0 - N\lambda X \ln \left\{ \frac{\lambda X}{v_1 - v_0} (1 - e^{-\frac{v_1 - v_0}{\lambda X}}) \right\} \tag{15}$$

$$V = N(v_0 + \lambda X) - \frac{N(v_1 - v_0)}{e^{\frac{v_1 - v_0}{\lambda X}} - 1}. \tag{16}$$

Thus we have the same limits as for volume function (12): $V = Nv_0$, when $X \to 0$ and $V = N(v_0 + v_1)/2$ when $X \to \infty$.

## 4. The Fokker-Planck Equation

The main physical idea of our approach is the following: all grains in the bulk experience the external perturbation as a random force with zero correlation time so that the process of compaction can be seen as the Ornstein-Uhlenbeck process for the degrees of freedom $\mu_i$, $i = 1, 2$. Therefore we write the Langevin equation:

$$\frac{d\mu_i}{dt} + \frac{1}{\nu} \frac{\partial W}{\partial \mu_i} = \sqrt{D} \, f_i(t) \tag{17}$$

where $\langle f_i(t) f_j(t') \rangle = 2\delta_{ij}\delta(t-t')$ and $\nu$ characterises the frictional resistance imposed on the grain by its nearest neighbours. The term $f_i(t)$ on the RHS of (17) represents the random force generated by a tap. The terms "shaken" or "tapped" have been used above and we have to make them more precise. The derivation gives the analogue of the Einstein relation that $\nu = (\lambda X)/D$. If we identify $f$ with the amplitude of the force $a$ used in the tapping, the natural way to make this dimensionless is to write the "diffusion" coefficient as :

$$D = \left(\frac{a}{g}\right)^2 \frac{\nu \omega^2}{v}, \tag{18}$$

that is we have a simplest guess for a fluctuation-dissipation relation:

$$\lambda X = \left(\frac{a}{g}\right)^2 \frac{\nu^2 \omega^2}{v} \tag{19}$$

where $v$ is the volume of a grain, $\omega$ the frequency of a tap and $g$ the gravitational acceleration. Use of the Langevin equation (17) is of course a crude simplification as it does not explicitly take into account the presence of boundaries and topological constraints. Generally speaking one would have to use the integro-differential Langevin equation with the memory kernel:

$$\frac{d\mu_i}{dt} + \int_0^t K(t - t') \, \mu_i(t') \, dt' = \sqrt{D} \, f_i(t), \tag{20}$$

as one sees in experiment that the final density depends sensitively on the history of vibration intensities. Clearly to solve such an equation is not a trivial task although the solution could give us the better understanding of many interesting features of granular compaction. The problem of how to choose the initial values of $\mu$ is in reality the deposition problem. We discuss it later.

The Langevin equation can be easily solved for $W$ quadratic in $\mu$:

$$\mu_i(t) = \mu_i(0)e^{-\gamma t} + \sqrt{D}e^{-\gamma t} \int_0^t f(t')e^{\gamma t'}\,dt' \tag{21}$$

Averaging over the ensemble we get:

$$\langle \mu_i(t) \rangle = \mu_i(0)e^{-\gamma t} \tag{22}$$

where $\mu_i(0) = 1$ is the initial value of $\mu_i, \gamma = 2\frac{(v_1 - v_0)}{\nu}$ has the meaning of relaxation time of the degree of freedom $\mu$. As $t \to \infty$ $\mu$ goes to $\mu_f = 0$ which corresponds to the random close packing limit. The Fokker-Planck equation seems to be quite generic in modelling the response of granular materials to an externally applied shear although in that problem it is more convenient to use the volume "Hamiltonian" $W$ as a function of the coordination number of each grain [14]. The standard treatment of the Langevin equation (17) is to use it to derive the Fokker-Planck equation:

$$\frac{\partial P}{\partial t} = \left(D_{ij}\frac{\partial^2}{\partial \mu_i \partial \mu_j} + \gamma_{ij}\frac{\partial}{\partial \mu_i}\mu_j\right)P = 0 \tag{23}$$

where $D_{ij} = D\delta_{ij}$ and $\gamma_{ij} = \gamma\delta_{ij}$. Equation (23) can be solved explicitly. It has right- and left-hand eigenfunctions $P_n$ and $Q_n$ and eigenvalues $\omega_n$ such that:

$$\omega_n P_n = \frac{\partial}{\partial \mu_j}\left(D_{ij}\frac{\partial}{\partial \mu_i} + \gamma_{ij}\mu_j\right)P_n \tag{24}$$

$$\omega_n Q_n = \left(-D_{ij}\frac{\partial}{\partial \mu_i} + \gamma_{ij}\mu_j\right)\frac{\partial}{\partial \mu_j}Q_n \tag{25}$$

or equivalently a Green function:

$$G = \sum_n P_n(\mu)Q_n(\mu)e^{-\omega_n t}. \tag{26}$$

It follows that if we start with a non-equilibrium distribution:

$$P^{(0)}(t = 0) = \sum_{n=0}^{\infty} A_n P_n, \qquad A_n = \int Q_n P^{(0)}\,d\mu_1\,d\mu_2 \tag{27}$$

and it will develop in time as:

$$P^{(0)}(t) = A_0 P_0 + \sum_{n \neq 0}^{\infty} A_n P_n e^{-\omega_n t} \qquad (28)$$

where $\int P^{(0)} \, d\mu_1 d\mu_2 = A_0$. This coefficient is determined by a number of grains present in the powder, hence must be a constant. The steady-state distribution function is:

$$P^{(0)}(t \to \infty) = \frac{e^{-\frac{(v_1 - v_0)(\mu_1^2 + \mu_2^2)}{\lambda X}}}{\int_0^1 e^{-\frac{(v_1 - v_0)(\mu_1^2 + \mu_2^2)}{\lambda X}} \, d\mu_1 \, d\mu_2} \qquad (29)$$

The Fokker-Planck operator (23) has a complete orthogonal set of eigenfunctions:

$$P_n = H_n e^{-\frac{(v_1 - v_0)(\mu_1^2 + \mu_2^2)}{\lambda X}} \qquad (30)$$

where $H_n$ are Hermite polynomials and $\mu_i \in (0, \infty)$. In our case $\mu_i \in (0, 1)$. One can avoid this mathematical difficulty, taking into account the crudety of our model and constructing the "first excited state":
$P_2 = (a(\mu_1^2 + \mu_1^2) + b) \, e^{-\frac{(v_1 - v_0)(\mu_1^2 + \mu_1^2)}{\lambda X}}$ orthogonal to the ground-state eigenfunction $P_0$. This eigenfunction describes the initial state of our system i.e. loosely packed deposited powder. Therefore it is easy to see the initial nonequilibrium distribution (27) depends on how the the powder is deposited. Constants $a$ and $b$ can be defined from the orthonormality relations. By using:

$$P_n = Q_n P^{(0)}(t \to \infty), \quad Q_0 = 1 \qquad (31)$$

and:

$$\int_0^1 Q_2 \hat{L}_{FP} P_2 \, d\mu_1 \, d\mu_2 = \omega_2, \qquad (32)$$

one can easily verify that the eigenvalue $\omega_2$ (which corresponds to $P_2$ and gives us the decay rate of our nonequilibrium distribution) is a constant dimensionless number.

Suppose now that deposition produces a highly improbable configuration, indeed the most improbable configuration with: $\mu_1^2 + \mu_2^2 = 2$ and the mean volume function is $\bar{W} = 2v_1 - v_0 \approx v_1$, where:

$$\bar{W}(X, t) = \int P^{(0)}(X, t) W \, d\mu_1 d\mu_2. \qquad (33)$$

It is possible to imagine a state where all the grains are improbably placed, i.e. where each grain has its maximum volume $v_1$. In a thermal analogy this would be like fully magnetised magnetic array of spins where the magnetic field is suddenly reversed. Such a system is highly unstable and equilibrium statistical mechanics does not cover this case at all. It will thermalize consuming the very high energy whilst establishing the appropriate temperature. Powders however are dominated by friction, so if one could put together a powder where the grains were placed in high volume configuration, it will just sit there until shaken; when shaken it will find its way to the distribution (7). It is possible to identify physical states of the powder with characteristic values of volume in our model. The value $V = Nv_1$ corresponds to the "deposited" powder, i.e. the powder is put into the most unstable condition possible, but friction holds it. When $V = Nv_0$ the powder is shaken into closest packing. The intermediate value of $V = (v_0 + v_1)/2$ corresponds to the minimum density of the reversible curve. Thus we can offer an interpretation of three values of density presented in the experimental data [6].

The general solution of the Fokker-Planck equation (23) goes to its steady-state value when $t \to \infty$ so we can expect $\bar{W}(X, t)$ to diminish (as the amplitude of tapping increases) until one reaches the steady-state value $\bar{W}(X)$. The formula (12) can be obtained using (33) when $t \to \infty$ and represents a reversible curve in experimental data of [6]: altering $a$ moves one along the curve $\rho = \frac{v}{\bar{W}(X)} = \rho(a)$. We can identify time with the number of taps, so wherever we start with any initial $\rho_{(0)}$ and $a$, successive tapping takes one to reversible curve $\rho(a)$. Or, if one decides on a certain number of taps, $t \neq \infty$ , one will traverse a curve $\rho_t(a)$, where $\rho_\infty(a) = \rho(a)$. Notice that the simple result lies within the crudety of our model. The general problem will not allow us to think of $X$ as $X(a)$ independent of the development of the system. The thermal analogy is this: if the Brownian motion in an ensemble of particles is controlled by a random force $f$ which is defined in terms of its amplitude and time profile, this random force defines the temperature in the system. Our problem is like a magnetic system where magnetic dipoles are affected by a constant magnetic field, being random at high temperature , and increasingly oriented by the external field as the temperature falls.

## 5. Discussion

The physical picture presented in this paper is consistent with everyday knowledge of granular materials: when poured they take up a low density but when shaken settle down, unless shaken violently when they return to low density. These effects are much more pronounced in systems with irregularly shaped grains then with fairly smooth uniform spheres, indeed the more irregular a grain is, the more the discussion above describes big differences between $\rho_{(0)}$ and $\rho$. The experimental data of [6] show the packing density dependence on parameter $\Gamma = a/g$ for a fixed number of taps. A loosely packed bead assembly first undergoes irreversible compaction corresponding to the lower branch of $\rho(\Gamma)$. The settling behavior becomes reversible only once a characteristic acceleration has been exceeded. Our theory gives three points $\rho(X = 0), \rho(X = \infty)$ and $\rho(t = 0)$ which are in the ratio: $v_0^{-1}$, $\frac{2}{(v_0+v_1)}$, $v_1^{-1}$ and these are in reasonable agreement with experimental data: $\rho(X = 0) = \frac{1}{v_0} \approx 0.64$, $\rho_0 = \frac{1}{v_1} \approx 0.58$ and $\rho(X = \infty) = \frac{2}{(v_0+v_1)} \approx 0.62$. Another important issue is the validity of the compactivity concept for a "fluffy" but still mechanically stable granular arrays e.g. for those composed of spheres with $\rho \leq 0.58$. In our theory $\rho(X = \infty)$ corresponds to the beginning of the reversible branch (see Fig.1) and using our analogy with a magnetic system is analogous to dipoles at a high temperature. The irreversible branch has an analogue in the behaviour of the magnetic system where initially the dipoles can, by orienting them in a strong field, then suddenly reversing the field, be put in a configuration inaccessible in any equilibrium temperature regime, butonce thermalised from then on behave in the usual way.

The fluffy powder is a very complicated object as it has plenty of topological defects and stress arches. Throughout the paper we assumed that our granular array is spatially homogeneous which is the case for densities of the reversible curve. However this is a very subtle problem which will be a subject of a future paper. It is a difficult problem to decide whether embarking on a vast amount of algebraic work that a superior mode would entail is worthwhile. But our simple model is quite physical and can be extended when experiments would justify the work.

A final point is that we find the lower (irreversible) curve build up to the upper (reversible) curve exponentially in time:

$$V(t) = V_i e^{-\omega t} + V_f(1 - e^{-\omega t}) \tag{34}$$

while one can expect the logarithmic in time approach to the steady state density e.g. the Vogel-Fulcher type curve which is typical of disordered thermal systems such as spin and structural glasses [15]

$$V(t) = V_f + (V_i - V_f)e^{-\omega t} + Tt^{-\epsilon} \tag{35}$$

where $\epsilon$ is large.

The function which has been found to fit the ensemble averaged density $\rho(t)$ better than other functional forms, is [4]:

$$\rho(t) = \rho_f - \frac{\Delta \rho_f}{1 + B \log(1 + \frac{t}{\tau})} \tag{36}$$

where the parameters $\rho_f$, $\Delta \rho_f$ and $\tau$ depend only on $\Gamma$.

In fact, identifying $t$ with the number of taps $n$, the law seems to be even slower at $(\ln t)^{-1}$. Our simple analysis is clearly inadequate to obtain such a result which is quite outside the straightforward method of expansion in the present set of eigenfunctions. However, there is an argument by de Gennes [16] which says that a Poisson distribution of voids can provide this logarithmic behaviour.

In statistical mechanics, the fluctuation-dissipation theorem relates the dissipative response of a system to an external perturbation, with the microscopic dynamics of the system in a state of equilibrium. The configurational space for the granular medium is explored by fluctuations induced by external perturbation sources, such as vibrations. In a vibrated granular material, density fluctuations from the steady-state represent the different volume configurations accessible to grains. In thermal statistical mechanics we can write the specific heat as

$$C_V = \left(\frac{d\,E_0}{d\,T}\right)_V = \frac{\langle (E - E_0)^2 \rangle}{k_B\,T^2} \tag{37}$$

where $E_0$ is the equilibrium average of the energy of the system, $k_B$ is Boltzmann's constant, $T$ is the absolute temperature and $\langle \dots \rangle$ represents the time average. In our theory, the analogous quantity to the specific heat (37) is

$$C = \frac{d\,V_{ss}}{d\,X} = \frac{\langle (V - V_{ss})^2 \rangle}{\lambda\,X^2} \tag{38}$$

where $V_{ss}$ is the steady-state volume.

$$\int_{V_1}^{V_2} \frac{d\,V_{ss}}{\langle (V - V_{ss})^2 \rangle} = \frac{1}{\lambda\,X_2} - \frac{1}{\lambda\,X_1} \tag{39}$$

The equation (39) has been used [7] to calculate the compactivity difference for any two volumes by measuring the fluctuations of the densities in the steady-state. Given a suitable point of reference, this equation allows the determination of an absolute value for compactivity. Despite all its merits, this is an indirect method of measuring $X$. However, how can one then measure $X$ directly? We argue that a direct method should be based on an analog the Zeroth Law of thermodynamics: no net volume will be transferred between the two granular materials when they are placed in contact with each other if they have the same compactivity. This means that we

156

should be able to define a state of equilibrium such that two systems in equilibrium with a third system are also in equilibrium with each other. To conclude, the applicability of the compactivity concept in the description of quasistatic granular materials requires further experimental investigation. In particular, it would be interesting to examine the properties of granular media comprised of irregular particles and subject to isotropic vibration.

## 6. Acknowledgements

We acknowledge financial support from Leverhulme Foundation (S. F. E.), Shell (Amsterdam) and Gonville & Caius College (Cambridge) (D. V. G.).

## References

1. Evans, P. E., Millman, R. E. (1967) The vibratory packing of powders, *Vibratory Compacting: Principles and Methods*, Hausner, H. H., Roll, K. H., Johnson, P. K. (eds.) **V.2**, pp. 237, Plenum Press, New York.
2. Rosato, A., Prinz, F., Standburg, K.J. and Swendsen, R. H. (1987) Why the brazil nuts are on top - size segregation of particulate matter by shaking, *Phys. Rev. Lett.*, **58**, pp. 1038–1042.
3. Jullien, R., Meakin, P., Pavlovitch, A. (1992) 3-D model for particle-size segregation by shaking, *Physical Review Letters*, **69**, pp. 640–643.
4. Knight, J. B., Fandrich, C. G., Lau, C. N., Jaeger, H. M., Nagel, S. R. (1995) Density relaxation in a vibrated granular material, *Phys. Rev. E*, **51**, pp. 3957–3963.
5. Knight, J. B., Ehrichs, E.E., Kuperman, V. Y., Flint, J. K., Jaeger, H.M. and Nagel, S. R. (1996) Experimental study of granular convection, *Phys. Rev. E*, **54**, pp. 5726–5738.
6. Nowak, E. R., Knight, J. B., BenNaim, E., Jaeger, H. M. and Nagel, S. R. (1997) Reversibility and irreversibility in the packing of vibrated granular material, *Powder Technology*, **94**, pp. 79–83.
7. Nowak, E. R., Knight, J. B., BenNaim, E., Jaeger, H. M. and Nagel, S. R. (1998) Density fluctuations in vibrated granular materials, *Phys. Rev. E*, **57**, pp. 1971–1982.
8. Pouliquen, O.Nicolas, M. and P. D. Weidman, P. D. (1997) Crystallization of non-brownian spheres under horizontal shaking, *Phys. Rev. Lett.*, **79**, pp. 3640–3644.
9. Edwards, S.F. and Oakeshott, R. B. S. (1989) Theory of powders, *Physica A*, **157**, pp. 1080–1090.
10. Mounfield, C. C. and Edwards, S. F. (1994) The statistical mechanics of granular systems composed of elongated grains, *Physica A*, **210**, pp. 279–289.
11. Mounfield, C. C. and Edwards, S. F. (1994) The statistical mechanics of granular systems composed of spheres and elongated grains, *Physica A*, **210**, pp. 290–300.
12. Mounfield, C. C. and Edwards, S. F. (1994) The statistical mechanics of granular systems composed of irregularly shaped grains, *Physica A*, **210**, 301–316.
13. Monasson, R. Pouliquen, O. (1997) Entropy of particle packings: An illustration on a toy model, *Physica A*, **236**, pp. 395–410.
14. Higgins, A. and Edwards, S. F. (1992) A theoretical approach to the dynamics of granular materials, *Physica A*, **189**, pp. 127–132.
15. Mezard, M. Parisi, G. and Virasoro, M. A. (1987) *Spin Glasses and Beyond*, World Scientific, Singapore.
16. Boutreux, T. and de Gennes, P. G. (1997) Compaction of granular mixtures: a free volume model, *Physica A*, **244**, pp. 59–67.

# MODELING GRANULAR FLOWS

E G. FLEKKØY[1], S. MCNAMARA[2] AND K. J. MÅLØY[1]

[1] *Department of Physics, University of Oslo*
*P.O. Box 1048 Blindern, 0316 Oslo 3, Norway*
[2] *City College, New York, USA*

**Abstract.** We study a variety of granular flows and some models by means of a combination of simulations, experiments and analytic considerations. The models as well as the physical systems are governed by the presence of an interstitial gas. Through experimental comparison and validation of the models their applicability is determined. In particular we study bubbles of air in tubes (by experiment and simulation) the formation of stagnant zones in grain flow from a hopper (by experiment and simulation), fluidized beds (by simulation) and the intermittent flow in an hourglass (by experiment and simulation).

## 1. Introduction

The modeling of granular flows is still, in many respects, in its infancy. In hydrodynamics a wide variety of phenomena are described by the Navier Stokes equations for incompressible fluid flow. No analogous unified descriptions for granular flows exists, and for every attempt at modeling it is important to examine carefully the scope and validity of the corresponding description. In most cases of interest, this cannot be done without direct experimental verification.

Recently much effort has been devoted to such modeling (Jenkins, 1983; Goldhirsch, 1993; Peng, 1994; Mc Namara, 1999; Flekkøy, 1996). Like all dynamic systems, granular flows obey the basic conservation laws of mass, momentum and energy. These conservation laws can be cast in the form of differential equations. However, in order to make such a description useful, constitutive equations are needed, in particular relations between strain and stress (Jenkins, 1983), and an equation of state. Since the validity of such

157

*A.T. Skjeltorp and S.F. Edwards (eds.), Soft Condensed Matter: Configurations,*
*Dynamics and Functionality, 157-184.*

constitutive relations depend strongly on the granular material at hand as well as its dynamical state, general continuum descriptions of granular flows are difficult, if not impossible. While it is true that existing descriptions may work well for strongly excited granular media, they are not applicable when the medium is less excited.

In granular flows many distinct phenomena governed by gas–grain interactions are known: Even in a stationary settled bed of grains a bubble of air will rise and dissolve as it moves (Flekkøy, 1996). In gas-filled hourglasses intermittent flow as well as bubble formation will occur due to the volume exchange between the gas and grain phases (see Refs. (Wu, 1993) and (lePennec, 1996; Veje, 1997) which introduces the "ticking hour glass"). The main purpose of the present paper is to study two different models for such granular flows. The first model is conventional in the sense that it describes both the gas and the grains in terms of density fields. The second model evolved from the first largely from the requirement of small scale particle resolution and a microscopic interpretation of the granular phase.

Among the models that focus on dynamic aspects of granular flows there exists various continuum descriptions (Hayakawa, 1995; Tan, 1995; Manger, 1994; Beverloo, 1961) as well as cellular automata models that seek to capture some essential part of the physics (Peng, 1994). For the case of fluidized beds (Davidson, 1995; Davidson, 1971; Gidaspau, 1994) continuum descriptions that describe the state of granular material excited by the flow of gas as an in-viscid liquid, have been studied (Davidson, 1995). In general these models describe granular flows that resemble the flow of fluids.

However, many granular flows are not in a strongly excited, fluid-like state. A salient feature of such 'cooler' flows is the transition from a solid-like behavior, where the material is kept in place by the walls of its container, to a fluid-like behavior where there are continuous internal deformations. This behavior has not been previously studied by continuum models— although it has frequently been observed in experimental studies, for instance in Refs. (Wu, 1993; Baxter, 1989). The solid-fluid transition is the main focus of the continuum model and it is shown that it captures the experimentally observed stagnant zones that form under hopper flow.

Both the fundamental and industrial interest in gas governed granular flows call for reliable and efficient numerical modeling. Among existing, particle based models the hydrodynamic description of the inter-granular fluid has represented a bottle neck. In these models the interacting particles are taken to define moving boundary conditions for the Navier Stokes equations (vanderHoof, 1991; Ladd, 1994; Ladd, 1994; Kalthoff, 1997), and the fluid flow field thus represents the finest level of resolution. This implies that the computational effort required per particle is too large for many applications. Moreover, for many of the large scale phenomena of interest this level of detail in the description is not needed. In the present particle based model the particles constitute the finest level of detail and are described by Molecular Dynamics (MD), and the hydrodynamics is described only on a coarse grained level. The main advantage of this scheme is to keep the microscopic description of the particles with the ability this gives to describe strong density variations like shock fronts as well as friction and stagnant particle regions, while avoiding the large computational cost of obtaining the interparticle fluid flow field. The fact that fluid inertia is neglected and the fluid described in terms of the pressure only gives a simpler and computationally more economical model.

The 'holy grail' of the present modeling efforts is the 'ticking hourglass' (Wu, 1993). In this system all the uniquely granular phenomena; arch formation, dilatancy (Reynolds, 1885), interaction with the interstitial fluid and the formation of heaps and stagnant zones combine. However, in the quest for the successful modeling of the ticking hourglass we exploit the opportunity to study intermediate and simpler systems.

These include sedimentation processes and fluidized beds (Lim, 1995). The fluidized bed in particular provides a natural testing ground on which to start the validation process. By means of the particle based model it is shown that the qualitative aspects of this phenomena is correctly captured: When external pressure gradients are applied the particle bed fluidizes and spontaneously produces bubbles of the shape observed experimentally. Moreover, these bubbles are seen to merge as is experimentally observed. Finally we simulate the ticking hourglass in the absence of interparticle friction (though particle collisions still dissipate energy) and show qualitatively that the intermittent flow and bubble formation at the orifice is captured in the model.

## 2. An inertialess continuum model

As a starting point of the description we define the local densities of sand and air. The sand volume fraction is $\rho_s = 1 - \phi$, where $\phi$ is the porosity and $\rho_g$ is the density of the material that makes up the grains. Likewise we define the dimension-less density of air $\rho$ as the mass of air per unit volume normalized by the density of air at atmospheric pressure in a random loose packing. We shall use the result given by Scott (Scott, 1960) for the porosity of a random loose packing of spheres. The value he obtained by extrapolating data obtained by measurements on finite size container to an infinite container size was 0.40. Hence, in a random loose packing at atmospheric gas pressure $\rho_s = 0.60$ and $\rho = 1.0$.

We shall take the threshold density $\rho_0$ below which the sand is allowed to move as that of the random loose packing. When $\rho_s$ exceeds $\rho_0$ the motion of the sand is frozen relative to the walls. We shall refer to this state of the sand as *solidified*, and we shall call the state where $\rho_s \leq \rho_0$ and the sand can move as *fluidized*.

The conservation of sand and air is described by the continuity equations

$$\partial_t \rho + \nabla \cdot \mathbf{j}_a = 0 \tag{1}$$

$$\partial_t \rho_s + \nabla \cdot \mathbf{j}_s = 0 \tag{2}$$

where the (normalized) mass currents of air and sand, $\mathbf{j}_a$ $\mathbf{j}_s$, have been introduced. These flow velocities are partly determined by the local Darcy law

$$\mathbf{j}_a = \rho(\mathbf{u}_s - \frac{\kappa}{\mu}\nabla P) \tag{3}$$

where the sand flow velocity $\mathbf{u}_s = \mathbf{j}_s/\rho_s$ and $\kappa = \kappa(\rho_s)$ is the local permeability of the sand matrix, $\mu$ the dynamic viscosity of air and $P$ the interstitial air pressure. Equation (3) says that in the local rest frame of reference for the sand the air flow is proportional to the pressure gradient. We shall take the local permeability as that given by the Carman-Kozeny expression (Carman, 1937) for packings of spheres. It reads

$$\kappa(\rho_s) = \frac{a^2}{9K}\frac{(1-\rho_s)^3}{\rho_s^2} \tag{4}$$

where $a$ is the spheres radius and the constant $K \simeq 5$ is obtained experimentally for a random packing of spheres. The above Darcy law plays the role of a constitutive equation for the mass currents and relies on the flow to be governed by viscous– rather than inertial forces, i.e. the pertinent Reynolds number must be small (Landau, 1959). In the experiments particles of diameter 50 $\mu$m and 65 $\mu$m are used. At terminal velocity for single particles in air the Reynolds number is less than 0.8, which in the present context will be considered marginally small. In the modeling of fluidized beds Eq. (3) is employed with an extra term that is second order in velocity (Homsy, 1988). This term can be neglected at small Reynolds numbers. For fluidized beds momentum equations both for the fluid and granular phases are used. The neglect of inertial terms constitute the main simplification in the present model. While inertial forces may be negligible for single particles they may be important for the motion of clusters of particles.

In Ref. (Flekkøy, 1996) we argue that the interstitial air will be isothermal when particles are sufficiently small. In that case the pressure is easily written in terms of the densities $\rho$ and $\rho_s$. It follows from the isothermal equation of state for an ideal gas that $\phi P \propto \rho$ where $\phi = 1 - \rho_s$ is the porosity. By taking the gradient on both sides of this equation and then dividing by the equation itself we obtain

$$\frac{\nabla P}{P} = \frac{\nabla \rho}{\rho} + \frac{\nabla \rho_s}{1 - \rho_s} . \tag{5}$$

Substituting this equation of state in Eq. (3) we get

$$\mathbf{j}_a = \rho \left[ \mathbf{u}_s - \frac{\kappa P}{\mu} \left( \frac{\nabla \rho}{\rho} + \frac{\nabla \rho_s}{1 - \rho_s} \right) \right] \tag{6}$$

as the governing equation for the flow of air.

The equations (1)–(6) do not fully determine the dynamic evolution of the conserved densities. The physical assumption we add to complete the description is that the sand flow immediately reaches steady state, i.e.we neglect the acceleration of the sand. This assumption is justified for the case of single small grains: The distance a spherical grain of glass of radius $a$ needs to fall to reach terminal velocity is approximately $(a/(50\mu m))^4$ 3cm if the particle is falling freely. In a dense packing where there is also an initial upwards flow of air this distance is likely to be significantly smaller. If grain accelerations thus are neglected the net force (per unit volume) acting on the sand must vanish, i.e. $\nabla P = \rho_s \mathbf{g}$, where $\mathbf{g}$ is the acceleration of gravity. We will substitute this relation in Equation (3) to get an expression for $\mathbf{u}_s$. However, since we are neglecting inertial effects, we will have problems when the falling sand causes the the air to move downwards, and, in the next time step, the sand again acquires terminal velocity in the moving air. This downwind instability may be avoided by simply imposing that the sand only feels air motion that reduces the sand velocity. The sand velocity is then prescribed by the following relation

$$\mathbf{u}_s = \begin{cases} \frac{\mathbf{j}_a}{\rho} + \frac{\kappa}{\mu}\rho_s \mathbf{g} & \text{when} & |\mathbf{u}_s - \frac{\mathbf{j}_a}{\rho}| \leq \mathbf{u}_s \\ \frac{\kappa}{\mu}\rho_s \mathbf{g} & \text{when} & |\mathbf{u}_s - \frac{\mathbf{j}_a}{\rho}| > \mathbf{u}_s \\ \mathbf{0} & \text{when} & \rho_s \geq \rho_0 \end{cases} . \tag{7}$$

Note that while the approximate Eq. (7) depends on the relative velocity $\mathbf{j}_a/\rho - \mathbf{u}_s$ only, the $\nabla \rho/\rho$ term in Eq. (6) depends on $\mathbf{j}_a/\rho$ through Eq. (1). Equations (7) and (6) are thus independent and fully determine the flow velocities $\mathbf{u}_a$ and $\mathbf{u}_s$. The physical reason for this is the approximation of vanishing accelerations in Eq. (7). By the above assumptions we have obtained a closed set of equations on the level of mass conservation, thus avoiding the complications of finding the correct constitutive equations that describe the flow of momentum or energy.

How do the density fields relax to a state as described by Eqs. (7) and (3)? This question is less obvious than in the case where a momentum

equation is employed. When the first line of Eq. (7) is used to get the
the current value of $\mathbf{u}_s$ at time $t_n$, the previous value of $\mathbf{j}_a$ at time $t_{n-1}$ is
required. By combining the first line of Eqs. (7) and Eq. (3) to eliminate $\mathbf{j}_a$
we obtain the recursion relation

$$
\begin{aligned}
\mathbf{u}_s^{(n+1)} &= \mathbf{u}_s^{(n)} + \left[ \frac{\kappa^{(n)}}{\mu} (\rho_s^{(n)} \mathbf{g} - \nabla P^{(n)}) \right] \\
&= \sum_{k=0} \left[ \frac{\kappa^{(k)}}{\mu} (\rho_s^{(k)} \mathbf{g} - \nabla P^{(k)}) \right]
\end{aligned}
$$

where the subscripts $n$ and $k$ denote the time at which the fields are evaluated and the last line follows by induction from the first. Upon convergence
of $\mathbf{u}_s$ the densities must organize to give $\nabla P = \rho_s \mathbf{g}$. This happens when
the downwind instability is handled as in Eq. (7).

Even though the description of the mass flows given by Eqs. (6) and
(7) neglect all inertial effects, they will deal correctly with the time dependence caused by the variation of permeability and air pressure. Hence, when
accelerations can be neglected the model will describe both the transient
pressure relaxation and the overall evolution of the bubbles following from
the semi-steady velocities.

By rewriting Eqs. (7) and (6) in terms of dimension less quantities,
obtained by rescaling of the velocities by $U_0 = \kappa(\rho_0)\rho_g g/\mu$, the densities by
their stationary values, the distance by some characteristic length $h$, the
pressure by $P_0$, and the time with $h/U_0$ it is seen that the physical system is
characterized by the Péclet number $\mathrm{Pe} = U_0 h/D_0$ where $D_0 = \kappa(\rho_0)P_0/\mu$.
When the above expressions for $U_0$ and $D_0$ are used, Pe takes the form

$$
\mathrm{Pe} = \rho_g g h/P_0 . \tag{8}
$$

The Peclet number derives its name from the fact that it may be interpreted
as the ratio between a diffusive and an advective $(l/U_0)$ time scale. Note
that it reduces to the ratio between the hydrostatic pressure $\rho_g l g$ caused
by the grains and the background pressure $P_0$. A bubble of air rising in a
tube full of sand is described by the Peclet number $\mathrm{Pe} = 0.125$, when $P_0$
is taken as the atmospheric pressure, $h = 0.5\mathrm{m}$ and the density of glass
$\rho_g = 2.5 \ 10^3 \mathrm{kg/m}^3$. Note that Pe depends on the background pressure $P_0$
but not on the permeability (particle radius) or viscosity. This is because
the sand flow rate in still air and the air diffusivity depend on $\kappa/\mu$ in a way
such that $\kappa/\mu$ cancels in Pe.

Among the other simplifications made in the model, the use of the approximate Carman Kozeny equation (4) for the permeability is important.
It can be shown to hold reasonably well for packings which are sufficiently

dense. Zick and Homsy (Zick, 1982) have demonstrated that there is agreement within 25 % between the predictions of Eq. (4) and simulations of flow in periodic arrays of spheres when $\rho_s > 0.3$. For lower densities Eq. (4) becomes less accurate. Because of the stability criteria of the Boltzmann model (Flekkøy, 1993), we have further introduced the cut-off condition on the permeability that $\kappa(\rho_s < 0.3) = \kappa(0.3) \equiv \kappa_{max}$.

## 2.1. THE NUMERICS: A LATTICE BOLTZMANN MODEL

The lattice Boltzmann model that we introduce for the present purposes represents a generalization of a model for advection diffusion phenomena (Flekkøy, 1993). These phenomena are described by the advection diffusion equation

$$\partial_t \rho + \nabla \cdot (\rho \mathbf{u} - D\nabla\rho) = 0 \qquad (9)$$

where $\mathbf{u} = \mathbf{u}(\mathbf{x}, t)$ is an arbitrary vector field and $D = D(\mathbf{x}, t)$ a diffusivity, which may also depend on space and time. The equation describes the conservation of the density $\rho$ which is transported by the combined action of advection and diffusion. We will exploit the fact that Eqs. (1) and (2) can be cast in the above form. In the following we briefly review how the lattice Boltzmann model works.

In general a Lattice Boltzmann (LB) model (mcNamara, 1988) describes a fluid by a large number of particle populations (Frisch, 1987) that move from site to site on a regular lattice, where they interact in collisions according to certain conservation laws. In the present case the lattice will be triangular. The lattice unit vectors connecting neighboring sites are $\mathbf{c}_i$, $i = 1, \ldots, 6$. While a finite difference approach would give a numerical solution to the conservation equations, a LB model is based directly on a conservative process, which in turn is described by the conservation equations. This is useful in implementing the boundary conditions.

The particle populations are denoted $N_i(\mathbf{x}, t)$. A common interpretation is to think of $N_i(\mathbf{x}, t)$ as the probability of finding a particle at $\mathbf{x}$ at time $t$ moving moving with unit velocity in one of the six lattice directions $i = 1, \ldots, 6$. However, we may also think of the $N_i$'s as actual masses. The density $\rho$ is defined as

$$\rho = \sum_{i=1}^{6} N_i \qquad (10)$$

and the algorithm consist of a two-step procedure: First the particle probabilities are propagated to their neighboring sites according to their associated velocities, i.e. $N_i(\mathbf{x} + \mathbf{c}_i)$ is given the value of $N(\mathbf{x}, t)$ for every $\mathbf{x}$. Second, the 6 $N_i$'s undergo a local interaction that conserves the value of $\rho$. These two steps constitute the basic simulation time step. The details

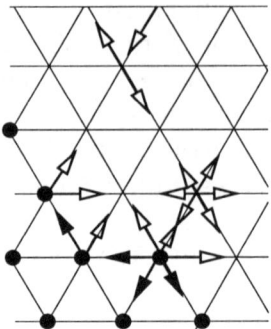

*Figure 1.* Particle populations in the Boltzmann model. The sites marked with a •
are solidified sites. Filled arrows represent stationary particle populations while empty
arrows represent moving particles. All directions on every site has an associated particle
population. Only selected ones are shown.

are described in Ref. (Flekkøy, 1996) where it is also shown that this algorithm solves Eqs.(1) and (2) with the addition of a small diffusive current of
sand. In order to describe the solid fluid boundary a Boolean field $C(\mathbf{x}, t)$
is introduced. At 'solidified' sites where $\rho_s \geq \rho_0$ $C(\mathbf{x}, t) = 1$, otherwise it is
0. The interaction between the fluid and solid phases which is responsible
for the dynamic transition between the two, takes place along the bonds
that connect solid and fluid sites. The mass exchange along these bonds
is free, just as in the fluid phase. This is illustrated in Fig. 1. In the solid
region the dynamics of the sand densities is otherwise frozen in the sense
that the particle populations are simply not touched. Eq. (6) which governs
the air dynamics then reduces to the simple diffusion equation. As the sand
mass transport between the solid and fluid phases is proportional to the
density difference between the adjacent solid and fluid sites, the solid–fluid
interaction is diffusive in nature. The existence of the granular diffusivity
$D_s$ adds to this picture. The discontinuous transition between the solidified
and fluidized phase is based on the physical approximation that all grain–
grain contacts are instantly replaced by hydrodynamic grain interactions
when the density falls below the threshold density.

It is a well established fact that a granular medium must dilate (expand) in order to move relative to itself or its surroundings (Reynolds, 1885;
lePennec, 1997). The dilatency necessary for motion relative to a smooth
wall might be significantly smaller than the dilatancy needed for internal
shear. In the present model however, we make no such distinction between
small and large dilatancies. In the simulations discussed in sequel the dilatancy needed for the motion relative to the walls to begin is determined

by the initial excess density $\Delta\rho_s = \rho_s^0 - \rho_0$ where $\rho_s^0$ is the initial granular density.

## 3. Experiments and Simulations using the continuum model

In order to see what can be learned from the present model we now turn to the direct comparison of it by two selected experiments.

### 3.1. A BUBBLE IN A TUBE

One of the simplest possible applications of the model to a case where gas interactions are strongly governing the flow, is a simple one dimensional flow through a tube. In the experiments, the flow is of course, three dimensional.

Bubble dynamics have been studied extensively in the case of fluidized beds, and an approximate analytic solution of corresponding continuum equations for mass– and momentum conservation, has been obtained by Davis and Taylor (Davis, 1950). While this solution neglects boundary effects and depends on inertia, the present numerical study focuses on a drag dominated bubble in an initially close packing in the presence of walls. The experimental setup here is similar to that used in the work of Raafat et al.(Raafat, 1995) who also introduce an analytic description of the non-linear essentially one-dimensional motion of granular plugs. However, while their work deals with the dynamics of a finite 'bubble' of sand moving through air, the present setup deals with the complimentary state where a bubble of air creates a local motion in a bed of stationary sand. The present description and the corresponding simulations are two-dimensional and can easily be extended to three dimensions.

In the experiment, illustrated in Fig. 2 the bubble was released at the bottom of a long vertical glass tube. The tube, which is closed on both sides, has an internal diameter of 5mm, and a total internal length of 105.1cm. The tube has two parts of length 5cm and 100.1cm respectively. The upper part of length 100.1cm was filled with small glass beads and the lower part 5cm was filled with air at the atmospheric pressure. Two types of glass beads of diameter $d = 51 \pm 8\mu m$ and $d = 65 \pm 9\mu m$ were used. A density of $\rho_s = 1.38g/cm^3$ was obtained by simply pouring the beads into the tube. To perform an experiment with an increased density, the $d = 65\mu m$ particle packing was further compactified by tapping uniformly on the side-walls. This gave a density $\rho_s = 1.44g/cm^3$. A shutter mechanism is placed between the upper and the lower tube. The shutter consists of a 0.5mm thick aluminum plate with a hole of diameter 5mm. Two small permanent magnets are mounted on each side of the plate, and an external permanent magnet was used to move the plate. The bubble propagation starts when the hole in the aluminum plate is in position with the tube. To prevent

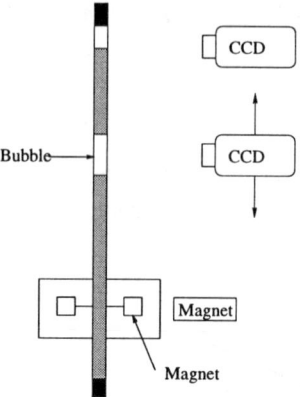

*Figure 2.* The experimental bubble setup. A plate with a hole is pulled to the right by an external magnet, thus allowing the bubble to form within the sealed system. The bubble motion and the top interface are recorded by two video cameras.

air leakage and to make the system transparent, the shutter mechanism is sealed in a piece of Plexiglas. Both the upper and the lower parts were initially kept at atmospheric pressure. To visualize the motion of the bubble and the top level of the sand we used two video cameras. The camera which recorded the bubble motion was mounted on a vertical translation stage to follow the bubble. The second camera recorded the motion of the top interface of the sand.

The bubble has a rather sharply defined top and bottom. The position of the upper and lower interface of the bubble as functions of time were easily determined from video recodings of the bubble motion. Also the position of the free surface on the top was easily defined.

In these experiments one should really consider two values of the dilation needed for the granular packing to start its motion. To move relative to the walls the static wall–particle friction must be exceeded. Due to the slight elasticity of the particles this requires a small dilation of the packing. This dilation will be larger if the walls are rough. To move relative to each other particles must be able to pass by each other. This internal shear motion requires a much larger dilatancy, which may be of the order 10 % . In the case of the flow in a smooth tube like the present one, an initial compactification of the packing may survive throughout the experiment (lePennec, 1997). This confirms that in order to move relative to the walls the packing need only reduce its density by a small fraction of the initial density.

In the simulations no wall friction exists, and the wall interactions are either on- or off according to the value of $\hat{\rho}_s$. The bubble was initialized as a depletion of Gaussian shape in an otherwise constant density profile

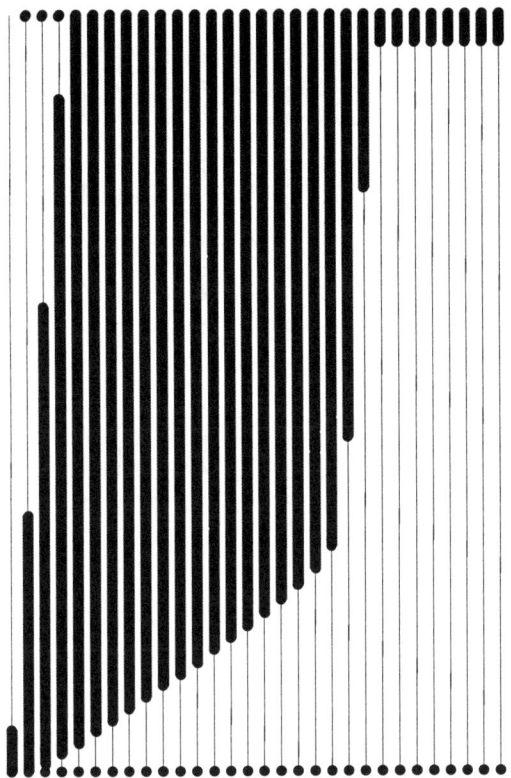

*Figure 3.* Simulation: The fluidized region shown at a series of different times. The system size is 512 by 1, $D_s = 0.001$ and the compactification $\Delta\rho_s = 0.0025$

$\hat{\rho}_s = (1 + q)\rho_0$. Initially the solidification field is therefore on outside the depletion. The term $\Delta\rho_s$ is the relative excess compactification. Initially in the simulation a fluidization front propagates through the system, as illustrated in Fig. 3, thus reducing the density below $\rho_0$. For moderate values of $\Delta\rho_s$ and significant values of the terminal velocity, which the particles are assumed to reach instantly, the layer of sites that become fluidized will have a sufficient mass transport away from the front to allow the next layer above to fluidize in the next time step. In this case the speed of the front will be between $\sqrt{3}/2$ and 1 lattice unit per time step, depending on the orientation of the lattice.

This rapid fluidization front, although realistic in appearance, only captures part of the physics in the experiment. In the experiments the packing

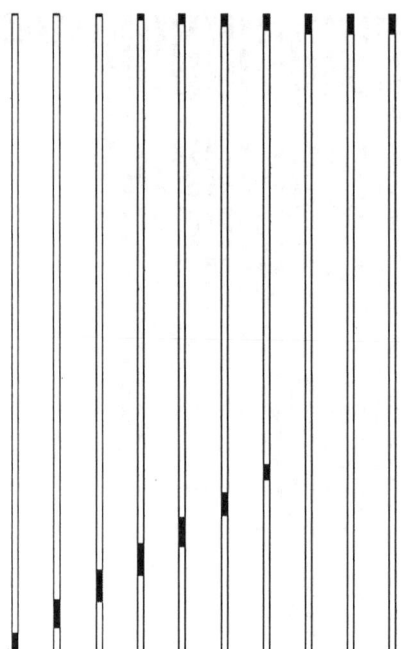

*Figure 4.*  The bubble and the top regions shown at a series of different times.

must also loosen from the walls by the action of a rapidly propagating front. But here inertia, wall friction and long range force network will all be potentially active mechanisms. Several questions regarding the fallout process from a region of grains supported by the walls are still open, both in this and related works (Raafat, 1995; lePennec, 1997).

In the simulations the delay of the free surface motion relative to the initialization of the bubble is sensitive to the value of $\Delta\rho_s$. On the other hand, the motion of the bubble itself, which takes place on a larger time scale than the front propagation, depends only weakly on $\Delta\rho_s$.

The location of the bubble bottom and top as well as the position of the free surface was defined in the simulations as the position where the density was half way between its local minimum and maximum values. These

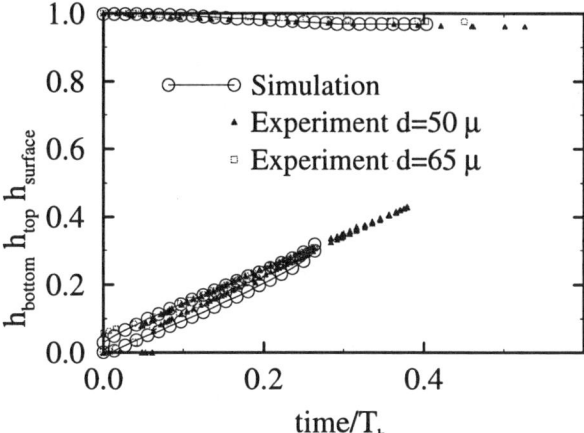

*Figure 5.* The top and bottom positions of the bubble and the position of the top surface for the simulation and experiment. The system size is 512 by 1, $D_s = 0.001$ and the compactification $\Delta\rho_s = 0.0025$

positions are illustrated in a graphical way in Fig. 4, which also closely resembles the visual appearance of the experiment. The final comparison between experimental and simulation results is shown in Fig. 5. In the experiment the bubble location is determined by visual inspection. In the simulation, where the bubble undergoes some diffusive smearing, its extent is defined by the region where the density is below $(\rho_0 + \rho_{min})/2$ where $\rho_{min}$ is the minimum value of the density in the bubble. The position of the free surface is defined similarly, $\rho_{min}$ is in this case the minimum value of the density above the surface. The time is normalized by the time $T_0$, which is the time the (extrapolated) bubble needs to reach the top of the sand packing. The positions are normalized with the position of the top surface $h$ before the bubble was released. Both experiments and simulation agree that the velocity of the bubble is independent of its position and size. The agreement is fairly good both for the position where the bubble finally disappears, and for the top motion. In view of the simplifications introduced in the simulations this agreement is rather encouraging. However, it should be noted that both the interior shape of the bubble and its exact boundary positions are not captured by the simulations.

## 3.2. STAGNANT REGIONS AND HOPPER FLOW

There are two main aspects of the modeling, the gas–grain interaction described by the co-moving Darcy law and the grain–grain interactions introduced by the solid fluid transition rules. Here we investigate the latter in a

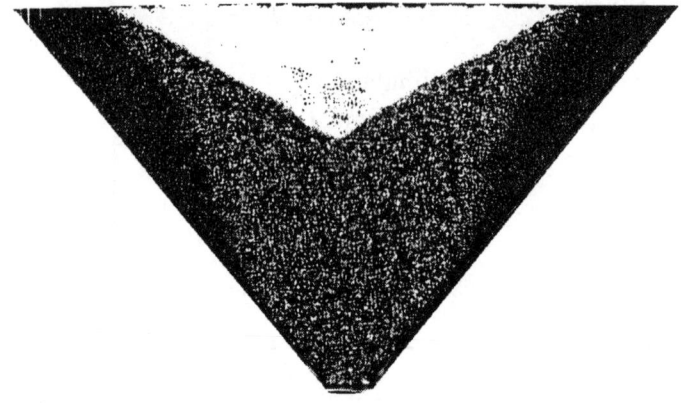

*Figure 6.* Picture of the hopper flow experiment. This is a difference image obtained by subtracting an image before– from an image during the flow. The stagnation zones, which are regions of no displacement, are made visible as distinct dark zones with very little noise.

simple hopper flow experiment. The solid–fluid like transition is a uniquely granular phenomenon that is linked both to the static friction present between grains and the dilatancy which is a prerequisite for granular motion.

In the experiment a quasi two-dimensional hopper consists of two glass plates separated a distance $5mm$ apart. The inclined hopper side-walls makes an angle $\Theta/2$ with the vertical. These side-walls are made rough by absorbing glass beads onto double sided tapes, glued on the side-walls. The width of the orifice, initially closed with a piece of tape, is $D = 10.1mm$. The hopper was filled with glass beads of $d = 1.0mm$, and the flow was initiated by removing the tape at the orifice. To visualize the flow, pictures were taken with a high resolution (1500 times 1200 pixels) kodak DCS 420 CCD camera. The stagnant zones become apparent by subtraction of the pictures with a picture before the flow started. As seen in Fig. 6 the stagnation zones will then appear as regions with much less noise than in regions with particle movement. Note that on account of the gluing of particles to the walls, a layer or two along the sides will always be stagnant.

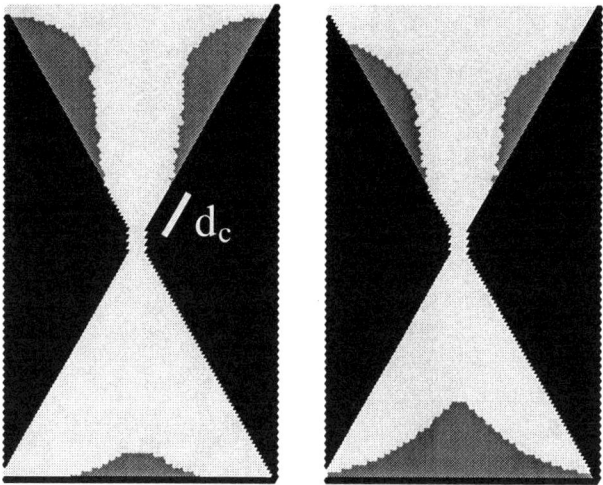

*Figure 7.* The solidified region shown as dark gray, at two consecutive instants in time. The fluidized region is shown as light grey. Here $d_c$ is the distance between the hopper opening and the stagnation zone.

Fig. 7 shows the corresponding simulations. These were carried out in an hourglass geometry in order to see the pile formation in the bottom as well. This pile has the same qualitative shape that is observed in experiments (Alonso, 1996). The distance $d_c$ to the stagnant region from the orifice, as indicated in Fig. 6, was measured as function of the hopper angle $\Theta$. In the simulations the stagnant regions are directly available as the regions where the solidification field is on. Figure 7 shows this field as a function of time. In both experiment and simulations the lower position of the stagnant region was time independent to a good approximation, so that a single $d_c$ could be defined. In the simulations the field $\hat{\rho}_s$ interacts with solidified sites in the same way as wall sites. Hence, the simulations model an experimental situation where the wall has the property of a collection of fixed grains. For that reason smooth walls were inadequate for comparison with the present simulations. This corresponds to walls with grains glued onto them as in the experiments. Figure 8 shows $1/d_c$ as a function of $\Theta$ both for the simulations and experiments. Both graphs are consistent with a linear dependence. This is in a striking contrast to the smooth wall experiments by Baxter and Behringer (Baxter, 1989) who obtained an exponent of $2.2 \pm 0.1$. The prefactors are seen to differ in the experiment and the simulations

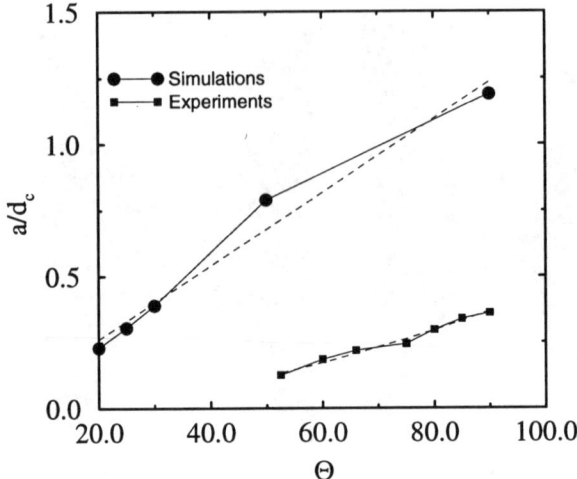

*Figure 8.* The ratio $a/d_c$, where $d_c$ is the distance between the hopper opening and the stagnation zone, and $a$ is the diameter of the hopper opening as a function of the hopper opening angle $\Theta$. Both experiments and simulations are shown.

though. These are expected to depend on grain-grain friction as well as the grain geometry and possibly size.

While the simulations do not include inertia, the experiment clearly does as the particles in the hopper opening are large and fall freely. The flow rate close to the stagnant zones, on the other hand, is very slow. However, it may be still be affected by the inertia controlled central flow. Hence, it is not obvious *a priori* that the simulations are suited to capture the experimental behavior, and the results must be judged in view of that. It is therefore an encouraging observation that the stagnant zones indeed appear to be predictable by the present, inertia-less model.

## 4. A hybrid model combining particles and continuum fields

In order to obtain both the resolution required to describe the shock like density contrasts typical for granular flows as well as the inertia of the granular mass we have in the following chosen a full particle description of the grains while keeping a continuum description of the fluid. This strategy combines the advantage of a microscopic interpretation of the particles with the numerical economy of a Darcy-level description of the fluid. The model is described in more detail in Ref. (Mc Namara, 1999).

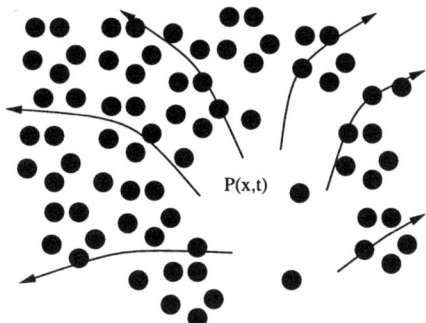

*Figure 9.* A conceptual picture of the dynamics. The MD particles move according to Newton's laws while the pressure evolves according to a local Darcy law for which the particles form a porous medium.

Figure 9 illustrates conceptually the basis for the model. The gas flow, shown by the arrows, takes place between grains that define a local permeability $\kappa(\rho_s)$.

## 4.1. A PRESSURE EVOLUTION EQUATION

In the pressure evolution equation derived in the following the air current has both an advective term caused by the motion of the grains, and a diffusive term describing the Darcy flow in the local rest frame of reference for the sand. Substituting $\rho_s = 1 - \phi$ in Eq. (2) we get

$$-\frac{\partial \phi}{\partial t} + \nabla \cdot ((1 - \phi)\mathbf{u}) = 0 . \tag{11}$$

Using the isothermal equation of state for an ideal gas, $\rho_a \propto \phi P$, we can write Eq. (1) in the form

$$\frac{\partial(\phi P)}{\partial t} + \nabla \cdot \left( \phi P \left[ \mathbf{u} - \frac{\kappa}{\mu} \nabla P \right] \right) = 0 . \tag{12}$$

By eliminating $\partial \phi / \partial t$ between Eq. (12) and Eq. (11) a small manipulation gives

$$\phi \left( \frac{\partial P}{\partial t} + \mathbf{u} \cdot \nabla P \right) = \nabla \cdot \left( \phi P \frac{\kappa}{\mu} \nabla P \right) - P \nabla \cdot \mathbf{u} . \tag{13}$$

In this equation the left hand side is just the substantial derivative of the pressure. The first term on the right hand side describes the Darcy flow in the local rest frame of reference of the grains. The last term describes pressure changes due to changes in the grain density. See Ref. (Gidaspau, 1994) for a more elaborate discussion of continuum equations like Eq. (13).

Finally, the numerical solution of Eq. (13) can be simplified by dividing the pressure into an average and fluctuating part: $P = P_0 + P'$. In the experiments we wish to study, the changes in pressure are only a small fraction of atmospheric pressure, so $P' \ll P_0$. Neglecting terms of order $O(P'/P_0)$ in Eq. (13) leads to

$$\phi \frac{\partial P'}{\partial t} = P_0 \nabla \cdot \left( \phi \frac{\kappa}{\mu} \nabla P' \right) - P_0 \nabla \cdot \mathbf{u} . \tag{14}$$

The simplifications leading to this equations are by no means crucial. In applications where it is needed the neglected terms may well be re-inserted.

For the numerical implementation it is convenient to non-dimensionalize Eq. (14). Writing the characteristic magnitude of the permeability $\kappa_0 = a^2/45$ we may introduce the characteristic grain velocity $U_0 = (\kappa_0/\mu)\rho_g g$, where $\rho_g$ is the mass density of the material that makes up the grains. Introducing the characteristic length scale $l$ a characteristic time scale $l/U_0$ follows. The dimensional quantities may then be written in terms of non-dimensional (primed) quantities as $P = P_0 P'$, $\mathbf{u} = U_0 \mathbf{u}'$, $x = lx'$ and $t = (l/U_0)t'$. Substituting these relations in Eq. (14) we get

$$\phi \frac{\partial P'}{\partial t'} = \mathrm{Pe}^{-1} \nabla' \cdot \left( \frac{\phi^4}{(1-\phi)^2} \nabla' P' \right) - \nabla' \cdot \mathbf{u}' . \tag{15}$$

where the Peclet number is defined in Eq. (8). In the simulations that follow we use Pe = 0.029.

Equations (13) and (4) describe fluid flow where the fluid inertia may be neglected, as discussed in Sec. 2. However, when particles become larger the description does not break down in a dramatic way as the first corrections in the Reynolds number amounts to small corrections in the pressure forces on the particles. As fluid inertia becomes increasingly important, however, it affects not only the fluid–particle coupling but also the fluid dynamics itself, and in these cases an equation describing the flow of fluid momentum, like the Euler equation (Kawaguchi, 1998) is needed.

## 4.2. PARTICLE DYNAMICS

The particles evolve according to Newton's second law:

$$m \frac{d\mathbf{v}}{dt} = m\mathbf{g} + \mathbf{F}_I - \frac{\nabla P}{\rho} \tag{16}$$

where $\rho$ is the number density of particles, $\mathbf{g}$ is the gravity, $m$ the particle mass and $\mathbf{F}_I$ the inter-particle force. The number density $\rho = \rho_s/(\pi d^3/6)$ where $d$ is the particle diameter. What distinguishes the present model from conventional models of granular materials is the pressure force per particle $\nabla P/\rho$. It is the pressure gradient obtained from the continuum equation (13), distributed over the particles present in that volume. In this paper, we use a version of the "TC model"(Luding, 1998), which is an event-driven algorithm to solve Eq. (16). Soft sphere molecular dynamics(Brendel, 1998) and contact dynamics(Radjai, 1998) could be used instead. In the event driven method, the output velocities after a collision are computed directly in terms of the input velocities. If $\mathbf{v}$ is the relative velocity between two particles, and $\hat{\mathbf{n}}$ is a normal vector pointing along the line of centers, the normal component of the output velocity $\mathbf{v}'$ is $\mathbf{v}' \cdot \hat{\mathbf{n}} = -r\mathbf{v} \cdot \hat{\mathbf{n}}$ where $r \leq 1$ is the restitution coefficient. When $r < 1$ the collisions are dissipative, and setting $r = 1$ conserves energy. The velocities perpendicular to $\hat{\mathbf{n}}$ are left unchanged: the particles are perfectly smooth, so we can ignore their rotation. Collisions between a particle and the walls can be considered in the same way, except that the wall has infinite mass.

The algorithm for computing the grain trajectories is outlined below:

- 1. Advance all particles by the time step $\Delta t$, assuming the particles do not interact. Since $\Delta t$ must be chosen so that particles move only a small fraction of their diameter each timestep, the pressure force can be taken as a constant during the short time $\Delta t$. Therefore, the particles are advanced along parabolic line segments, corresponding to free flight in a constant force field.
- 2. Compile a list of all overlapping particles.
- 3. Scan through the list of overlapping particles. If any pair has relative velocities such that the two particles are approaching each other, implement a collision between these two particles. There are two types of collisions: energy conserving and dissipative. If a particle has already suffered a collision in the current or preceding time step, then all collisions involving that particle are energy conserving. Otherwise, they are dissipative. This rule is necessary to avoid inelastic collapse (an infinite number of collisions in finite time) (Luding, 1998), a feature which is characteristic for event driven methods. Repeat this step until all pairs of particles are separating. Then go to step 1.

Sometimes it is undesirable to have perfectly smooth walls, because these introduce unrealistic slip-planes, and a more complicated particle-wall interaction must be used. The present simulations have "rough walls", which are implemented by particle-wall collisions that derive from the particle–particle collisions with tangential friction. For this we use a model which has been experimentally verified(Foerster, 1994). In this model the tangential unit vector $\hat{\mathbf{t}}$, defined to satisfy $\hat{\mathbf{n}} \cdot \hat{\mathbf{t}} = 0$, is needed. Then $\mathbf{v}' \cdot \hat{\mathbf{t}} = -\beta \mathbf{v} \cdot \hat{\mathbf{t}}$, where $\beta$ is the tangential restitution coefficient and $\beta = -1$ corresponds to perfectly smooth particles ($\mathbf{v}' \cdot \hat{\mathbf{t}} = \mathbf{v} \cdot \hat{\mathbf{t}}$). For energy dissipating collisions, $\beta$ is chosen to be the largest value which satisfies both the following conditions:

$$(1 + \beta) \left| \mathbf{v} \cdot \hat{\mathbf{t}} \right| \leq \mu_f (1 + r) \left| \mathbf{v} \cdot \hat{\mathbf{n}} \right| \text{ and } \beta \leq \beta_0. \tag{17}$$

The first condition says that Coulomb friction acts during the collision. The impulse transferred between the particles in the normal direction is $(1 + r)|\mathbf{v} \cdot \hat{\mathbf{n}}|$, and the impulse in the tangential direction is $(1 + \beta) \left| \mathbf{v} \cdot \hat{\mathbf{t}} \right|$. The ratio between these must not exceed the friction coefficient $\mu_f$. The second condition requires $\beta$ to be less than some limit $\beta_0$. If equality held in the Coulomb friction condition during all collisions, we would have $\beta \gg 1$ when $|\mathbf{v} \cdot \hat{\mathbf{n}}| \gg |\mathbf{v} \cdot \hat{\mathbf{t}}|$, leading to energy creation during collisions. Experimentally, $\beta_0 \approx +0.4$ (Foerster, 1994). The use of this model in simulations is extensively discussed in (Luding, 1995). For the energy conserving collisions we must have $|\beta| = 1$, and we first try to set $\beta = 1$, which corresponds to a bounce-back. If this leads to a violation of the first condition in Eq. (17), then $\beta = -1$ which corresponds to smooth particles. We now need to define $\rho$ and $\mathbf{u}$ in terms of the particle positions and velocities $\mathbf{v}_i$ ($i$ labels individual particles). In order to obtain a continuous density field we will distribute the particle mass in a halo which goes continuously to zero around the particle. This is done by introducing the halo function described in Ref.(Mc Namara, 1999), which has the virtue of smearing out the particle density so that it does not vary discontinuously as particles pass in and out of their coarse graining boxes. The divergence $\nabla \cdot \mathbf{u}$ is evaluated as a finite difference.

Just as the halo function may be used to obtain smooth particle input to Eq. (13), it may be used the other way, to distribute the pressure forces on the particles. The effect of this is to give each particle a pressure force which is an average of the pressure force computed on the four nearest neighboring sites.

## 4.3. IMPLEMENTATION

For the model to work in practice it is necessary to introduce a cutoff $\rho_{\min}$ on the density. This has both physical and numerical reasons. Physically there is no sense in defining a permeability field if the particle density is too low. The Carman Kozeny permeability gives a reasonable prediction only when $\rho_s > 0.2$(Zick, 1982). The pressure computations may become unstable both when the permeability becomes too high and when the source term becomes too erratic, as will happen when $\rho_s \to 0$. Therefore we shall take $\rho_s = \rho_{\min}$ wherever the measured density is less than $\rho_{\min}$. This introduces a cutoff on the permeability, $\kappa < \kappa(\rho_{\min})$. Likewise, when the pressure force on the dilute particles are computed we shall use $\rho_{\min}$ in place of the actual density when it is too small. This implies that the pressure feels a permeability corresponding to a higher than actual particle density. Correspondingly, the particles are subjected to the force $\nabla P/\rho = \nabla P/\rho_{min}$, when $\rho_s < \rho_{min}$. This means that the particles in the volume cell $\Delta V$ corresponding to a lattice site will not absorb the entire force $\nabla P \Delta V$ when $\rho_s < \rho_{min}$. However, due to the over estimate made by $\kappa(\rho_s)$ in dilute regions, the force per particle will still be larger than the single particle Stokes drag (Landau, 1959). This means that the error made by introducing the cut-off is mainly that dilute particles fall somewhat more slowly than they should.

Although the practical implementation of the present model in three dimensions is not significantly harder than in two dimensions we wish to simulate a two-dimensional system because it is numerically much less expensive. However, the Carman-Kozeny equation (4) is a three dimensional relation as it gives the permeability in terms of the volume fraction of spheres $\rho_s$, and we wish in the end to compare our results to real three dimensional experiments. Consequently we need to transform the area fraction of grains in the simulations $\rho_s^{(2D)}$, to the volume fraction $\rho_s$ in such a way that the closed packed value of $\rho_s^{(2D)}$ corresponds to the closed packed value of $\rho_s$. This is approximately achieved by the transformation $\rho_s = (2/3)\rho_s^{(2D)}$, which we use in the following.

In the simulations we use a distribution of particle sizes to avoid the 2 dimensional hexagonal ordering. To improve the relation between the 2 and 3 dimensional packing densities the closed packed value of $\rho_s^{(2D)}$ will eventually be measured and compared to the three dimensional random closed packed values. For the present validation process however, this is not needed and we use the 2/3 factor.

## 5. Applications of the particle model

As a test of the model we apply a constant pressure drop from the bottom to the top of a container in order to produce the bubbling behavior as observed experimentally in gas fluidized beds (Davidson, 1995; Lim, 1995; Rowe, 1985).

### 5.1. FLUIDIZED BEDS

While incompressible fluids (liquids) do not produce bubbles, compressible ones spontaneously do. The formation of bubbles is thus a salient phenomenon in gas–grain flow, of which the model captures the main features.

Figure 10 shows a time series of a fluidization simulation with an initial layer of colored particles at the bottom of the system. These particles are different from the above particles by their color only. The computation was

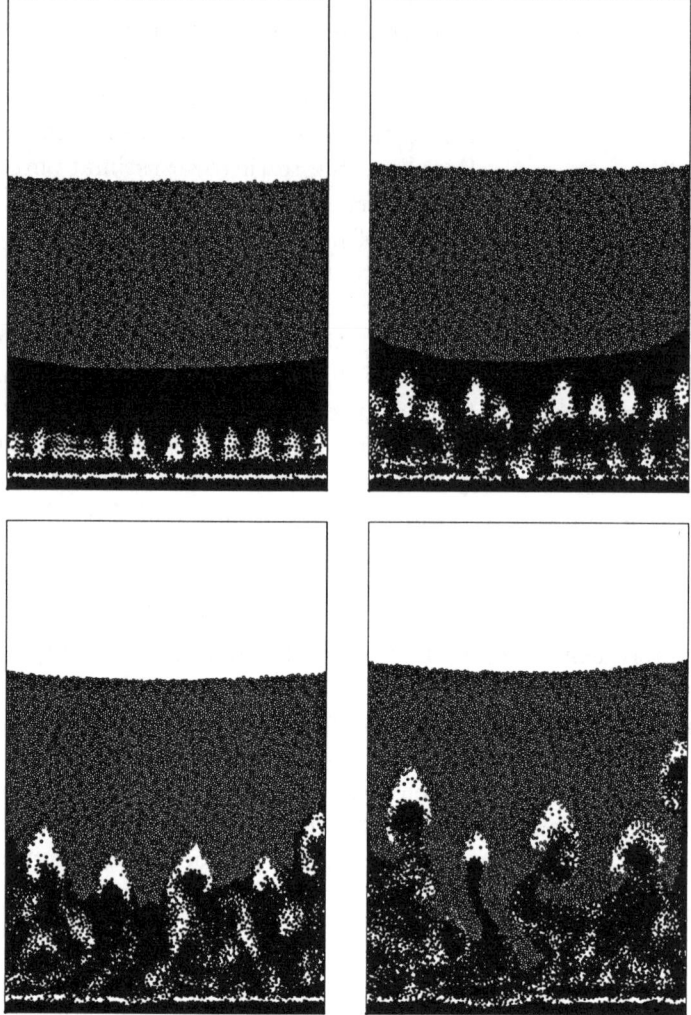

*Figure 10.* A time sequence showing the simulation of fluidization. The walls are rough with $\mu_f = 0.3$ and $\beta_0 = 0$.

carried out with 20.000 particles and the size of the domain is $62\ell \times 94\ell$. Here $\ell$ is the lattice constant for the grid of the pressure computations and $\tau = \ell/U_0$. The entire simulation took less than 10 hours on an aging workstation. An air flux of $1860\ell^2/\tau$ was injected uniformly across the bottom , and gravity was $3333\ell/\tau^2$.

Three main features of this simulation should be noted. First the bubbles form spontaneously when the pressure difference is turned on. Second, small bubbles coalesce as they move upwards forming larger bubbles. And, finally the shape of the bubbles have the the same qualitative shape that is observed experimentally(Rowe, 1985). The two first features are also seen experimentally (Davidson, 1995; Rowe, 1985). The first and last of the above features were also observed in the simulations of Kawaguchi et al. (Kawaguchi, 1998). We note that due to the lack of sliding friction between the particles in the present simulations the system may start forming bubbles at a denser state than in real systems. The stationary layer of particles at the very bottom is caused by the discretization of the pressure field, and is a numerical rather than physical effect.

Note the shape of the colored particle tail with the accumulation of black particles at the bottom of the bubble. This is a striking feature also in experiments where there is a bottom layer of colored particles (Rowe, 1985).

## 5.2. THE TICKING HOURGLASS

The description of the original experiment is given in Ref. (Wu, 1993) and illustrated in Fig. 11. When sand flows from the upper to the lower compartment in an hourglass, the volume carried by the grains increase the pressure enough to stop the flow. The reason for this is that the pressure need only balance the weight of the grains near the orifice. The remaining weight is carried but the walls via force arches (Wu, 1993). Simulations of this system have been carried out by Manger et al.(Manger, 1994) who used finite differenced 2d equations for a *fluidized* granular medium. Being intrinsically simulations of liquid– rather than granular flow and neglecting the nature of the internal force propagation in granular media these simulations could not describe the experimental fact that wall forces prevent the granular pressure from building up with depth, and thus could not reproduce the height independence of the ticking frequency. The present simulations will contain this effect, even without inter-particle friction as along as the walls are not vertical. With vertical walls friction is needed to screen the pressure at the bottom from forces on the top.

## 5.3. CONNECTING MODEL AND REALITY

Approaching the modeling problem along the route of maximum simplicity we neglect particle-particle friction entirely. Moreover, since experiments show that $\delta P/P_0 \approx 0.001$, where $\delta P$ is the pressure variation and $P_0$ the

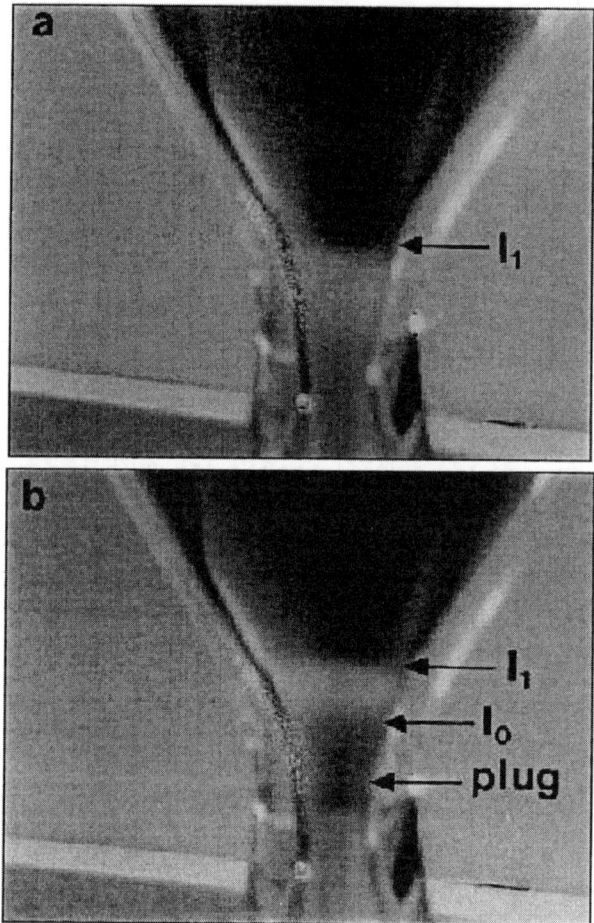

*Figure 11.* Picture of the ticking hourglass. The sand-air interface ($I_1$) moves upwards with time and eventually forms the upper surface of an airbubble, with lower interface $I_0$, that disappears into the sand. The bubble forms due to a plug formation below it.

background pressure (Wu, 1993), the simplifications leading to Eq. (14) hold.

The ticking frequency will be determined by the ratio between the granular volume flux $\rho u$ through the orifice and the volume $V$ available to the air in the lower or upper compartment. For a given value of the volume flux, which is driven by a combination of gravity and wall forces (and produces the pressure forces as a response), this ratio may be estimated as

$$\frac{\dot{V}}{V_0} = \frac{\rho u}{V_0} \left\{ \begin{array}{ll} \frac{D}{h} & \text{in 2d} \\ (\frac{D}{h})^2 & \text{in 3d} \end{array} \right. , \tag{18}$$

*Figure 12.*   Six subsequent snapshots of a simulation of the ticking hourglass: The parameters are as in Fig. 10, there are 1580 particles and the square grid on which the pressure is evolved is 60 × 60.

where $D$ is the orifice diameter, $h$ is the typical linear size of one of the hour glass chambers ($h^d$ is roughly the hour glass volume), $u$ is evaluated in the orifice. The key difference between the 2d and 3d equations is the extra factor $D/h$ in 3d. This suggests that the ticking frequency should be roughly a factor $h/D$ larger in 2d than 3d. However, the force networks that support the main part of the granular weight will also be different in 2d and 3d. Since the forces propagate along convoluted lines, a single arch may completely screen the underlying medium in 2d while having a more marginal effect in 3d. Finally, the pressure gradients will be more localized in 3d than in 2d so that the pressure force may fall over a slightly taller part of the packing in 2d than in 3d. Since our interest is mainly directed towards the qualitative aspects of the system — as a first step, the 'ticking' phenomenon itself — we perform the simulations in 2d in spite of the significant differences between 2d and 3d.

### 5.3.1. *Simulations*

Figure 12 shows the present simulations As the characteristic scales we have taken the orifice diameter $D = 0.3$ cm, the experimentally observed ticking period $T = 1$s and the measured pressure difference $\delta P = 0.001$ atm.

The result is shown in Fig. 12. In these 2d simulations the bubble that forms at the orifice is clearly seen. This should be compared to the experimental image of Fig. 11 where the bubble forms but disappears inside the granular packing. Note in the simulations how the bubble is decreased in size. As the bubble is being formed the flow rate of grains through the orifice is decreased by roughly a factor two. The fact that this decrease is smaller than in the experiments is most likely due to the lack of sliding friction between the simulated grains.

## 6. Conclusions

We have introduced two models for granular flow under the influence of an interstitial gas. We have tested the models by comparing their main features directly with experiments. In the present context we have emphasized the qualitative aspects leaving out some quantitative results on both bubble propagation and sedimentation which is discussed elsewhere (Mc Namara, 1999; Flekkøy, 1996).

In designing the models we have sought to maximize the conceptual and numerical simplicity, rather than the accuracy. For that reason the continuum model has been simplified to the level of leaving out inertia altogether, and in the second model particles move without rotation and friction and fluid (though not particle-) inertia is neglected. This allows in both cases for relatively simple overall description as the fluid is described in terms of a Darcy law, and the particles in terms of event driven molecular dynamics or a simple continuum equation.

From a physical point of view the main results are

1. Bubble propagation in granular columns: On a qualitative and on a rough quantitative level simulations and experiments agree, though the simulations lack the ability to resolve the sharp density contrasts observed experimentally. They also lack inertia.
2. Formation of stagnant regions: Simulations and experiments agree on the linearity of the relation between the inverse hopper opening angle $\Theta$ and the distance $d_c$, a result that differs from earlier measurements (Baxter, 1989).
3. Fluidized beds: The particle based model reproduces the key experimental aspects of these systems, such as the fluidization, bubble formation, coalescence of bubbles and the characteristic particle tails.
4. The ticking hourglass: The phenomenological aspects of the intermittency, i.e. the formation of bubbles and force arches at the hourglass orifice is observed. This effect is correlated with a marked decrease in the granular flow rate.

# References

J. T. Jenkins and S. B. Savage. A theory for the rapid flow of identical, smooth, nearly elastic, spherical particles. *J. Fluid Mech.*, **Vol. 130**, pp. 187, 1983.

I. Goldhirsch and G. Zanetti. Clustering instability in dissipative gases. *Phys. Rev. Lett.*, **Vol. 70**, pp. 1619, 1993.

G. Pengand H. J. Herrmann. Density waves of granular flow using lattice gas automata. *Phys. Rev. E*, **Vol. 47**, pp. 1796, 1994.

S. Mc Namara, E. G. Flekkøy, and K. J. Måløy. Grains and gas flow. Molecular dynamics with hydrodynamic interactions. preprint, 1999.

E. G. Flekkøy and K. J. Måløy. A continuum description of dry granular flows: Experiment and simulation. *Phys. Rev. E*, **Vol. 57**, pp. 6962, 1998.

X. l. Wu, K. J. Måløy, A. Hansen, M. Ammi, and D. Bideau. Why hourglasses tick. *Phys. Rev. Lett.*, **Vol. 71**, pp. 1363, 1993.

T. Le Pennec, K. J. Måløy, A. Hansen, M. Ammi, D. Bideau, and X l. Wu. Ticking hour glasses: Experimental analysis of intermittent flow. *Phys. Rev. E*, **Vol. 53**, pp. 2257–2264, 1996.

C. T. Veje and P. Dimon. Power spectra of flow in an hourglass. *Phys. Rev. E*, **Vol. 56**, pp. 4376, 1997.

H. Hayakawa, S. Yue, and D. C. Hong. Hydrodynamic description of granular convection. *Phys. Rev. Lett.*, **Vol. 75**, pp. 2328, 1995.

M-L. Tan, Y. H. Qian, I. Goldhirsch, and S. A. Orzaag. Lattice bgk approach to simulating granular flows. *J. Stat. Phys.*, **Vol. 81**, 1995.

E. Manger, T. Solberg, and B. H. Hjertager. Numerical simulation of the ticking hourglass. preprint, 1994.

W. A. Beverloo, H. A. Leniger, and J. van de Velde. The flow of granular solids through orifices. *Chem. Eng. Science*, **Vol. 15**, pp. 260, 1961.

J. F. Davidson. Bubbles in fluidized beds. In E. Guazzelli and L. Oger, editors, *Mobile Particulate Systems*, page 197, New York, 1995. Kluwer Academic Publisher.

J. F. Davidson and D. Harrison. *Fluidization*. Academic Press, New York, 1971.

D. Gidaspau. *Multiphase flow and fluidization*. Academic Press, San Diego California, 1994.

G. W. Baxter and R. P. Behringer. Pattern formation in flowing sand. *Phys. Rev Lett.*, **Vol. 62**, pp. 2825, 1989.

D. Frenkel M. A. van der Hoef and A. J. C. Ladd. Self-diffusion in colloidal particles in a two-dimensional suspension: Are deviations from Fick's law experimentally observable? *Phys. Rev Lett.*, **Vol. 67**, pp. 3459, 1991.

A. J. C. Ladd. Numerical simulations of particulate suspensions via a discretized boltzmann equation. part 1. *J. Fluid Mech.*, **Vol. 271**, pp. 285–309, 1994.

A. J. C. Ladd. Numerical simulations of particulate suspensions via a discretized boltzmann equation. part 2. *J. Fluid Mech.*, **Vol. 271**, pp. 309–339, 1994.

W. Kalthoff, S. Schwarzer, and H. J. Herrmann. Simulations of particle suspensions with inertial effects. *Phys. Rev. E*, **Vol. 56**, pp. 2234, 1997.

O. Reynolds. On the dilatancy of media composed of rigid particles in contact. *Phil. Mag.*, **Vol. 20**, pp. 469–482, 1885.

K. S. Lim, J. X. Zhu, and J. R. Grace. Hydrodynamics of gas-solid fluidization. *Int. J. Multiphase Flow*, **Vol. 21**, pp. 141—193, 1995.

G. D. Scott. Packing of equal spheres. *Nature*, **Vol. 188**, pp. 908–909, 1960.

P. Carman. Fluid flow through granular beds. *Trans. Inst. Chem. Eng. Lond.*, **Vol. 15**, pp. 150–160, 1937.

L. D. Landau and E. M. Lifshitz. *Fluid Mechanics*. Pergamon Press, New York, 1959.

G. M. Homsy. Aspects of flow and mixing in fluidized beds. In *Disorder and Mixing*, pages 185–201, Dordrecht, 1988. Kluwer academic publisher.

A. Zick and G. Homsy. Stokes flow through periodic arrays of spheres. *J. Fluid Mech.*, **Vol. 115**, pp. 13–26, 1982.

E. G. Flekkøy. Lattice bgk models for miscible fluids. *Phys. Rev. E*, **Vol. 47**, pp. 4247, 1993.

G. McNamara and G. Zanetti. Use of the Boltzmann equation to simulate lattice-gas automata. *Phys. Rev. Lett.*, **Vol. 61**, pp. 2332, 1988.

U. Frisch, D. d'Humières, B. Hasslacher, P. Lallemand, Y. Pomeau, and J.-P. Rivet. Lattice gas hydrodynamics in two and three dimensions. *Complex Systems*, **Vol. 1**, pp. 648, 1987.

T. Le Pennec, K. J. Måløy, E. G. Flekkøy, J. C. Messager, and M. Ammi. Silo hiccups. dynamic effects of dilatancy in granular flows. *Phys. of Fluids*, **Vol. 10**, pp. 3072, 1998.

R. M. Davies and G. I. Taylor. Bubbles in fluidized bed. *Proc. Roy. Soc.*, **Vol. A200**, pp. 375, 1950.

T. Raafat, H. J. Herrmann, and J. P. Hulin. The pneumatic instability of granular materials. preprint, 1995.

J. J. Alonso and H. J. Herrmann. Shape of the tail of a two-dimensional sandpile. *Phys. Rev. Lett.*, **Vol. 76**, pp. 4911–4915, 1996.

T. Kawaguchi, T. Tanaka, and Y. Tsuji. Numerical simulation of two-dimensional fluidized beds using the discrete element method (comparision between two- and three dimensional models). *Powder technology*, **Vol. 96**, pp. 129–138, 1998.

S. Luding and S. McNamara. Tcmodel. *cond-mat/9810009*, **Vol. 1**, pp. 111, 1998.

L. Brendel and S. Dippel. Physics of dry granular media. In H. J. Herrmann, J.-P. Hovi, and S. Luding, editors, *Physics of Dry Granular Media*, page 313, Dordrecht, 1998. Kluwer Academic Publishers.

F. Radjai and D. Wolf. Contact dynamics. *Granular Matter*, **Vol. 1**, pp. 3, 1998.

S. Foerster, M. Louge, H. Chang, and K. Allia. Measurements of the collision properties of small spheres. *Phys. Fluids*, **Vol. 6**, pp. 1108, 1994.

S. Luding. Granular materials under vibration–simulations of rotating spheres. *Phys. Rev. E*, **Vol. 52**, pp. 4442, 1995.

P. N. Rowe. *Fluidization*, chapter 4, page 121. Academic Press, London and New York, 1971.

# CRYSTALLINE ARCHITECTURES AT THE AIR-LIQUID INTERFACE: FROM NUCLEATION TO ENGINEERING

IVAN KUZMENKO[†], HANNA RAPAPORT[†], KRISTIAN KJAER[¥],
JENS ALS-NIELSEN[◊], ISABELLE WEISSBUCH[†], MEIR LAHAV[†],
LESLIE LEISEROWITZ[†]

[†]*Department of Materials and Interfaces, The Weizmann Institute of Science, 76100 Rehovot, Israel;* [¥]*Condensed Matter Physics and Chemistry Department, Risø National Laboratory, DK 4000, Roskilde, Denmark;* [◊]*Niels Bohr Institute, H. C. Ørsted Laboratory, DK 2100, Copenhagen, Denmark.*

**KEYWORDS/ABSTRACT:** air-liquid interface/crystalline molecular films/ structure/surface x-ray diffraction

The air-liquid interface is a convenient medium to obtain monolayers or multilayers. The molecular structure of such crystalline films can be established by grazing incidence X-ray diffraction (GIXD), complemented by spectroscopic techniques and lattice energy calculations. Formation of two-dimensional clusters using GIXD may provide insight into the first stages of crystal nucleation. Langmuir monolayer can serve as a template for induced nucleation of bulk crystals or to be a model system of natural membranes. A host of architectures composed of water-insoluble molecules can be formed on liquid surfaces. The number of layers and the polymorphism can be controlled. The concept of supramolecular design was applied in two-dimensions with the use of water-insoluble and water-soluble components.

*A.T. Skjeltorp and S.F. Edwards (eds.), Soft Condensed Matter: Configurations,*
*Dynamics and Functionality, 185-217.*

## 1. Introduction

Ordered molecular clusters are currently attracting wide attention in the physical and biological sciences. The design and preparation of functional materials such as thin-layered microstructures, reagent films for biosensors, devices for optoelectronics requires knowledge and control of nanoarchitectures from the very early stages of self-organization. This requirement touches upon the control of nucleation, growth, morphology and structure of crystals, particularly at interfaces.

Over the past decade it has become possible to glean information on the growth and dissolution of molecular crystals on the nanometer scale by a variety of experimental and theoretical techniques. By comparison, our understanding of and ability to monitor and control crystal nucleation is still rudimentary, even though the process is of central importance to living systems and to the pure and applied sciences.

Crystal nucleation in solution has been described classically in terms of two distinct steps. The first involves aggregation of the molecules in the supersaturated solution into organized nuclei, thus developing a surface which separates them from the environment. The free energy of this surface, which is positive, is proportional to the square of the radius $r$ of the nucleus. On the other hand, the free energy arising from the drive towards molecular aggregation within the nucleus is negative and proportional to the volume of the nucleus and so to $r^3$. Thus, for a small radius $r$, where the positive surface energy predominates, the nucleus is metastable and may disintegrate. But, in the second step, once the nucleus has crossed a critical size, beyond which the total free energy begins to fall eventually becoming negative, continued growth will be a stable process [1].

This theory provides guidelines for crystallization studies, but yields little structural information on the nanometer scale. None the less, theoretical methods to glean information on molecular assembly in the early stages in solution are currently being developed, such as molecular dynamics [2]. To date, reported experimental studies of structures of supersaturated solutions at the onset of crystallization, have provided experimental proof for the existence of clusters [3] but little information on their structures and on their role as intermediates in the crystallization process.

Crystal nucleation occurs either via a homogeneous or a heterogeneous process. The former implies an isotropic environment about the nucleus, for example, liquid solvent, melt solute, or a gas. Such a process has not yet been monitored by X-ray diffraction because of experimental difficulties, although experiments have been performed in which clusters of small molecules formed on expansion through a nozzle were studied by electron diffraction. Such experiments have yielded information on cluster size and structure [4], and have provided the incentive for studies on the dynamics of such cluster formation by molecular simulations [4b].

It is possible to glean information on crystal nucleation of polymorphic systems by the affect of tailor-made inhibitors thereon. It is generally assumed that as the molecules start to associate in supersaturated solutions, they form nuclei with structures resembling those of the to-be-grown crystals. For polymorphic systems, we may query whether nuclei of all the observed polymorphs are formed; clearly the structure of each type of nucleus would resemble the crystal into which it will eventually develop. If this hypothesis is correct, one may use the structural information of the macroscopic crystals to design inhibitors of a particular crystalline phase. The overall consequence of this inhibition process can be that the unaffected phase, if less stable, will grow by kinetic control, the logic of which has been applied to several systems [5].

Heterogeneous crystal nucleation is a frequently observed process that takes place at an interface, thus lowering the activation barrier for nucleation. The interaction between the nucleus and the interface may occur at different levels of specificity, ranging from weak adsorption to strong binding to epitaxial nucleation. The recent development of various methods for the elucidation of molecular ordering at interfaces [6] has provided the means to probe the early stages of molecular assembly at various levels of interfacial specificity. Here much use is made of grazing incidence X-ray diffraction (GIXD) using synchrotron radiation to extract structural information at the subnanometer scale of crystalline films at the air-liquid interface [7]. A brief description of the principles of GIXD is given in an appendix (§ 8).

Recent experiments have shown that certain families of water-insoluble molecules, that include amphiphiles, bolaform amphiphiles, chainlike hydrocarbons and fluorocarbons, spontaneously form crystalline films at the air-water interface. These crystallites are oriented vis-a-vis the water surface and can be studied in situ by GIXD. Such studies can shed light onto the crystal nucleation process.

Another approach to probe crystal nucleation at interfaces involves the use of crystalline monolayers on water which induce nucleation at the air-solution interface of three-dimensional crystals composed of solute molecules from the aqueous subphase. The monolayer is designed to provide a complementarity between the molecular arrangement within a particular plane in the three-dimensional crystal and that of the hydrophilic head groups in the monolayer. We shall describe the use of this method to induce nucleation of ice by long-chain alcohols and to get an estimate of the critical size of the ice nucleus. We shall also discuss how ice may be nucleated via the surfaces of amino acid crystals and by frost bacteria.

The tendency of water-insoluble molecules to aggregate into crystalline multilayers also provides a means for engineering various packing arrangements; for example, multi-component thin film crystallites of supramolecular architecture may be prepared via water-insoluble and solute molecules interacting at the solution surface.

188

This work shall be presented in three different regimes of molecular organization at the air-solution interface (Scheme 1): firstly, monolayer arrangements of amphiphilic molecules at the air-liquid interface and interaction between the monolayer and the solution subphase (Scheme 1, central and left columns). We next describe crystallite multilayers composed of water-insoluble molecules only, and finally multilayers generated by reaction between the water-insoluble species with solutes from the subphase to form complex architectures (Scheme 1, right column). The topics described here shall encompass the spontaneous separation of racemates of amphiphilic molecules into enantiomorphous two-dimensional (2D) monolayer domains, followed by the induced nucleation of ice via alcohol monolayers.

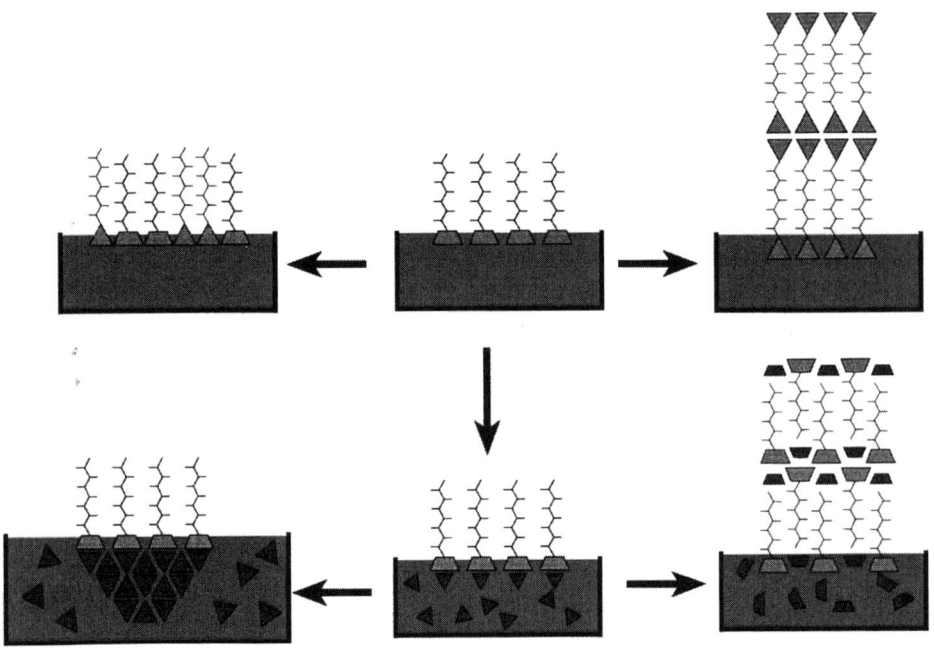

*Scheme 1*

We then discuss the formation on the water surface of alkane multilayers, the early stages of crystalline aggregation of cholesterol, and the organization of channel-forming molecules. The interplay between water-soluble and insoluble components, resulting in the assembly of trilayers containing interdigitated molecules, and the self-organization of supramolecular thin film architectures, is the last topic covered.

## 2. Separation of Racemic Mixtures into Two-Dimensional Crystals

The spontaneous separation of left-and right-handed molecules (*i.e.* enantiomers) from a racemic mixture has intrigued scientists since it was discovered by Pasteur a century and half ago. This phenomenon might have played a significant role in an abiotic process proposed to explain the transformation from a racemic chemistry to a chiral biology, in particular the separation of enantiomers in two dimensions may have relevance to the transfer of chiral information within or across an interface. This latter possibility prompted us to focus on the requirements for separation at the air-water interface of left- and right-handed α-amino acids into 2D arrays.

Unlike in 3D crystals, the detection of spontaneous separation in two-dimensions is not straightforward. Chiral segregation into 2D crystals has been detected via techniques that probe the structure at the nanometer scale, such as scanning force [8] and scanning tunneling [9] microscopy, and GIXD [10,11].

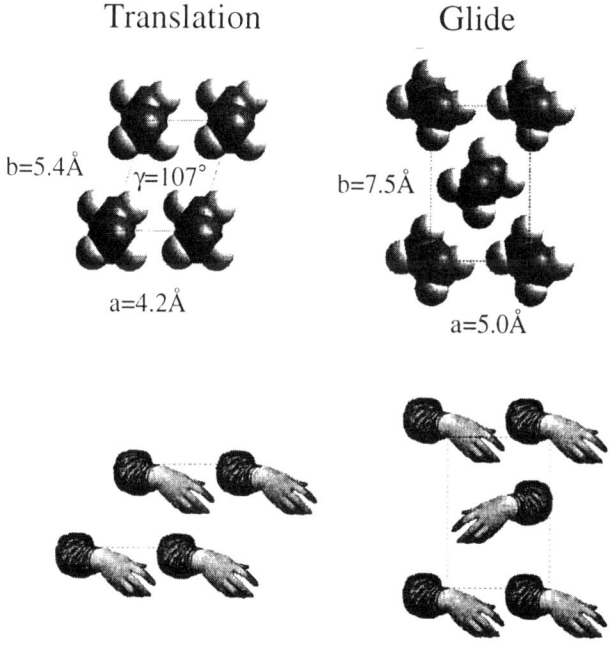

*Scheme 2*

Chiral discrimination of molecules in solution and in Langmuir films has been examined theoretically by Andelman, de Gennes and Orland [12]. Their overall conclusion that heterochiral pair interactions should be generally favoured should be relevant to dilute fluid phases but hardly to crystalline systems where multiple molecular interactions and crystal symmetry play a dominating role. We shall see that the occurrence of spontaneous resolution in 2D crystals may depend upon subtle differences in molecular structure.

Racemic mixtures of molecules tend to form 3D crystals which embody centers of inversion or glide planes that relate chiral molecules of opposite handedness. Only translation (Scheme 2, left) and a glide (Scheme 2, right) whose plane is perpendicular to the water surface are the symmetry elements common for 2D crystals composed of hydrocarbon chain-like molecules but which tend to pack in the herring-bone motif generated by glide symmetry. Spontaneous segregation of enantiomeric territories of chiral amphiphiles at the solution surface as depicted for the set of left and right hands in Scheme 3, can be more easily induced if the glide can be prevented.

*Scheme 3*

## 2.1. AMINO ACIDS

Reasoning along the above lines was applied to achieve chiral segregation for long chain α-amino acids $^+H_3NCHXCOO^-$ [11]. According to the GIXD pattern (Fig. 1a), when X was chosen to be $C_nH_{2n+1}$, $n = 12,16$, the racemic mixture on water crystallized into heterochiral monolayer domains in a rectangular unit cell containing molecules related by glide symmetry in a herring-bone arrangement of the chains (Fig. 1c). By comparison the enantiomer crystallized in an oblique cell (Fig.1d) as determined by the GIXD pattern (Fig. 1b). Enantiomeric separation was achieved by introducing into the chain an amide group to form $X = C_nH_{2n+1}CONH(CH_2)_4$, $n = 15, 17, 21$.

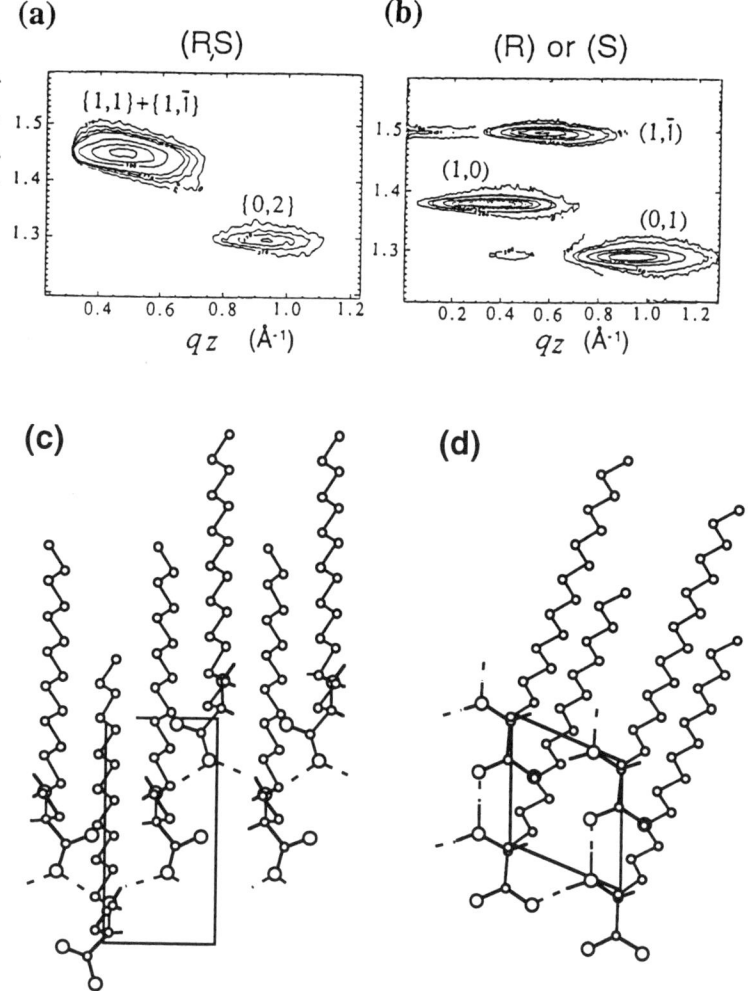

*Figure* 1. The monolayer crystals of racemic, (R,S), and enantiomeric, (R) or (S), $\alpha$-amino acid amphiphile $C_{16}H_{33}CH(NH_3^+)CO_2^-$ on water. (a,b) Their corresponding GIXD patterns are presented as two-dimensional intensity contour plots $I(q_{xy}, q_z)$. (c,d) Their packing arrangements are viewed perpendicular to the water surface. The aliphatic hydrogen atoms have been omitted.

The GIXD patterns of the monolayers of the enantiomeric and racemic mixtures, which were similar to each other (Fig. 2a,b), indicated an oblique unit cell with translation symmetry only, induced by N-H···O hydrogen-bonding of the amide groups along a 5Å axis, complemented by the translation-related N-H···O network of the $^+H_3NCHCOO^-$ moieties (Fig. 2c). This spontaneous separation into domains of opposite handedness was substantiated by X-ray structure factor computations. The subtle effect of change in molecular structure came into play once again on increasing the length of X to $C_{29}H_{59}CONH(CH_2)_4$; the herring-bone motif dominated the packing arrangement even in the enantiomeric 2D crystal structure in which molecules were related by pseudo-glide symmetry.

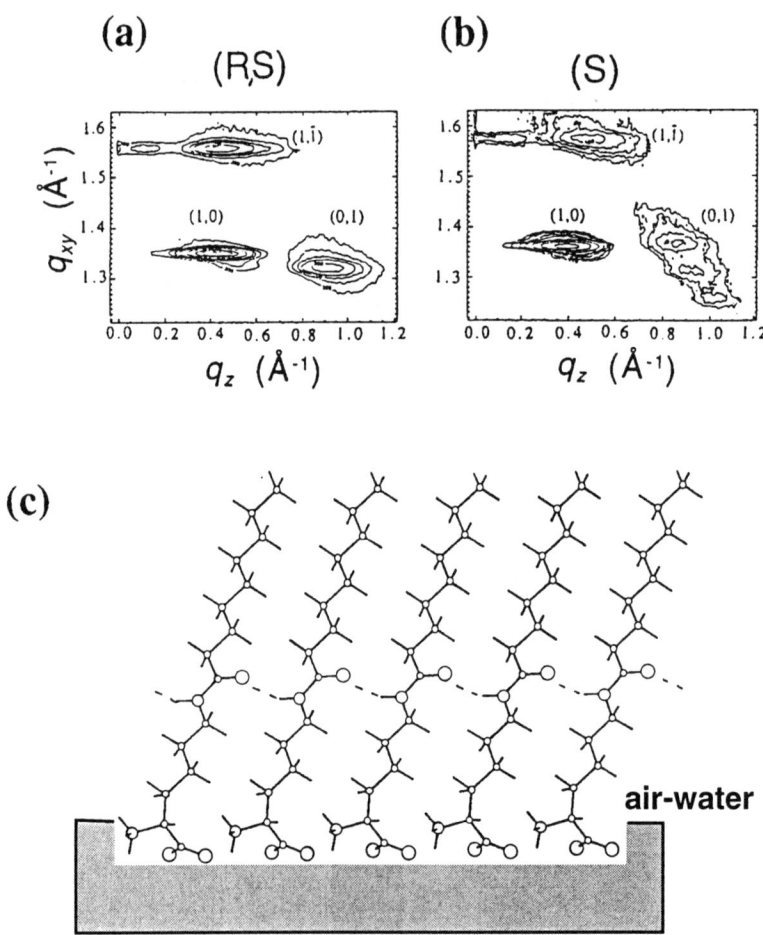

*Figure 2.* The monolayer crystals of racemic, (R,S), and enantiomeric, (R) or (S), $\alpha$-amino acid amphiphile $C_nH_{2n+1}CONHC_4H_8CH(NH_3^+)CO_2^-$ $n = 15, 17, 21$ on water. (a,b) The GIXD patterns for the amino acid $n = 17$ in the form of two-dimensional intensity contour plots $I(q_{xy},q_z)$. (c) The packing arrangement of the racemic mixture $n = 17$ viewed parallel to the water surface. Upper chain section is not shown.

## 2.2. AMPHIPHILIC ACID-BASE COMPLEXES

Spontaneous separation of enantiomers of more complex monolayer systems has been achieved making use of acid-base interactions in mixtures, taking advantage of the 3D crystal structures [13] composed of $R$ or $S$ mandelic acid, $C_6H_5CH(OH)CO_2H$, and $R'$ or $S'$ phenylethylamine, $C_6H_5CH(CH_3)NH_2$. These crystal structures suggested that if both the acid and amine molecules would be modified by attaching a hydrocarbon chain in the *para* position of the phenyl ring, a spontaneous separation of their enantiomers on water could be achieved from a monolayer containing the $(R,S,R',S')$ mixture. The 1:1 mixtures of $R$ or $S$ $C_{15}H_{31}$-$C_6H_5$-$CH(OH)COOH$ and $R'$ or $S'$ $C_{14}H_{29}$-$C_6H_4$-$CH(CH_3)NH_2$, form on water crystalline monolayers [14] depicted schematically in the central column of Scheme 4.

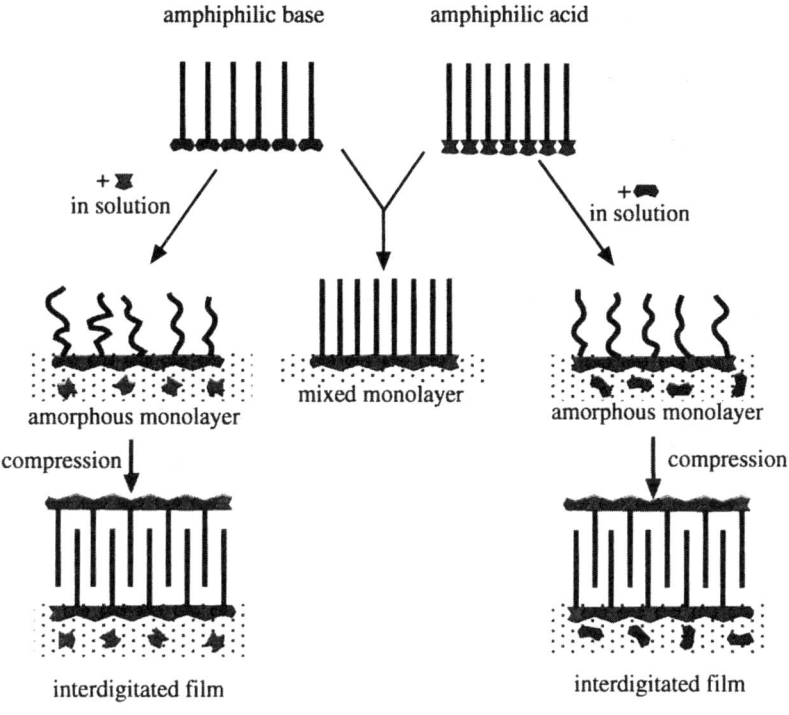

amphiphilic base          amphiphilic acid

+ in solution          + in solution

amorphous monolayer          mixed monolayer          amorphous monolayer

compression          compression

interdigitated film          interdigitated film

*Scheme 4*

The GIXD patterns of the $(R,R')$ and $(R,S')$ films (Fig. 3a,c,d) are distinctly different. An equimolar mixture of the four components $(R,S,R',S')$ gave rise to a diffraction pattern (Fig. 3b) almost identical to that of the $(R,R')$ mixture, signifying a separation of $(R,R')$ and $(S,S')$ territories depicted in Fig, 3, right.

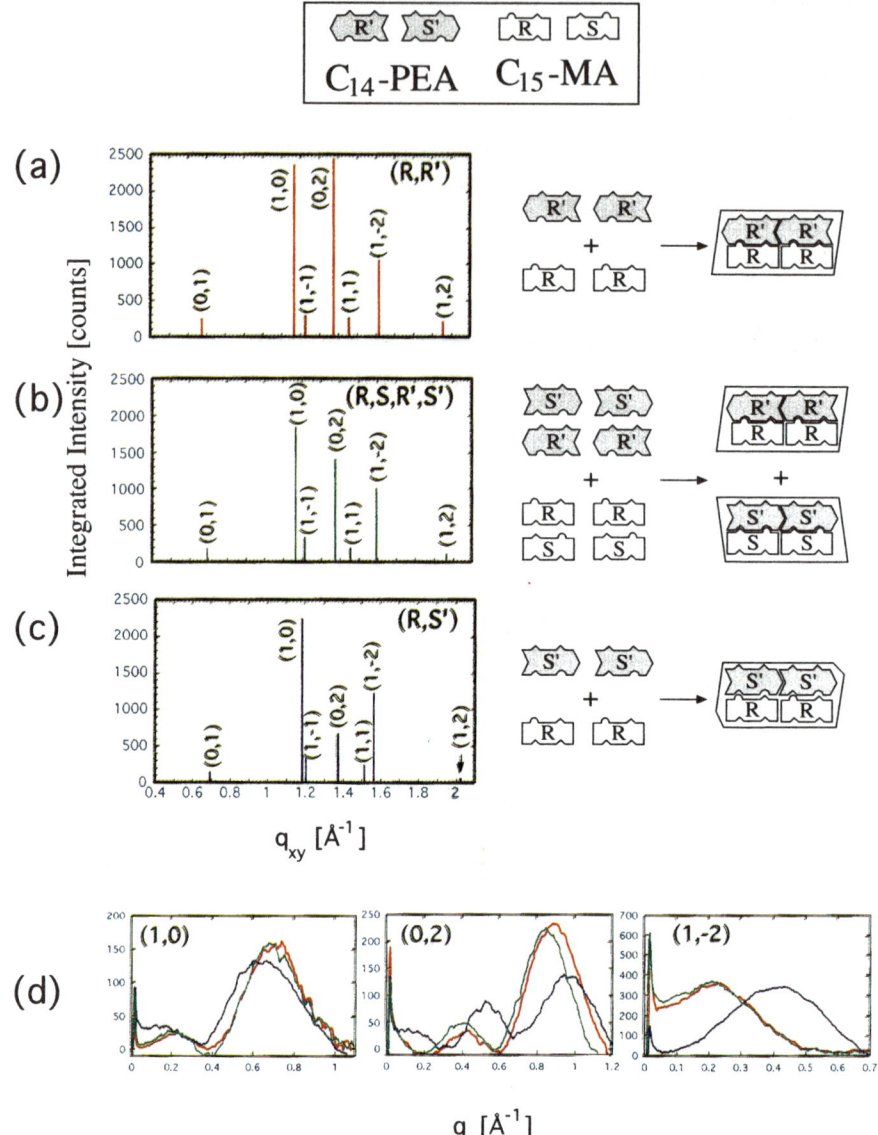

*Figure* 3. The GIXD intensities of the crystalline monolayers on water of the three acid-base mixtures, $(R,R')$, $(R,S')$, and $(R,S,R',S')$ are depicted as red, blue and green lines respectively. (a-c, left) GIXD patterns of the mixtures showing Bragg peaks $I(q_{xy})$ with their $(h,k)$ indices. (d) Comparison of the Bragg rod profiles $I(q_z)$ of the (1,0), (0,2), and (1,-2) reflections of the $(R,R')$, $(R,S')$, and $(R,S,R',S')$ mixtures. The spontaneous separation of left- and right-handed molecules in the $(R,S,R',S')$ mixture into $(R,R')$ and $(S,S')$ territories is depicted schematically in (b, right). Also shown in (a,c, right) is a schematic packing of the molecules in the $(R,R')$ and $(R,S')$ mixtures.

Studies involving nonracemic mixtures demonstrated, however, that mixing of the enantiomers can take place. Thus the formation of an oblique unit cell containing one molecule only as in the systems studied by GIXD [10,11] does not rule out the presence of some enantiomeric disorder in the monolayer crystals of opposite handedness.

## 3. Induced Nucleation of Ice

Pure water can be supercooled to temperatures as low as -20°C to -40°C. Therefore the inhibition or induction of the freezing point of ice, in particular through the role of auxiliaries, has far-reaching ramifications for the living and the non-living world. An intriguing example of inhibition involves the activity of antifreeze proteins in the blood serum of fish in polar seas; these proteins lower the freezing point by no more than 2°C, but this small difference is vital for their survival. Promotion of ice nucleation, has been exploited in the induced precipitation of rain by silver iodide seeded in clouds. On the other hand, it can result in wide scale damage to non-coniferous plants in temperate climates by frost bacteria which nucleate ice at temperatures as high as -2°C.

Ice nucleation has been widely studied, yet the role played on the atomic level by auxiliaries involved in the promotion of ice nucleation, requires further clarification. To help unravel the complexity of this process, we designed experiments where mechanisms such as a structural match and the possible effect of an electric field can be studied somewhat systematically.

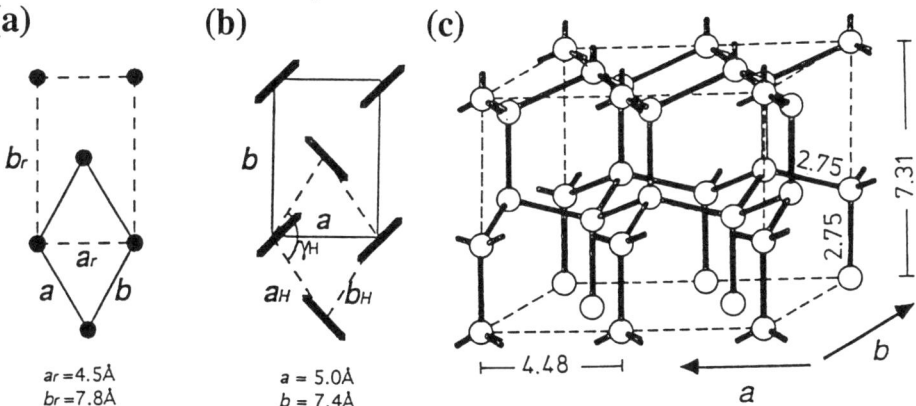

*Figure 4.* (a) Schematic representation of $a,b$ layer of hexagonal ice in terms of the hexagonal cell $(a,b)$ and of a $c$-centered rectangular cell $(a_r,b_r)$. (b) Schematic representation of the herring-bone packing of hydrocarbon chains $C_nH_{2n+2}$ in a rectangular cell $(a,b)$, viewed along the chain axis. The distorted hexagonal representation of the cell $(a_h,b_h)$ is also depicted. (c) Three-dimensional structure of hexagonal ice showing oxygen atoms interlinked by hydrogen bonds (bold lines).

## 3.1 INDUCED NUCLEATION OF ICE UNDER MONOLAYERS OF LONG-CHAIN ALCOHOLS

Various considerations suggested that water-insoluble long-chain aliphatic alcohols $C_nH_{2n+1}OH$ would promote ice nucleation. For example, there is a close lattice match between the arrangement of oxygen atoms in the $ab$ layer of hexagonal ice (Fig. 4a) and of the oxygen atoms in the alcohol molecules when packed in the herring-bone arrangement of hydrocarbon chains, of unit cell dimensions 5x7.5Å$^2$ (Fig. 4b). Indeed, alcohol films of the series $C_nH_{2n+1}OH$, $n$ = 13-31, were found to catalyze ice nucleation [15] (Fig. 5). The induced freezing into ice for $n$ odd reaches an asymptotic temperature just below $0°C$ for an upper value of $n = 31$; the freezing temperature for $n$ even reaches a plateau of -8$°$C for $n$ in the range 22-30.

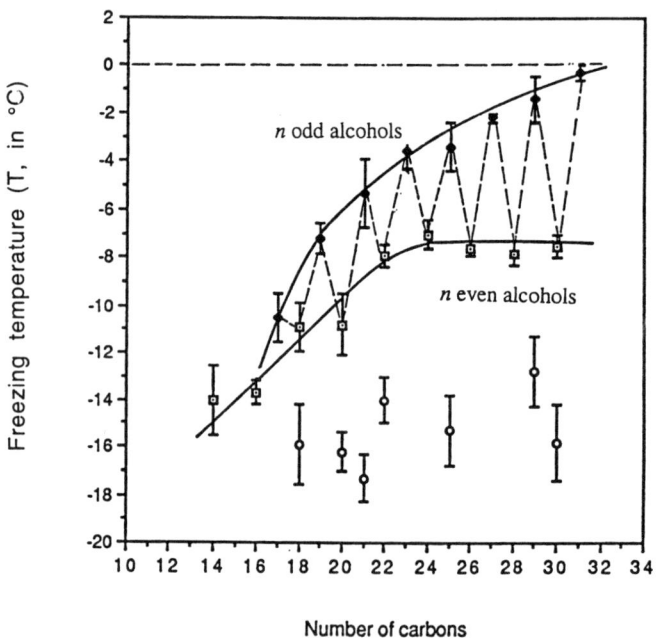

Figure 5. Freezing temperatures (T) of supercooled water drops covered by monolayers of amphiphilic alcohols $C_nH_{2n+1}OH$ (rhombs and squares) and carboxylic acids $C_nH_{2n+1}COOH$ (open circles). Curves are drawn separately for alcohols with $n$ odd (filled rhombs) and $n$ even (open squares).

The epitaxial nature of the induced ice nucleation from an essentially single crystal of the $C_{31}H_{63}OH$ monolayer was demonstrated by an electron diffraction pattern [16a] (Fig. 6c,d). GIXD studies of self-assembled monolayers of the alcohols $C_nH_{2n+1}OH$, $n$

=13-31, on water at 5°C revealed information that could be correlated with the ice freezing behavior [16b,c]. For example, the epitaxial relation between ice and the alcohol monolayers is obvious from the similarity between the arrangement of oxygen atoms in the (001) layer of ice and in the alcohol monolayers $C_n H_{2n+1}OH$, $n = 31, 30$ (Fig. 6a,b). The alcohols crystallize in a rectangular cell, with the chains adopting the herring-bone motif. There is a gradual change in crystal structure with chain shortening that may be correlated with poorer ice nucleation behavior; the molecules become more tilted (see Fig. 7a), there is an increase in lattice area mismatch between ice and the alcohol monolayer, a significant increase in chain librational motion and a reduction in the relative amount of alcohol monolayer crystallites formed and their extent of lateral coherence.

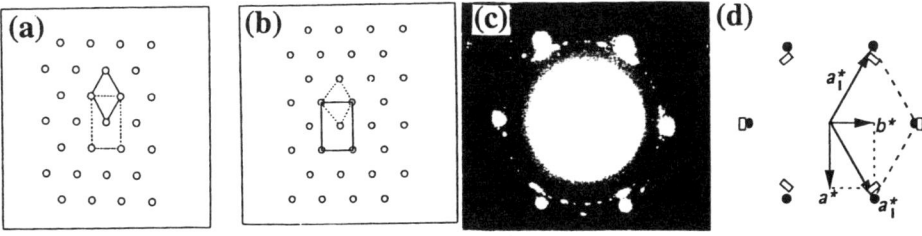

Figure 6. Layer arrangements of the O atoms in: (a) An (001) layer of hexagonal ice; (b) the alcohol $C_nH_{2n+1}OH$ ($n = 30, 31$) monolayers which have the same arrangement of O atoms. (c) Electron diffraction pattern from an essentially single crystallite of the $C_{31}H_{63}OH$ monolayer on a single crystal of hexagonal ice. (d) Relation between the reciprocal lattice of ice and of the monolayer, showing the six {1,0,0} reflections of ice (circles) and the four {1,1} and two {0,2} reflections of the monolayer (rectangles).

The odd-even effect in the ice-nucleating behavior of $C_nH_{2n+1}OH$, and the GIXD data implies that the absolute azimuthal orientation of the hydrocarbon chains differing in length by one $CH_2$ group are the same, leading to a difference in orientation of their $CH_2OH$ groups. The absolute orientation of the molecular chains, and thus of the $CH_2O(H)$ groups, was established for the alcohols $C_mH_{2m+1}CO_2C_nH_{2n}OH$, $m = 19$, $n=9,10$, (Fig. 7b), by GIXD complemented by lattice energy calculations [16c], and independently by sum frequency generation measurements [16d]. For $n=9$, which is the poorer ice nucleator, the O-H bond and the lone-pair electron lobes are equally exposed to water, for $n=10$ (Fig. 7b, left), the better ice nucleator the O-H bond can point vertically into the water subphase (Fig. 7b, right). This model was confirmed from an assignment of the absolute azimuthal orientation of the bromoalcohols $BrC_nH_{2n}OH$, $n = 21,22$ on water, determined from an analysis of the GIXD data [16d].

198

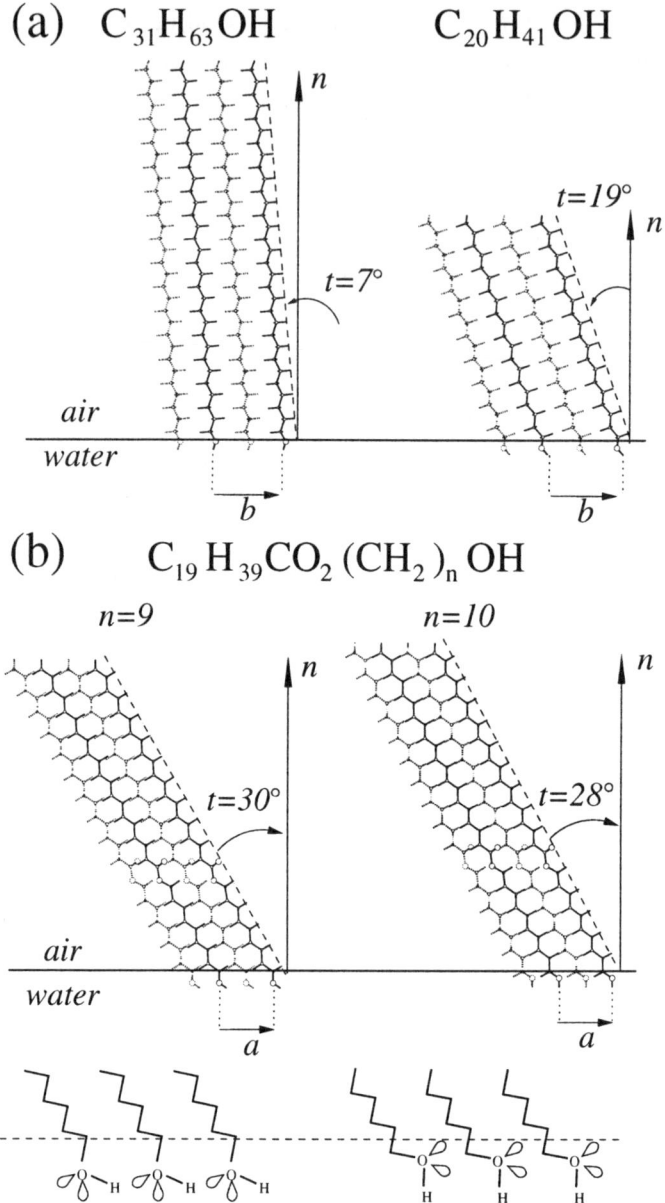

*Figure* 7. Monolayer crystalline arrangements of alcohols viewed perpendicular to the glide plane. Molecules related by glide symmetry are in full and dotted lines respectively. (a) Packing of the alcohols $C_nH_{2n+1}OH$, $n$ = 31, 20. (b) Packing of the hydroxyl alkyl esters $C_{19}H_{39}CO_2(CH_2)_nOH$ $n$ = 9,10, showing also blown-up views of the different orientations of the C-OH groups.

3.1.1 *Critical Size of the Ice Nucleus.*

An estimate of the critical size of ice nuclei as induced by an alcohol monolayer, just below $O^{\circ}C$, was gleaned from a GIXD study monitoring growth of (001) plates of ice by a $C_{31}H_{63}OH$ monolayer [16e] (Fig. 8a-d). Many of these ice crystals had a lateral coherence length of 25 Å, as determined from the width of the (100) reflection of ice (Fig. 6d). This value of 25 Å compares well with the domain diameter of about 30-35 Å, over which there is a match to within 0.5 Å between the oxygen positions of the $a,b$ lattice of ice and that of the $C_{31}H_{63}OH$ monolayer on water (Fig. 8e).

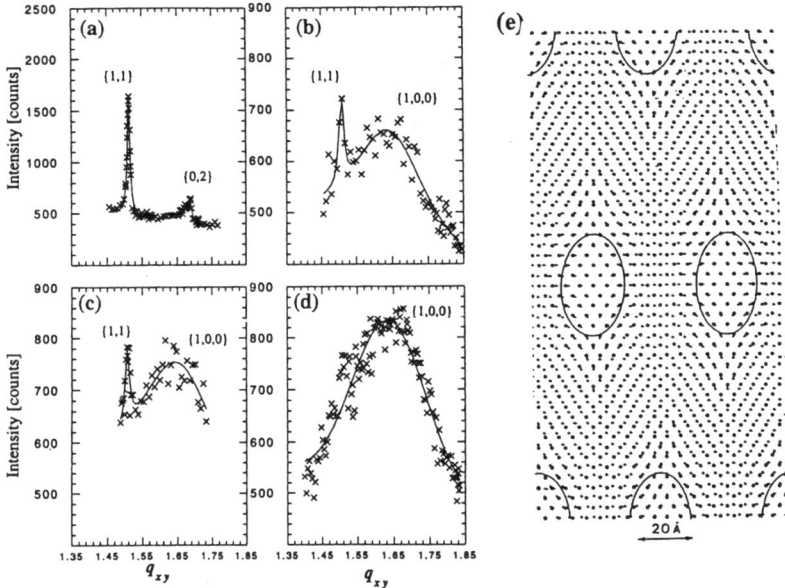

*Figure* 8. GIXD measurements made on a monolayer of $C_{31}H_{63}OH$ over pure water cooled to freezing. (a) The Bragg peaks {1,1} and {0,2} of the monolayer on water at 4°C. (b) First stage of ice nucleation just below 0°C under the monolayer. The {1,1} Bragg peak of the monolayer and the (100) Bragg peak of ice are visible. (c) Same as part (b) but after a time interval of 15 min. (d) After a further 15 min only the (100) Bragg peak of ice is visible. The disappearance of the Bragg peaks of the crystalline monolayer was not due to its destruction, but rather because the ice crystals bound to the monolayer were no longer parallel. The $q_{xy}$ values are in $Å^{-1}$ units. (e) A diagram of the oxygen positions within the $ab$ layer of hexagonal ice (triangles) superimposed on the oxygen positions in the monolayer of $C_{31}H_{63}OH$ (squares). Within each ellipse the oxygen atoms of ice and the monolayer match to within 0.5Å.

The value of 25 Å is also compatible with a model calculation which accounts for the lowering of ice nucleation temperature as a function of point defects concentration (tailored additives) in the $C_{29}H_{61}OH$ monolayer [16f]. Thus, assuming a maximum diameter of the ice critical nucleus of 30Å, would constitute a hemisphere containing about 250 molecules of water. This analysis also suggests that for a homogeneous ice nucleation just below 0°C, the number of water molecules in a critical nucleus is unlikely to be less than 500. According to electron diffraction studies of clusters of water molecules formed on expansion through a nozzle, the crystalline ice core formed in a water cluster had a diameter of about 45Å containing about 1200 water molecules [17]. These results suggest that the alcohol monolayer reduces the effective diameter of the critical ice nucleus.

## 3.2 INDUCED NUCLEATION OF ICE ON SURFACES OF α-AMINO ACID CRYSTALS AND BY FROST BACTERIA.

There are reports in the literature on induced ice nucleation that indicate mechanisms other than an epitaxial fit. For example, the onset temperature of ice nucleation by crystalline powders of chiral-resolved hydrophobic α-amino acids such as valine, leucine and isoleucine was higher than the temperature induced by their corresponding racemic counterparts [18]. These results were surprising because the chiral and racemic crystal forms of these amino acids resemble each other in molecular packing and crystal morphology. A comparative study of ice nucleation on isostructural faces of single crystals of chiral and racemic α-amino acids showed that crystal polarity plays an important role [19], as described below.

The hydrophobic α-amino acids pack in hydrogen-bonded layers. In the chiral crystals, molecules within the layer are related by translation and interlinked by N-H···O bonds. In the corresponding racemates, the molecules appear in a similar layer motif. Neighboring layers in both the chiral and racemic crystals are interlinked on one side by N-H···O bonds to form bilayers. These bilayers are generated in the chiral crystals by twofold symmetry and in the racemates by centers of inversion. The chiral crystals are polar by virtue of the twofold axis parallel to the plane of the bilayer as depicted in Fig. 9c. The racemic crystals are centrosymmetric and so nonpolar as depicted in Fig.9d. The plate-like faces of the hydrophobic α-amino acid crystals expose hydrocarbon residues. Water vapor condensed on these faces at 0°C froze, on cooling, at temperatures higher by 3°C to 5°C on the polar chiral amino acid crystals than on the corresponding centrosymmetric racemates. In a different type of experiment, the substrate crystals were cooled to -15°C before water vapor deposition. Ice crystals hexagonal in shape, and with the $c$ axis tending to be parallel to the face of the substrate, were found to emerge from cracks at their surfaces (Fig. 9a).

The above results were explained as follows: the structure of the plate faces of the polar and non polar α-amino acid crystals, depicted schematically in Fig. 9c,d, are similar. In a polar crystal the structure of the opposite surfaces within a crack, perpendicular to the polar axis, are different. Were such a surface to carry a net charge, the two opposite faces would carry charges of opposite sign. In contrast, the opposite faces within a crack of a centrosymmetric crystal have the same structures with no net charge on each of the two faces. Therefore it was deduced that the opposite faces of cracks along the polar axis produced an electric field strong enough for the alignment of water molecules into proton-ordered ice clusters which would be polar along its hexagonal axis (Fig. 9b).

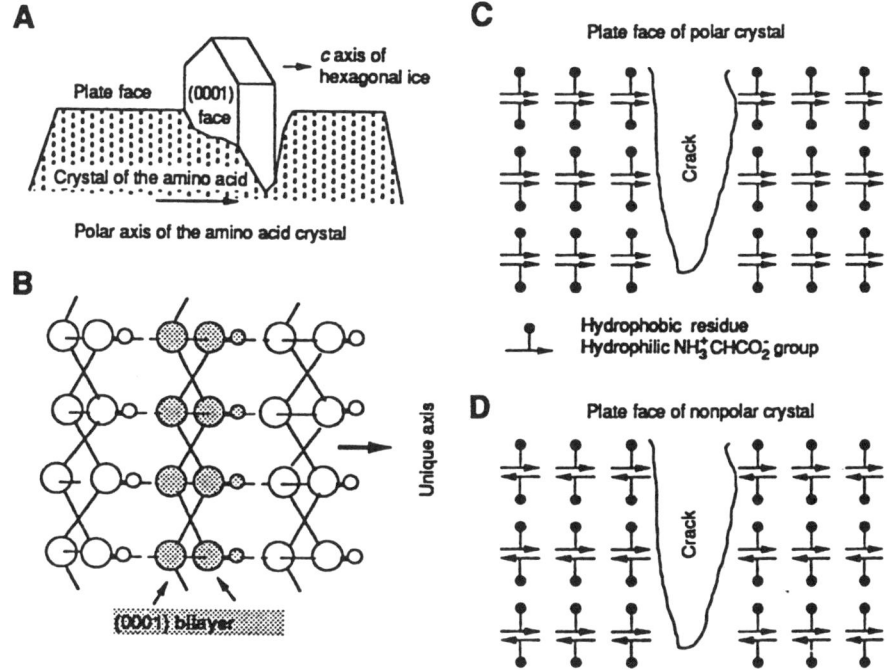

*Figure* 9. (a) Schematic view of hexagonal ice crystals (exaggerated for clarity) emerging from the plate face of an α-amino acid crystal. (b) Molecular packing of a hexagonal ice from which the O-H⋯O hydrogen bonds along the unique axis are assumed to be proton-ordered and so the structure is polar along this direction. The remaining O-H⋯O bonds per molecule within each (0001) bilayer are assumed to be proton-disordered. (c) Schematic view of a polar α-amino acid crystal composed of hydrogen-bonded bilayers. The polar axis is along the horizontal direction. The opposite faces within the cracks expose different groups. (d) Schematic view of the centrosymmetric α-amino acid crystal. The opposite faces within the crack are equivalent.

According to this model it was predicted and shown for alanine and tyrosine, whose racemic crystals have a polar axis and whose chiral counterparts do not, that the racemic crystal induces ice nucleation at temperatures higher by ~5°C. This model is in agreement with calculations of interaction energy between polar and non-polar hexagonal ice-like clusters contained within the crevice of both (R,S) and (S) of alanine [19c]. The ice freezing point results were explained on a more macroscopic level by Wilen [19d], who invoked the role of the difference in dielectric constants of water and ice.

We have already alluded to the fact that several strains of bacterial species, such as *pseudomonas syringae*, can catalyze ice nucleation at temperatures as high as -2°C. The membrane proteins responsible for ice nucleation have been sequenced [20] It was recognized by Warren and coworkers [20b] that the protein probably consists largely of β-strands. They proposed a schematic arrangement having a triangular shape similar to that of the *ab* layer of hexagonal ice. Mizuno proposed a more detailed model [21]. Kajava and Lindow [22] pointed out some disadvantages of these models: the lack of close packing of the protein structures as well as the inability of the water molecules at the ice-like template to form all necessary hydrogen bonds. They also presented a structural model which consists of a largely planar extended macromolecule with one side serving as a template for orienting water into an ice lattice and the other side interacting with the membrane.

## 4.   Multilayer Crystallites on Liquid Surfaces

Amphiphilic molecules almost invariably form monolayers on the water surface. Certain amphiphiles spontaneously form multilayers on liquid formamide [23]. A variety of other molecules were found to form crystalline multilayers on water, such as bolaform amphiphiles containing hydrogen-bonding groups at both ends of the chain [24], pure long-chain alkanes [25] and heterocyclic aromatics [26]. Their packing arrangements could be determined to almost atomic resolution by means of GIXD. Some of these multilayer systems exhibited polymorphic behavior whose interlayer growth and polymorphism could be controlled with the use of tailor-made inhibitors.

The formation of crystalline multilayers of the bolaform amphiphiles on water led to the realization that long-chain alkanes $C_n H_{2n+2}$ should also self organize on the water surface to form multilayer crystallites as indeed was demonstrated by GIXD for the alkanes $20 < n < 30$ [26]. For $n > 31$, monolayers were formed, and as the chain length decreased from $C_{29}H_{61}$ to $C_{23}H_{48}$, the number of layers increased from 2 to about 20. Multilayer growth of the alkanes $C_{23}H_{48}$ and $C_{24}H_{50}$ was inhibited on addition of 5-10% of alcohol additives (see Fig. 10). Not only was film thickness reduced, but a change was induced in the packing arrangement of $C_{24}H_{50}$ from triclinic $P$-1 to orthorhombic $Pbc2_1$ according to their GIXD patterns (Fig. 10), that could be explained by lattice energy calculations.

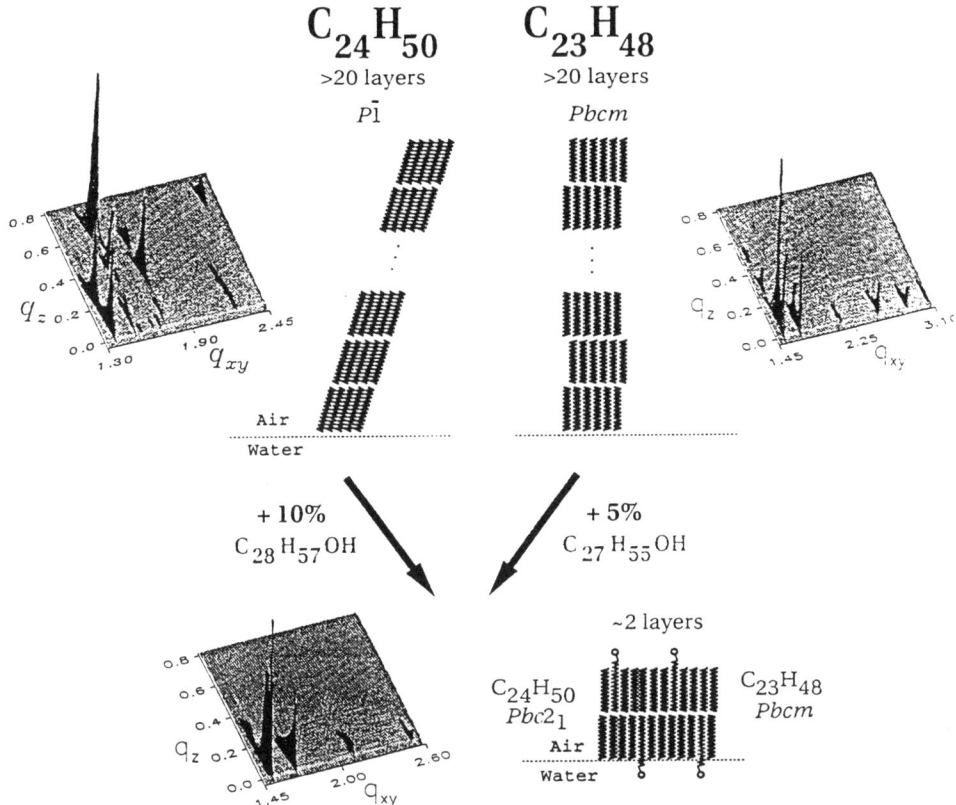

$C_{24}H_{50}$
>20 layers
$P\bar{1}$

$C_{23}H_{48}$
>20 layers
$Pbcm$

Air
Water

+ 10%
$C_{28}H_{57}OH$

+ 5%
$C_{27}H_{55}OH$

~2 layers

$C_{24}H_{50}$
$Pbc2_1$

$C_{23}H_{48}$
$Pbcm$

Air
Water

*Figure* 10. Top: Schematic representation of the multilayer crystalline packing arrangements on water of the alkanes $C_{24}H_{50}$ of triclinic symmetry $P$-1 and $C_{23}H_{48}$ of orthorhombic symmetry $Pbcm$, together with the two-dimensional intensity distributions $I(q_{xy}, q_z)$ of their GIXD patterns. Bilayers of orthorhombic symmetry for $C_{24}H_{50}$ and $C_{23}H_{48}$ were formed on addition of small amounts of alcohol additives. Their GIXD patterns are similar.

## 5. Ordered Assemblies of Membrane-Active Compounds at the Air-water Interface.

Natural biomembranes are systems of a variety of compounds such as phospholipids, cholesterol, proteins held together by non covalent bonds. Structural studies of model membrane films may provide knowledge on the tendency of the different components to form ordered domains.

### 5.1 CHOLESTEROL

Cholesterol is the most abundant sterol in animal tissues playing an important role in determining the rigidity of cell membranes. Faulty control of cholesterol levels may lead to it's precipitation, also in crystalline form, in human bodies, prevailing in the pathological processes of atherosclerosis and gallstone formation in the bile.

A monolayer film of cholesterol at the air-water interface in the uncompressed state exhibits one broad reflection (Fig. 11, curve A) corresponding to a crystalline lateral coherence length of about 70Å [27]. The Bragg rod profile of that peak indicates a monolayer, in which the long molecular axes are aligned normal to the water surface. A hexagonal unit cell $a = b = 6.6$Å containing one molecule is the most basic structure that can be assigned for this single reflection. Such a lattice, that implies high librational motion about the long molecular axis, is precluded since the intermolecular contacts would be too close. A trigonal crystal composed of a unit cell $a' = b' = \sqrt{3}a$ containing three cholesterol molecules related by threefold symmetry (Fig. 11) meets the condition of acceptable intermolecular contacts, and of calculated X-ray structure factors, that are in agreement with the experimental diffraction data [28]. High compression (inset to Fig. 5) gives rise to several sharp Bragg reflections (Fig. 11, curve C) corresponding to a rectangular unit cell $a=10.07$Å, $b=7.57$Å of bilayer thickness (Fig. 11). This rectangular unit cell, which contains two symmetry-unrelated molecules, is fingerprint evidence of a particular layer packing found in several cholesterol-like structures.

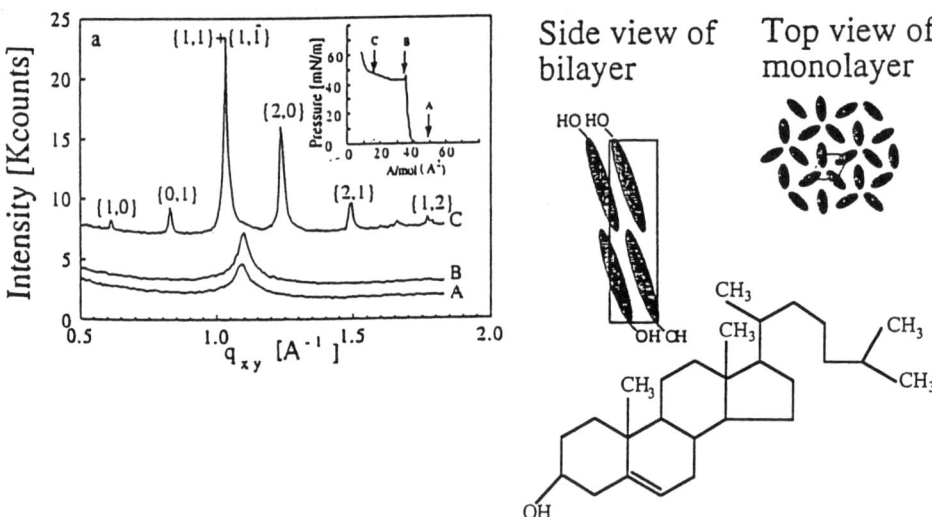

*Figure* 11. (Left): GIXD intensity patterns of a film of cholesterol on water at different compressions on water. (Inset): The surface pressure-area isotherm with arrows indicating the points at which the GIXD measurements were made. (Right): Schematic packing arrangements of the bilayer (side view) and of the monolayer (top view).

## 5.2 IONOPHORES

Ionophores are capable of selectively carrying ions across natural and artificial membranes. The ion transport is associated with structural reorganization that takes place inside the membrane and at its interface. Physico-chemical studies, aiming at revealing the mechanism of ion transport by ionophores, have been the major source of information regarding the conformations adopted by ionophores in different solutions. Structural studies of model membranes containing ionophores, as films at air-solution interfaces should contribute to the elucidation of the processes that take place at the hydrophilic-hydrophobic interface and inside the 2D environment of the long-chain lipids. Moreover, the design of ionoselective films for electrodes and sensors utilizing architectures containing ionophores, may also benefit by this type of study.

GIXD has been applied for the study of the structural characteristics of 2D crystallites of the ionophores valinomycin and nonactin when complexed with various cations at the air-solution interface [29]. Valinomycin complexed with KCl assumes a bracelet shape that packs in a crystalline monolayer in a hexagonal unit cell (Fig. 12, right) according to the GIXD pattern (Fig. 12, left). Other crystalline phases were formed on potassium iodide and barium perchlorate solutions. The presence of hydrophobic monolayers such as phospholipids or stearic acid, that provide a membrane-like environment, induced ordered stacking of valinomycin-potassium chloride complexes into three to four layers. Nonactin packs in a pseudo tetragonal unit cell on solutions of $NH_4SCN$ and KSCN. Upon compression of the nonactin-$NH_4SCN$ film, crystallites of seven to eight layers thick were detected.

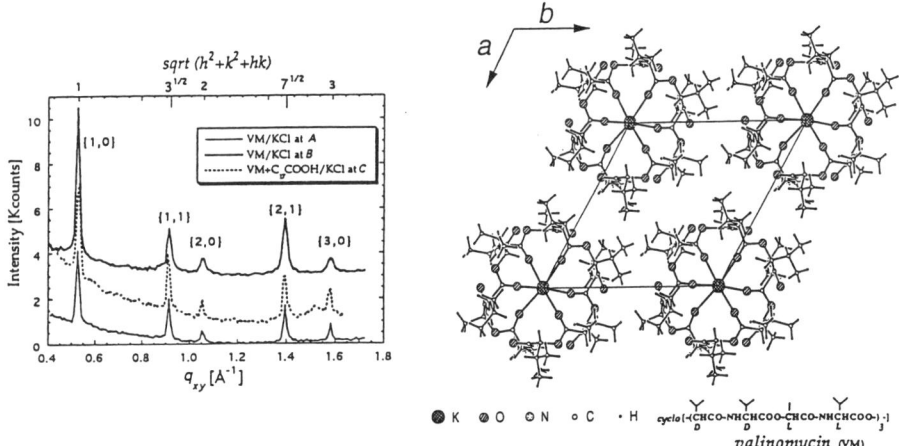

*Figure* 12. (Left) The GIXD patterns of pure valinomycin (VM) at increasing surface pressures (*A,B*), and when deposited with stearic acid, on a 1 M KCl solution. (Right) Molecular packing arrangement in the 2D unit cell of the valinomycin-potassium cation complex.

The high molecular symmetry characterizing these ionophores, reflected in their tendency to form 2D ordered arrays and the relative ease by which multilayers of these complexed ionophores are assembled at interfaces, forming channel-like architectures, may have bearing on ion transport through membranes, via stacking.

## 5.3. CYCLIC PEPTIDE NANOTUBES

Cyclic peptides of an even number of alternating $D$- and $L$- amino acids have been shown to self-assemble by N-H$\cdots$O hydrogen bonding into nanoscalar tubular structures [30]. As these architectures have a potential use as size-selective transporting components of molecular devices, biosensors, and drug-delivery systems it is advantageous to study the principles that govern the organization of these tubular component at interfaces.

The assembly, orientation, and the structural features of nanoscalar tubes composed of cyclic peptides, formed at the air-water interface were detected by GIXD, complemented by atomic force microscopy on transfer to mica support [31]. The eight residue peptide $cyclo$ $(L$-Phe-$D$-N-MeAla-$)_4$ exhibits 2D crystallinity in which the peptide ring lies parallel to the water interface in an arrangement similar to it's 3D crystal form. The peptide $cyclo$ $[(L$-Trp-$D$-Leu$)_3$-$L$-Ser-$D$-Leu$]$ forms mostly aggregates composed of several nanotubes in lateral registry assuming a raft-like structure parallel to the air-water interface (Fig. 13). The cyclic peptide $[(L$-Trp-$D$-Leu$)_4]$ has the lowest tendency, if at all, to form a crystallline 2D array of nanotubes.

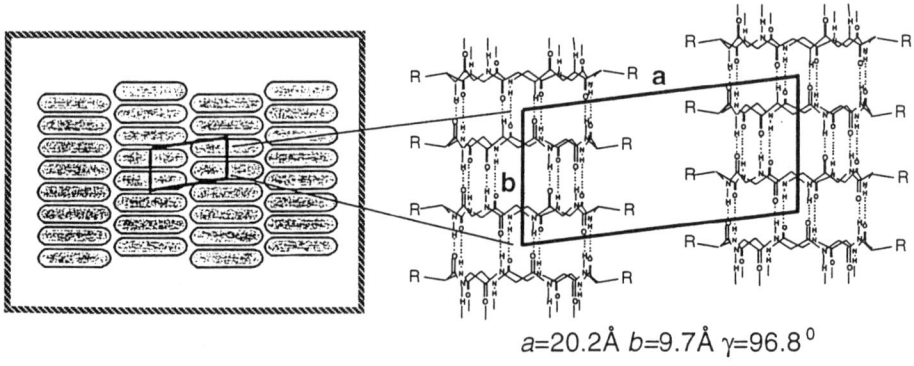

$$a=20.2Å\ b=9.7Å\ \gamma=96.8^\circ$$

*Figure* 13. Raft-like structure of the nanotobe composed of peptide molecules $cyclo$ $[(L$-Trp-$D$-Leu$)_3$-$L$-Ser-$D$-Leu$]$, viewed perpendicular to the water surface.

According to this study, a key factor for generating a continuous domain of nanotubes oriented normal to the interface resides in achieving a more equal balance between the strong hydrogen-bond energy of attachment of the cyclic peptide to an end of a growing tube as against the relatively weak lateral forces between tubes.

## 6.    Supramolecular Architectures Prepared *in-situ* on the Solution Surface

The self-assembly at the air-water interface of water-insoluble molecules into thin crystalline films was extended to the generation of crystalline films made up of two complementary molecules, one water-insoluble and the other water-soluble, extracted from the solution subphase. Different architectures may be prepared in this way as described below.

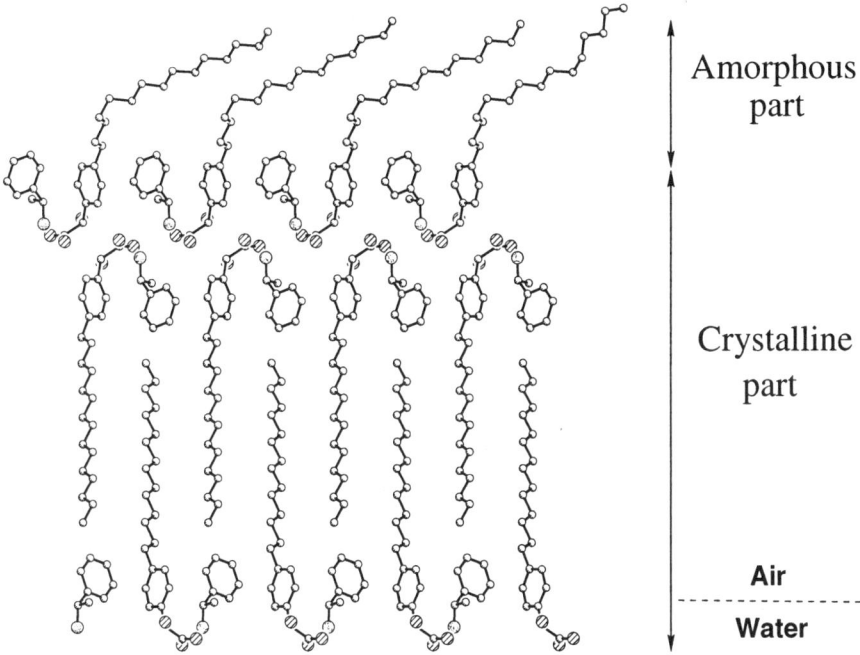

*Figure* 14. Packing arrangement of the interdigitated crystalline trilayer of the 1:1 molecular complex (*R*-$C_{15}H_{31}$-MA, *R'*-PEA) on the solution surface

## 6.1 INTERDIGITATED ARRANGEMENTS

It is possible to form at the air-solution interface bilayers composed of interdigitated molecules, of type $A$ containing a long aliphatic chain, and water-soluble type $B$, exhibiting acid-base complementarity, as depicted at the bottom of the left and right columns of Scheme 4. As appropriate systems use was made of the chiral mandelic acid (MA) derivative $R$ $C_{15}H_{31}$-$C_6H_5$-CH(OH)COOH, (labeled $C_{15}H_{31}$-MA) and $R'$ or $S'$ phenylethylamine, $C_6H_5CHCH_3NH_2$ (PEA), already alluded to in §2.2. A film of $R$-($C_{15}H_{31}$-MA) over a solution of $S'$-(PEA) did not diffract along the compression isotherm. Replacing the solution by $R'$-(PEA) yielded a GIXD pattern after high compression, that corresponded to a crystalline interdigitated trilayer film (Fig. 14) [32a]. The lack of interdigitation when the PEA units were of $S$ configuration was interpreted in terms of an incompatibility between the packing of the chains and the head groups.

Crystalline interdigitated films have been formed from a variety of water-insoluble and water-soluble molecules with acid-base complementarity [32b]. A basic requirement for interdigitation is the formation first of the mixed monolayer comprising the two acid-base components in registry, despite the loss in chain packing as shown in Scheme 4 (central row, left and right columns).

## 6.2. ORGANOMETALLIC FILMS

The formation of the interdigitated systems led to the idea that water-insoluble and water-soluble components can interact at the air-solution interface to spontaneously form oriented crystalline architectures. The dicarboxylic acid $HOOC$-$(CH_2)_{22}$-$COOH$ when deposited on water, form multilayer crystallites, with their long molecular axes aligned somewhat off the normal to the water surface [33]. On aqueous subphases containing divalent $Cd^{2+}$ and $Pb^{2+}$ ions[17], these diacid molecules organize differently, as determined by GIXD. The diacid reacts with the ions to form the corresponding salt, which then self assemble with their chains parallel to the water surface, forming crystalline films about 50Å thick. On transfer to hydrophobized glass, the films were reacted with $H_2S$, yielding quantum dots 2 to 4 nm in diameter [34].

More complex architectures could be self-assembled at the air-solution interface. Recently, oriented bilayer crystalline films (Fig. 15b) composed of 3x3 grids of nine silver ions bound to six ligand molecules (Fig. 15a), were prepared *in-situ* at the air-aqueous solution interface by interaction of the free ligand molecules spread on the aqueous solution, with silver ions from solution, as determined from the GIXD pattern (Fig. 15a) and other methods [26].

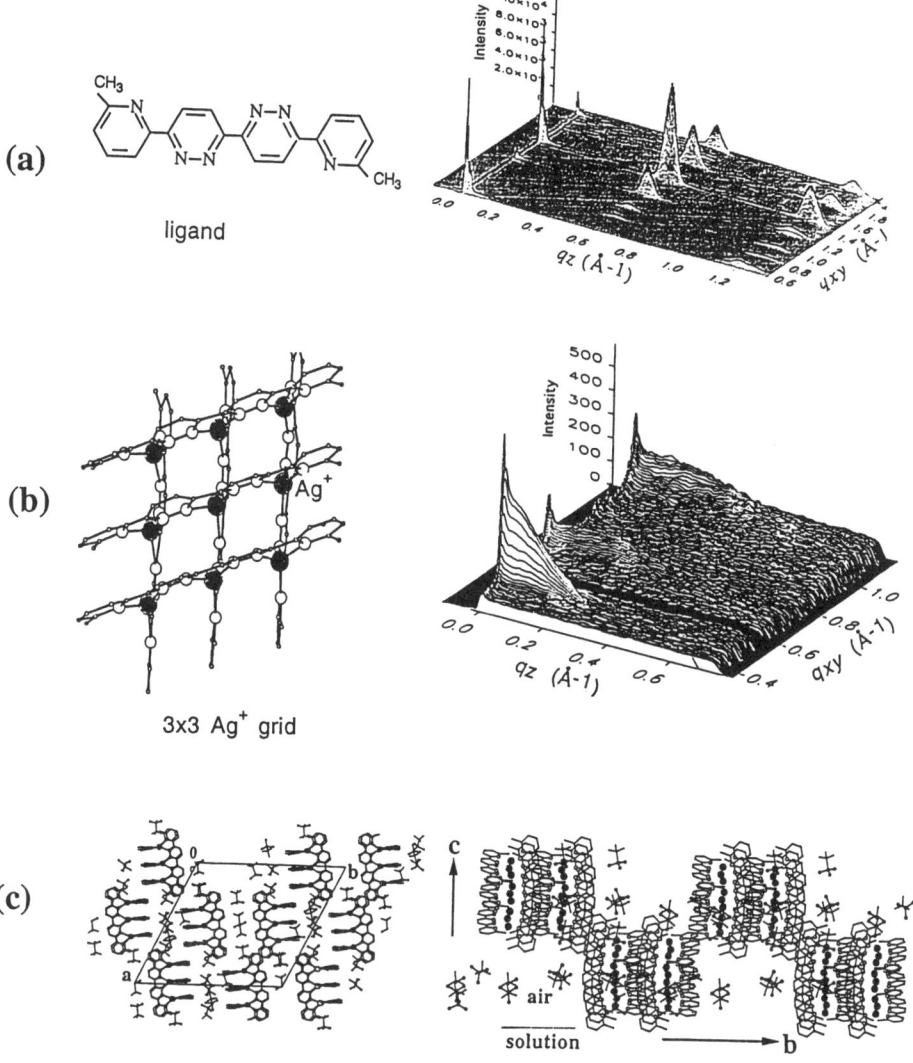

*Figure* 15. (a) The 3x3 silver grid, composed of six ligand molecules and nine silver ions (taken from a 3D crystal structure determination [35]), and the GIXD pattern obtained from an *in-situ* self assembly of the 3x3 silver grid after deposition of the ligand on a silver-containing solution. (b) Proposed packing arrangement of the bilayer film on solution.

## 7. Outlook

Over the past decade enormous strides have been made in the development of surface-sensitive methods for *in-situ* investigations of molecular organization at the air-liquid interface. These advances have led to an extension from early structural studies of monolayers involving simple chain-like amphiphilic molecules to the design and characterization of supramolecular thin film architectures and the examination of membraneous compounds at the air-liquid interface. We may anticipate in the next decade a more intense focus on chemical reactions and biological-type processes at interfaces.

## 8. Appendix: Principles of Grazing Incidence X-ray Diffraction

In the GIXD geometry (Fig. 16), the angle of incidence of the X-ray beam is kept just below the critical angle, $\alpha_i$, limiting the penetration of the beam to that of an evanescent wave with a depth in the range 50-100Å. X-ray scattering due to the subphase is thus efficiently eliminated, which allows an accurate measurement of the weak signals originating from the crystalline film. The evanescent wave is diffracted by the 2-D order in the film (Fig.17 bottom). If the order is crystalline, the evanescent wave may be Bragg scattered from a grain which is oriented so its $h,k$ lattice "planes", with a $d_{h,k}$-spacing, make an angle $\theta_{h,k}$ with the evanescent beam fulfilling the Bragg condition $\lambda = 2d_{h,k}\sin\theta_{h,k}$.

There is no restriction on the z-component of the Bragg scattered ray- it may go deeper into the liquid or it may go out of the liquid at any exit angle $\alpha_f$. In short, a 2-D crystalline lattice confines the X-ray scattering vector to Bragg rods along the $z$ direction, not Bragg points as for a 3-D crystal (Fig. 17 top). The grazing incidence beam illuminates a certain footprint of the surface. In general, the layerlike crystallites on the water surface are azimuthally randomly oriented, and so may be described as a "2-D powder".

The collection of the diffracted radiation by means of a position-sensitive detector (PSD) is made by scanning the detector over a range along the horizontal scattering vector $q_{xy}$ ($\approx 4\pi\sin\theta/\lambda$), where $2\theta$ is the angle between the incident and diffracted beams projected onto the horizontal plane, and integrating over the whole $q_z$ window of the PSD, to yield the Bragg peaks. Simultaneously, the scattered intensity, recorded in channels along the PSD, but integrated over the scattering vector $q_{xy}$ in the horizontal plane across a Bragg peak, produce $q_z$-resolved scans called Bragg-rods. For various purposes the scattered intensity may also be presented in 2-D contour plots as a function of $q_{xy}$ and $q_z$.

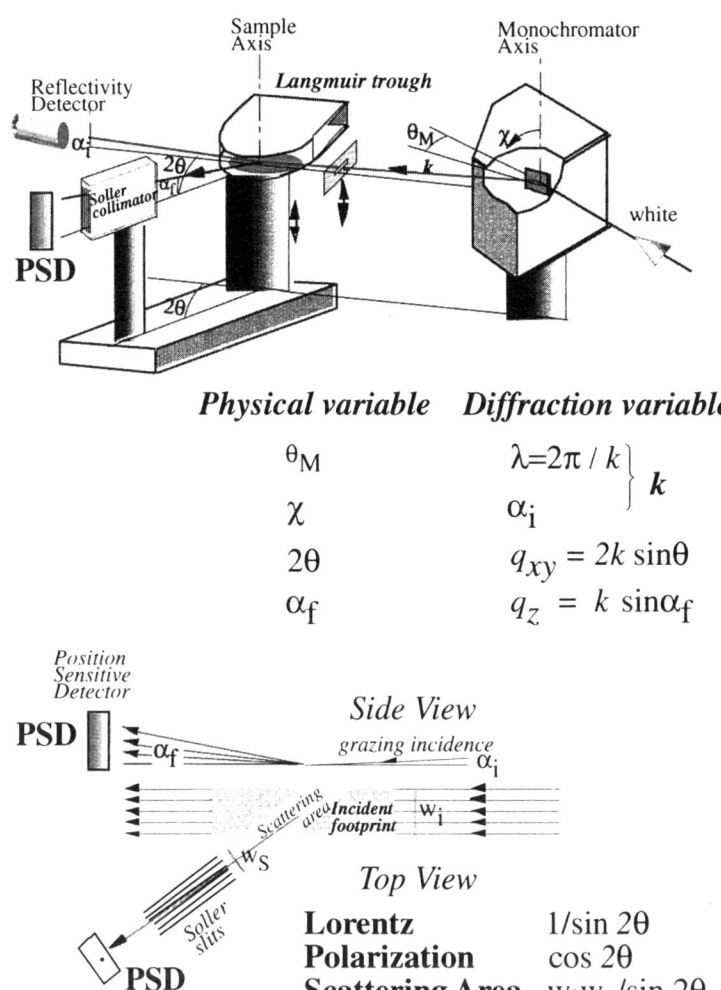

**Physical variable**    **Diffraction variable**

$\theta_M$          $\lambda = 2\pi / k$ ⎫

$\chi$          $\alpha_i$   ⎬ $k$

$2\theta$        $q_{xy} = 2k\sin\theta$

$\alpha_f$        $q_z = k\sin\alpha_f$

*Position Sensitive Detector*

**PSD**   $\alpha_f$     *Side View*

*grazing incidence*   $\alpha_i$

*Scattering area*   *Incident footprint*   $w_i$

$w_s$   *Top View*

*Soller slits*

**PSD**

| | |
|---|---|
| **Lorentz** | $1/\sin 2\theta$ |
| **Polarization** | $\cos 2\theta$ |
| **Scattering Area** | $w_i w_s / \sin 2\theta$ |

*Figure* 16. (*Top*) Experimental setup of the diffractometer at beamline BW1 (X-ray synchrotron source at Hasylab, Hamburg) for studying liquid surfaces. The white X-ray beam is monochromated and deflected down toward the sample by tilting the Be monochromator crystal. The specular reflected beam intensity is monitored by a NaI XR Detector. The diffracted beam is detected with a position-sensitive detector (PSD). (*Bottom*) Side and top views of the grazing incidence X-ray diffraction geometry. The PSD has its axis along the vertical. Only the cross-beam area contributes to the measured scattering. The Soller collimator defines the horizontal resolution of the detector, $k_i$ and $k_f$ are the wave vectors, of length $2\pi/\lambda$, of the incident and diffracted beams, respectively. The scattering vector $q$ is given by $k_f - k_i$ and has vertical and horizontal components $q_z \approx (2\pi/\lambda)\sin\alpha_f$ and $q_{xy} \approx (4\pi/\lambda)\sin\theta$.

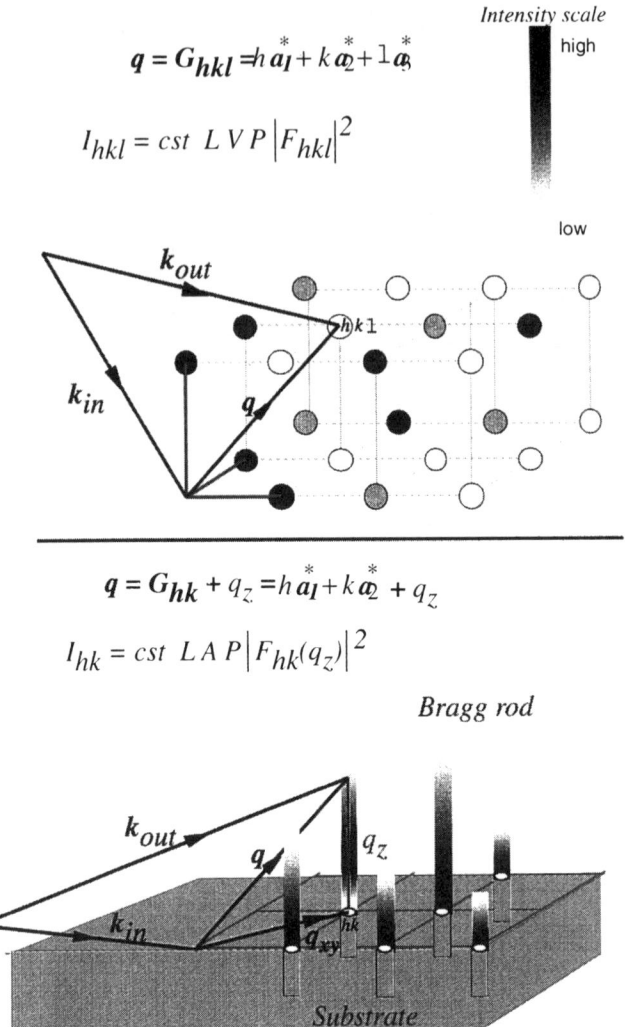

$$q = G_{hkl} = h\,a_1^* + k\,a_2^* + l\,a_3^*$$

$$I_{hkl} = cst\;L\,V\,P\left|F_{hkl}\right|^2$$

*Intensity scale*

high

low

$$q = G_{hk} + q_z = h\,a_1^* + k\,a_2^* + q_z$$

$$I_{hk} = cst\;L\,A\,P\left|F_{hk}(q_z)\right|^2$$

*Bragg rod*

*Substrate*

*Figure* 17. The conditions for diffraction from: (a) a three-dimensional crystal, (b) a two-dimensional crystal. Diffraction from a particular set of $h,k,l$ planes of a 3-D crystal will be achieved if the scattering vector $q$ is coincident with the reciprocal lattice vector $h a_1^* + k a_2^* + l a_3^*$. For a 2-D crystal the diffraction for a particular value of $q_z$ of an $(h,k)$ Bragg rod requires that the horizontal component $q_{xy}$ of the vector $q$, is coincident with the reciprocal lattice vector $h a_1^* + k a_2^*$.

Several different types of information may be extracted from the measured profiles. The angular $2\theta$ positions of the Bragg peaks yield the repeat distances $d = 2\pi/q_{xy}$, for the 2-

D lattice structure. The Bragg peaks may be indexed by the two Miller indices $h,k$ to yield the $a,b$ unit cell. The full width at half maximum (FWHM) of the Bragg peaks in $q_{xy}$ units yields the crystalline coherence length $L$ in the $a,b$ plane associated with the $h,k$ reflection. The intensity at a particular value of $q_z$ in a Bragg rod is determined by the square of the molecular structure factor $|F_{h,k}(q_z)|^2$. For chainlike molecules, precise information on the molecular chain orientation in a 2-D crystal may be obtained from the positions of the maxima of the Bragg rods, assuming the chains to be uniformly tilted.

## References

1. Mullin, J.W. (1993) *Crystallization;* Third edition, Butterworth-Heinemann: Oxford, Chapter 5, pp. 172-201.

2. Gavezzotti, A. (1998) The Crystal Packing of Organic Molecules: Challenge and Fascination Below 1000 Da, *Crystallography Rev. 7*, pp. 5-121.

3. Mullin, J.W. and Leci, C.L. (1969) Evidence of Molecular Cluster Formation in Supersaturated Solutions of Citric Acid, *Phil. Mag.* **19**, 1075-1078; Larson, M.A. and Garside, J. (1986) Solute Clustering in Supersaturated Solutions, *Chem. Eng. Sci.,* **41**, 1285-1289

4. (a) Stein, G.D. and Armstrong, J.A. (1973) Structure of water and carbon dioxide clusters formed via homogeneous nucleation in nozzle beams, *J. Chem. Phys.* **58**, 1999-2003. (b) Bartell, L.S. (1997) Nucleation and Phase Transitions in Molecular Clusters: Molecular Dynamics Simulation and Experiment, *Theoretical Aspects and Computer Modeling of the Molecular Solid State,* Ed. A. Gavezzotti, John Wiley & Sons, Chichester, Chapter 5, pp. 147-184, and references cited therein.

5. (a) Staab, E., Addadi, L., Leiserowitz, L. and Lahav, M. (1990) Control of Polymorphism by 'Tailor-Made' Polymeric Crystallization Auxiliaries, *Adv. Mater.* **2**, 40-43. (b) Weissbuch, I., Leiserowitz, L. and Lahav, M. (1994) "Tailor-Made" and Charge-Transfer Auxiliaries for the Control of the Crystal Polymorphism of Glycine, *Adv. Mater.* **6**, 952-956. (c) Weissbuch, I., Kuzmenko, I., Vaida, M., Zait, S., Leiserowitz, L. and Lahav, M. (1994) Twinned Crystals of Enantiomorphous Morphology of Racemic Alanine Induced by Optically Resolved α-Amino Acid; A Stereochemical Probe for the Early Stages of Crystal Nucleation, *Chem. Mat.* **6**, 1258-1268.

6. Ulman, A., *Ultrathin Organic Films, From Langmuir-Blodgett to Self-Assembly*; Academic Press, Inc.: San Diego, 1991.

7. (a) Als-Nielsen, J. and Kjaer, K. (1989) X-ray reflectivity and diffraction studies of liquid surfaces and surfactant monolayers, *Phase Transitions in Soft Condensed Matter* Ed.; Plenum Press, New York: Geilo, Norway, Vol. 211, Series B, pp. 113-138. (b)

Als-Nielsen, J., Möhwald, H. (1991) Synchrotron X-ray Scattering Studies of Langmuir Films, In *Handbook on Synchrotron Radiation*; S. Ebashi, E. Rubinstein and M. Koch, Ed.; North-Holland: Amsterdam, Vol. IV. pp. 1-55. (c) Jacquemain, D., Grayer Wolf, S., Leveiller, F., Deutsch, M., Kjaer, K., Als-Nielsen, J., Lahav, M. and Leiserowitz, L. (1992) *Angew. Chem. Inter. Ed. Engl.* **31**, 130-152. (d) Als-Nielsen, J., Jacquemain, D., Kjaer, K., Leveiller, F., Lahav, M. and Leiserowitz, L. (1994) Principles and Applications of Grazing-Incidence X-Ray and Neutron-Scattering from Ordered Molecular Monolayers at the Air-Water-Interface, *Physics Reports* **246**, 251-313. (e) Dutta, P. (1997) Studies of Monolayers Using Synchrotron X-ray Diffraction, *Curr. Opin. Solid St. M.* **2**, 557-562. (f) Berge, B., Lenne, P. F. and Renault, A. (1998) X-ray Grazing Incidence Diffraction on Monolayers at the Surface of Water, *Curr. Opin. Colloid. In.* **3**, 321-326.

8. Eckhardt, C.J., Peachey, N.M., Swanson, D.R., Takacs, J.M., Khan, M.A., Gong, X., Kim, J.-H., Wang, J. and Uphaus, R.A. (1993) Separation of Chiral Phases in Monolayer Crystals of Racemic Amphiphiles, *Nature* **362**, 614-616.

9. Stevens, F., Dyer, D.J. and Walba, D.M. (1996) Direct Observation of Enantiomorphous Monolayer Crystals from Enantiomers by Scanning-Tunneling-Microscopy, *Angew. Chem. Int. Ed. Engl.* **35**, 900-901.

10. Nassoy, P., Goldmann, M., Bouloussa, O. and Rondelez, F. (1995) Spontaneous Chiral Segregation in Bidimensional Films *Phys. Rev. Lett.* **75**, 457-460.

11. Weissbuch, I., Berfeld, M., Bouwman, W., Kjaer, K., Als-Nielsen, J., Lahav, M. and Leiserowitz, L. (1997) Separation of Enantiomers and Racemate Formation in Two-Dimensional Crystals at the Water Surface from Racemic α-Amino Acid Amphiphiles: Design and Structure, *J. Am. Chem. Soc.* **119**, 933-942.

12. Andelman, D. and de Gennes, P.G. (1988) Chiral Discrimination in a Langmuir Monolayer *C.R. Acad. Sc.* **307**, 233-237; Andelman, D. and Orland, H. (1993) Chiral Discrimination in Solutions and in Langmuir Monolayers, *J. Am. Chem. Soc.* **115**, 12322-12329.

13. Brianso, M.C., Leclercq, M. and Jacques, J. (1979) Mandélate de Phényl-1 Ethylamine *Acta Crystallogr.* **B35**, 2751-2753; Larsen, S. and Diego, H.L. (1993) A study of two polymorphic modifications of (S)-1-phenylethylammonium (S)-mandelate and the structural features of diastereomeric mandelate salts, *Acta Crystallogr.* **B49**, 303-309.

14. Kuzmenko, I., Kjaer, K, Als-Nielsen, J., Lahav, M. and Leiserowitz, L. (1999) Detection of Chiral Disorder in Langmuir Monolayers Undergoing Spontaneous Chiral Segregation, *J. Am. Chem. Soc.* , in press.

15. (a) Gavish, M., Popovitz-Biro, R., Lahav, M. and Leiserowitz, L. (1990) Ice Nucleation by Alcohols Arranged in Monolayers at the Surface of Water Drops, *Science* **250**, 973-975. (b) Popovitz-Biro, R., Wang, J.L., Majewski, J., Shavit, E.,

Leiserowitz, L. and Lahav, M. (1994) Induced Nucleation of Supercooled Water into Ice by Self-Assembled Crystalline Monolayers of Amphiphilic Alcohols at the Air-Water Interface, *J. Am. Chem. Soc.* **116**, 1179-1191.

16. (a) Majewski, J., Margulis, L., Jacquemain, D., Leveiller, F., Bohm, C., Arad, T., Talmon, Y., Lahav, M. and Leiserowitz, L. (1993) Electron Diffraction and Imaging of Uncompressed Monolayers of Amphiphilic Molecules on Vitreous and Hexagonal Ice, *Science* **261**, 899-902. (b) Majewski, J., Popovitz-Biro, R., Bouwman, W.G., Kjaer, K., Als-Nielsen, J., Lahav, M. and Leiserowitz, L. (1995) The structural Properties of Uncompressed Crystalline Monolayers of Alcohols $C_nH_{2n+1}OH$ (n=13-31) on Water and Their Role as Ice Nucleators, *Chem.Eur.J.* **1**, 304-311. (c) Wang, J.L., Leveiller, F., Jacquemain, D., Kjaer, K., Als-Nielsen, J., Lahav, M. and Leiserowitz, L. (1994) Self-Aggregates of Amphiphilic Alcohols at the Air-Water Interface as Studied by Grazing Incidence Synchrotron X-ray Diffraction, *J. Am. Chem. Soc.* **116**, 1192-1204. (d) Edgar, R.; Huang, J.Y.; Popovitz-Biro, R.; Kjaer, K.; Bouwman, W. G.; Howes, P. B.; Als-Nielsen, J.; Shen, Y.R.; Lahav, M., Leiserowitz, L. To be submitted for publication (1999). (e) Majewski, J., Popovitz-Biro, R., Kjaer, K., Als-Nielsen, J., Lahav, M. and Leiserowitz, L. (1994) Towards a Determination of the Critical Size of Ice Nuclei. A Demonstration by Grazing Incidence X-ray Diffraction of Epitaxial Growth of Ice under the $C_{31}H_{63}OH$ Alcohol Monolayer, *J. Phys. Chem.* **98**, 4087-4093. (f) Majewski, J., Popovitz-Biro, R., Edgar, R., Arbel-Haddad, M., Kjaer, K., Bouwman, W., Als-Nielsen, J., Lahav, M. and Leiserowitz, L. (1997) An Insight into the Ice Nucleation Process via Design of Crystalline Ice Nucleators of Variable Size, *J. Phys. Chem. B* **101**, 8874-8877.

17. Torchet, G., de Feraudy, M.F., Farges, J., Raoult and B. Schwartz, P.,(1983) Structure of Solid Water Clusters Formed in a Free Jet Expansion, *J. Chem. Phys.* **79**, 6196-6202.

18. (a) Power, B.A. and Power, R.F. (1962) Amino Acids as Ice Nucleators, *Nature* **194**, 1170-1171. (b) Barthakur, N, and Maybank, J. (1963) Anomalous Behavior of Some Amino Acids as Ice Nucleators, *Nature* **200**, 866-868. (c) Parungo, F.P. and Lodge, J.P.Jr. (1967) Amino Acids as Ice Nucleators, *J. Atm. Sci.* **24**, 274-277.

19. (a) Gavish, M., Wang, J.L., Eisenstein, M., Lahav, M. and Leiserowitz, L. (1992) The Role of Crystal Polarity in α-Amino Acid Crystals for Induced Nucleation of Ice, *Science* **256**, 815-818. (b) McBride J. M. (1992) Perspective. Crystal Polarity: A Window on Ice Nucleation, *Science* **256**, 814. (c) Lahav, M., Eisenstein, M., Leiserowitz, L. (1993) The Energy Density of Water and Ice Nucleation, *Science* **259**, 1469-1470. (d) Wilen, L. *ibid.*, 1469

20. (a) Green, R.L. and Warren, G.J. (1985) Physical and Functional Repetiotion in a Bacterial Ice, *Nature* **317**, 645-648. (b) Warren, G.J., Corotto, L.V. and Wolber, P.K. (1986) Conserved Repeats in Divreged Ice Nucleation Structural Genes from Two

Species of Pseudomonas *Nucl. Acids Res.* **14**, 8047-8060. (c) Warren, G.J. and Corotto, L.V. (1989) The Consensus Sequence of Ice Nucleation Proteins from Erwinia Herbicola, Pseudomonas Fluorescens and Pseudomonas Syringae, *Gene* **85**, 239-242.

21. Mizuno, H. (1989) Prediction of the Conformation of Ice-Nucleation Protein by Conformation Energy Calculation, *Proteins: Struct. Funct. Genet.* **5**, 47-65.

22. Kajava, A.V. and Lindow, S.E. (1993) A Model of the 3-Dimensional Structure of Ice Nucleation Proteins, *J. Mol. Biol.* **232**, 709-717.

23. Weinbach, S., Kjaer, K., Bouwman, W. G., Grubel, G., Legrand, J. F., Als-Nielsen, J., Lahav, M. and Leiserowitz, L. (1994) Control of Growth and Structure of Polymorphic Crystalline Thin Films of Amphiphilic Molecules on Liquid Surfaces, *Science* **264**, 1566-1570.

24. (a) Majewski, J., Edgar, R., Popovitz-Biro, R., Bouwman, W. G., Als-Nielsen, J., Lahav, M. and Leiserowitz, L. (1995) Structure Determination in the Twilight Region between Monolayers and 3-D Crystals; a Grazing Incidence X-ray Diffraction Study of Nanocrystalline Aggregates of $\alpha,\omega$-Docosanediol at the Air-Water Interface, *Angew. Chem. Int. Ed. Engl.* **34**, 649-652. (b) Popovitz-Biro, R., Majewski, J., Margulis, L., Cohen, S., Leiserowitz, L. and Lahav, M. (1994) Self-Aggregation of $\alpha,\omega$-Alkane-Diols into 3-D Crystallites as Studied at Interfaces; The System of $\alpha,\omega$-Docosanediol, *J. Phys. Chem.* **98**, 4970-4972.

25. Weinbach, S., Weissbuch, I., Kjaer, K., Bouwman, W. G., Als-Nielsen, J., Lahav, M. and Leiserowitz, L. (1995) Self-Assembled Crystalline Monolayers and Multilayers of *n*-Alkanes on the Water Surface, *Adv. Mater.* **7**, 857-862.

26. Weissbuch, I., Baxter, P.N.W., Cohen, S., Cohen, H., Kjaer, K., Howes, P.B., Als-Nielsen, J., Hanan, G.S., Schubert, U.S., Lehn, J-M., Leiserowitz, L. and Lahav, M. (1998) Self-Assembly at the Air-Water Interface. *In-situ* Preparation of Thin Films of Metal Ion Grid Architectures, *J. Am. Chem. Soc.* **120**, 4850-4860.

27. Lafont, S., Rapaport, H., Somjen, G.J., Renault, A., Howes, P.B., Kjaer, K., Als-Nielsen, J., Leiserowitz, L. and Lahav, M. (1998) Monitoring the Nucleation of Crystalline Films of Cholesterol on Water and    in the Presence of Phospholipids, *J. Phys. Chem.* **102**, 761-765.

28. (a) Rapaport, H., PH. D. Thesis, The Feinberg Graduate School, Weizmann Institute of Science, Rehovot, Israel, 1998. (b) Rapaport, H.; Kuzmenko, I.; Lafont, S.; Kjaer, K.; Howes, P. B.; Als-Nielsen, J.; Lahav, M., Leiserowitz, L. To be submitted for publication, (1999).

29. Rapaport, H., Kuzmenko, I., Kjaer, K., Howes, P., Bowman, W., Als-Nielsen, J., Leiserowitz, L. and Lahav, M. (1997) Structural Characterization of Valinomycin and Nonactin at the Air-Solution Interface by Grazing Incidence X-ray Diffraction, *J. Am. Chem. Soc.* **119**, 11211-11216.

30. Ghadiri, M.R., Granja, J.R., Milligan, R.A., McRee, D.E. and Khazanovich, N. (1993) Self-Assembling Organic Nanotubes Based on a Cyclic Peptide Architecture, *Nature* **366**, 324-327.

31. Rapaport, H. , Kim H.S., Kjaer, K., Howes, P.B., Als-Nielsen, J., Ghadiri, M. R., Leiserowitz, L. and Lahav, M. (1999) Crystalline Cyclic Peptide Nanotubes at Interfaces, *J. Am. Chem. Soc.* **121**, 1186-1191

32. (a) Kuzmenko, I., Buller, R., Bouwman, W.G., Kjaer, K., Als-Nielsen, J., Lahav, M. and Leiserowitz, L. (1996) Formation of Chiral Interdigitated Multilayers at the Air-lLiquid Interface Through Acid-Base Interactions, *Science* **274**, 2046. (b) Kuzmenko, I., Kindermann, M., Kjaer, K., Howes, P.B., Als-Nielsen, J., Kiedrowski, G. v., Leiserowitz, L. and Lahav, M. To be submitted for publication (1999).

33. Weissbuch, I., Guo, S., Cohen, S., Edgar, R., Howes, P., Kjaer, K., Als-Nielsen, J., Lahav, M. and Leiserowitz, L. (1998) Self-Assembly of Metal Salts of $\alpha,\omega$-Tetracosanedioic Acid into Oriented Crystalline Thin Films at the Air-Solution Interface, *Adv. Mat.* **10**, 117-121.

34. Guo, S., Popovitz-Biro, R., Arad, T., Hodes, G., Leiserowitz, L. and Lahav, M. (1998) Superlattices of Semiconductor Quantum Size Lead Sulfide Particles Prepared via a Topotactic Gas-Solid Reaction, *Adv. Mater.* **10**, 657-661.

35. Baxter, P.N. W., Lehn, J.-M., Fischer, J., Youinou, M.-T. (1994) Self-Assembly and Structure of a 3X3 Inorganic Grid from 9 Silver Ions and 6 Ligand Components, *Angew. Chem. Int. Ed. Engl.* **33**, 2284-2287.

# X-RAY AND NEUTRON SCATTERING STUDIES OF COMPLEX CONFINED FLUIDS

SUNIL K. SINHA
Advanced Photon Source
Argonne National Laboratory
Argonne, Illinois 60439

We review recent X-ray and neutron scattering studies of the structure and dynamics of confined complex fluids. This includes the study of polymer conformations and binary fluid phase transitions in porous media using Small Angle Neutron scattering, and the use of synchrotron radiation to study ordering and fluctuation phenomena at solid/liquid and liquid/air interfaces. Ordering of liquids near a solid surface or in confinement will be discussed, and the study, via specular and off-specular X-ray reflectivity, of capillary wave fluctuations on liquid polymer films. Finally, we shall discuss the use of high-brilliance beams from X-ray synchrotrons to study via photon correlation spectroscopy the slow dynamics of soft condensed matter systems.

## 1. Introduction

The behavior of fluids in confinement often differs from their bulk behavior in interesting ways. Thus fluids confined in a microporous medium for instance may have their phase behavior relative to melting or freezing modified. In other cases, the way in which a binary fluid system mixes or demixes may be modified. To some extent, a rather simple (and consequently less interesting) effect of confinement is simply a shift of various regions of the phase diagram with regard to temperature or concentration. However, more interestingly, in other cases the behavior itself may be modified. In particular, one may ask the question in the case of a second-order (continuous) phase transition, such as at a consolute point for a binary liquid mixture or a liquid/gas critical point, how is the critical behavior altered when the fluid is confined in micropores?

One may also ask whether the behavior is affected by the _confinement_ or by the _randomness_ of the pore structure. In both cases the effect is due to interaction with the pore walls, but considerations of randomness led a number of people a few years ago to propose the exciting idea that, for instance, a binary fluid demixing transition (which has a 3D-Ising-like critical behavior in bulk) maps on to the Random Field Ising Model (RFIM). The conformation of polymers confined in porous media also differs from that

219

_A.T. Skjeltorp and S.F. Edwards (eds.), Soft Condensed Matter: Configurations,_
_Dynamics and Functionality, 219-245._

in free solution in ways which depend on both the concentration and the ratio of polymer size to pore size, as has been discussed theoretically by DeGennes, Brochard and others. Important structural aspects of all of the above phenomena can be studied by small angle neutron scattering. This technique has the advantage that it can study concentration or density fluctuations inside media opaque to light and at short length scales which can vary from a few nanometers to microns. In addition, the techniques of contrast matching into selective deuteration of the sample can simplify the interpretation of the results.

This is because it is possible to fill the pores with a fluid whose scattering density exactly matches that of the solid, thus removing the scattering from the pore structure itself, thereby reducing the 2-phase system to a 1-phase system. The only fluctuations which will then scatter radiation are deviations from the <u>average</u> density, which will be fluctuations in the fluid itself.

Besides being of basic scientific interest, the behavior of fluids or polymers in porous materials is of considerable importance for processes such as enhanced oil recovery, chemical processing, and chemical or biological separation or filtration processes involving porous membranes or gels. The <u>transport</u> of fluids and polymers inside porous materials is also of considerable interest, but can only be indirectly studied by neutron or X-ray scattering. However, recent application of <u>coherent</u> X-ray beams from the new synchrotron radiation sources can make it possible to study the dynamics of such processes by using photon intensity correlation spectroscopy, as will be discussed at the end of these lectures.

One may of course also consider fluids confined at a surface or an interface or in a thin film. Here again, there are interesting predictions and observations regarding differences from bulk behavior. Scattering experiments such as specular and off-specular reflection of neutrons and X-rays (in particular the later, where one may now exploit the high brilliance of X-ray beams from synchrotron radiation sources) can elucidate effects such as crystallization or melting phenomena at interfaces, fluctuations due to capillary waves, etc., and we shall discuss some of these experiments also in these lectures.

## 2. Small Angle Scattering Studied of Fluids Confined in Porous Media

The basic methodology of small-angle scattering has been discussed in a variety of excellent reviews [1-4]. Neglecting any inelasticity in the scattering, the number of particles per second scattered by a sample into a detector is given by

$$I = S(\bar{q})(I_o/A)(\Delta\Omega) \tag{1}$$

where $I_o$ is the number of particles per second in the incident beam of cross-sectional area A, $(\Delta\Omega)$ is the solid angle subtended at the sample by the detector, $S(\bar{q})$ is the scattering function characterizing the sample and $\bar{q}$ is the so-called wavevector transfer defined by

$$\bar{q} = \bar{k}_1 - \bar{k}_0 \tag{2}$$

where $\bar{k}_0$, $\bar{k}_1$ are the wavevectors of the incident and scattered radiation respectively. The magnitude of q is given by $q = 2k_0 \mathrm{Sin}\,\vartheta$ where $2\vartheta$ is the angle of scattering (see Fig. 1), where $k_0 = 2\pi/\lambda$, $\lambda$ being the wavelength of the incodent radiation. In general the small angle regime is defined by $q \ll k_0$. $S(\bar{q})$ in general is a function of the average of the instantaneous positions of all the particles in the scattering system, but in the small-angle regime (defined roughly by the range $0 < q < 3$ nm$^{-1}$) we may ignore the atomic and molecular structure of the constituents and deal only with the spatial variations (on length scales from a few nm on up) of the scattering length density (SLD) $\rho(\bar{r})$ of the sample. For small-angle neutron scattering (SANS) experiments, $\rho(\bar{r})$ is defined by

$$\rho(\bar{r}) = \Sigma \; b_i \, n_i(\bar{r}) \tag{3}$$

where $b_i$ is the nuclear scattering length [5] of nucleus of type I, and $n_i(\bar{r})$ is the associated number density of such nuclei, while for small-angle X-ray scattering (SAXS) experiments, $\rho(\bar{r})$ is defined by

$$\rho(\bar{r}) = \left(e^2/mc^2\right)n_{el}(\bar{r}) \tag{4}$$

where the factor $(e^2/mc^2)$ is the Thompson scattering length of the electron, and $n_{el}(\bar{r})$ is the electron number density. Since a uniform scattering length density does not scatter radiation (except in the forward direction), $S(\bar{q})$ will depend only on the deviations of $\rho(\bar{r})$ about its mean, or what is referred to as the contrast. The Kinematic or Born approximation to the scattering [6], where multiple scattering effects are neglected, then yields for the scattering function $S(\bar{q})$ the expression

$$S(\bar{q}) = \int\!\int d\bar{r}d\bar{r}' \langle \delta\rho(\bar{r})\delta\rho(\bar{r}')\rangle e^{i\bar{q}(\bar{r}'-\bar{r})} \equiv \int\!\int d\bar{r}d\bar{r}' \gamma(\bar{r}-\bar{r}')e^{i\bar{q}(\bar{r}'-\bar{r})} \tag{5}$$

where the statistical average is taken over the whole system. (This statement has to be modified if the incident radiation is highly coherent. It is commonly assumed that such an average depends only on the magnitude of $\bar{R} \equiv \bar{r} - \bar{r}'$, for an isotropic and translationally invariant system. In Eq. (5), $\delta\rho(\bar{r})$ is defined as the fluctuation from the average SLD as explained above. If we are dealing with a particulate system (e.g., a dilute polymer solution) and these are far enough apart that the interference effects between the scattering from different particles (which will occur at values of q typically of order $(2\pi/d)$ where d is the average interparticle distance) is not important in the range of q studied in the experiment, then Eq.(5) simplifies to

$$S(q) = (\Delta\rho)^2 \sum_i \upsilon_i^2 f_i^2(\bar{q}) \tag{6}$$

where the sum is over all particles, $\upsilon_i$ is the volume of the $i^{th}$ particle, $\Delta\rho$ is its SLD contrast with the average medium (assumed uniform throughout the particle and the same for all particles) and $f_i$ is its form factor defined by

$$f_i(\bar{q}) = \frac{1}{\upsilon_i} \int_{\upsilon_i} d\bar{r} \; e^{-i\bar{q}.\bar{r}} \tag{7}$$

where the integral is over the particle volume. In general, one can make a spherical average of $f_i^2(q)$ and assume some law of polydispersity in size of the particles to carry out the weighted sum in Eq.(6). In such a restrictive case (dilute system of random particles), by making an expansion for small $\bar{q}$, one finds that

$$S(q) = S(0) - \frac{1}{3} R_g^2 q^2 + \dots \tag{8}$$

where

$$R^2g = R_g^2 \sum_i (1/\upsilon_i) \int_{\upsilon_i} d\bar{r} \cdot r^2 \tag{9}$$

is the average <u>radius of gyration</u> of the particles. Eq.(8) suggests (at least for small q) in this case the approximation

$$S(q) = S(0)\exp\left(-\frac{1}{3} R_g^2 q^2\right) \tag{10}$$

which is the famous Guinier approximation [1] for scaling from a dilute system of uniform particles. If the particles are polymer chains, and the SLD or contrast for SANS experiments is modified by selective deuleration of either the chains or the solvent, the above considerations form the basis of the method for determining the radius of gyration of polymer chains is dilute solution.

A problem which has been studied by several groups by both SANS and light scattering techniques is that of the phase separation of a binary fluid mixture confined in a porous medium, such as Vycor glass or an aerogel [7-16]. A convenient system for such studies is a mixture of water and (2,6) lutidine, which in bulk has an inverted phase diagram (with a homogeneous phase at low temperatures, and phase separation occurring at <u>higher</u> temperatures). The critical concentration in the bulk mixture is 31.2% lutidine, and the critical temperature is 33°C. In the vicinity of the critical point in the single phase region, a bulk mixture shows critical fluctuations obeying 3D Ising-like behavior. The behavior in Vycor glass as seen by light scattering [12,16] is very different and indicative of the effects of confinement and preferential wetting of the pores (typically 8 nm in diameter), with weak or non-existent critical fluctuations and long-time relaxation and hysteresis effects with temperature in the 2-phase region. SANS data from such a system [11,12] as a function of temperature is shown in Fig. 1. In order to study the concentration fluctuations in the fluid alone, without the complications from the scattering due to the solid/liquid contrast, the homogeneous phase was taken to be contrast-matched with the Vycor using the appropriate mixture of $H_2O$, $D_2O$ and lutidine. (The fact that Vycor preferentially absorbed lutidine from the supernatant solution complicated the task of ensuring that the final single-phase mixture inside the Vycor was nominally critical and contrast-matched, but this was achieved by a systematic study of different initial mixtures [11]). The "Vycor peak" at q~ 0.025 is absent due to the contrast- matching with the silica, but a peak at a larger value of q ($\sim 0.035 A^{-1}$) was observed in the data. This peak was identified as due to a "skin" of lutidine-rich liquid adsorbed on the internal pore surfaces, as proved by a complementary experiment, in which a similar layer of hydrocarbon chains was attached to the internal pore surface of Vycor inside a contrast-matching solution of H/D toluene in the pores. The peak at the same q was clearly observed (see Fig. 2) and was used to subtract off the "skin" scattering from the observed S(q) from the Vycor/water/lutidine mixture i.e. from the observed data in Fig. 1. The remaining scattering was fitted by the sum of a Lorentzian to represent the critical fluctuations (which turned out to have a very small amplitude) and a Lorentzian-squared term (with a different length-scale) to represent the formation of microdomains of phase-separated water-rich and lutidine-rich phases in the 2-phase region). The fitted curves are shown in Fig. 1. A good fit was obtained, with the domain size saturating in the 2-phase region at the 8 nm length-scale of the Vycor pores. This provided partial confirmation of the phase diagram obtained theoretically by Liu et al. [15] for a fluid mixture phase separating inside a finite tube. In their phase diagram (see Fig. 3), the system goes from a "tube" phase (lutidine-rich "skin" lining the pore walls) to a "capsule"-like phase ("skin" and capsules of water-rich phase within the tubes) as the temperature is raised into the 2-phase region.

A similar study was carried out by Frisken et al. [14] in the much more open structures of a series of silica gels using a $D_2O$ lutidine mixture. In this case, the scattering was interpreted as that from a dispersed fractal structure of the silica framework, together with the associated static concentration variations induced in the fluid by the preference of the silica for wetting itself with lutidine, in addition to spontaneous critical fluctuations in the fluid. Thus this model is similar in spirit to the well-known "Random Field Ising Model" which was initially proposed for such systems [7,8].

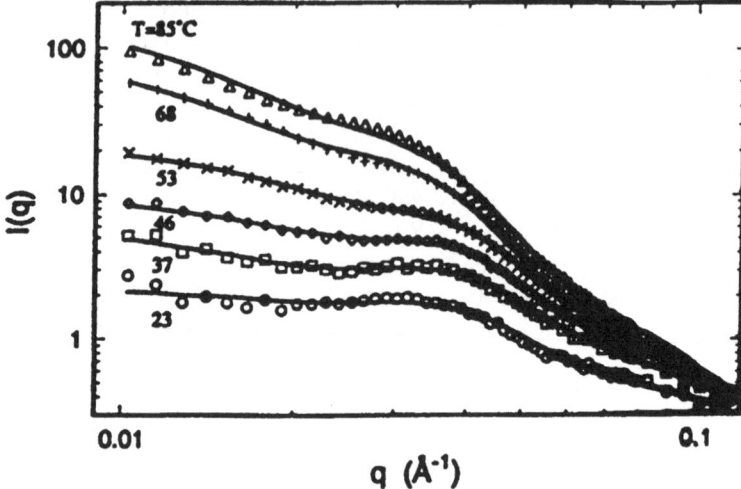

Figure 1. S(q) (arbitrary units) for a contrast-matching binary lutidine/water mixture near the critical concentration inside Vycor glass, as a function of temperature. (From Ref. [11]).

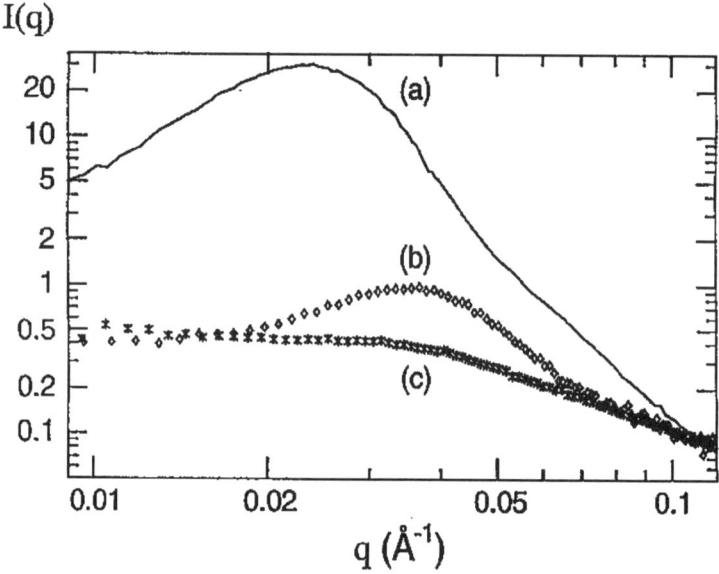

Figure 2. S(q) (log-log plot, arbitrary units) for (a) dry Vycor glass  (b) Vycor derivalized with $C_{18}$-alkylsiloxane surface layer and filled with contrast-matching hexane/d-hexane mixture and (c) Vycor with contrast-matching binary lutidine/water mixture. (From Ref. [11]).

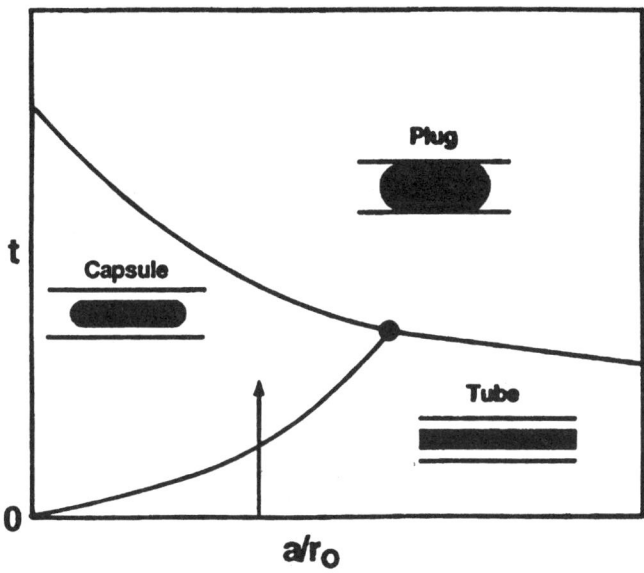

Figure 3. Phase diagram of the single-pore model of Liu, et al (Ref. [15]) for phase separation inside a confined pore geometry. The arrow indicates the observed behavior of the water/lutidine system in Vycor deduced from SANS data. t is the reduced temperature, $r_0$ is the tube radius and a is a molecular length.

Following Eq.(5) we may write

$$S(q) = \frac{1}{V} \langle \delta\rho(\bar{q})\delta\rho^*(\bar{q}) \rangle \tag{11}$$

where $\delta\rho(\bar{q})$ is the Fourier transform of the SLD fluctuation $\delta\rho(\bar{r})$. If $\phi_s(r), \phi_1(r), \phi_w(r)$ are respectively the local volume fractions of silica, lutidine and $D_2O$ and their fluctuations from the mean are $\delta\phi_s, \delta\phi_1, \delta\phi_w$, then

$$\rho(\bar{q}) = \rho_s\phi_s(\bar{q}) + \rho_1\phi_1(\bar{q}) + \rho_w\phi_w(\bar{q}) \tag{12}$$

where $\rho_s, \rho_1, \rho_w$ are the SLD's of silica, lutidine and water respectively.
Since

$$\delta\phi_S + \delta\phi_1 + \delta\phi_w = 0 \tag{13}$$

226

We obtain from Eq. (12)

$$\delta\rho(\bar{q}) = (\rho_s - \rho_w)\delta\phi_s(\bar{q}) + (\rho_1 - \rho_w)\delta\phi_1(\bar{q}) \tag{14}$$

Let us ignore spontaneous critical concentration fluctuations in the liquid for the moment and concentrate on the response of the lutidine concentration to the silica concentration via the "wetting" interaction. Assuming linear response, this may formally be written as

$$\delta\phi_1(\bar{q}) = \alpha(\bar{q}, T)\delta\phi_s(\bar{q}) \tag{15}$$

where $\alpha(\bar{q}, T)$ may be written in the Ornstein-Zernike form

$$\alpha(\bar{q}, T) = -\frac{\overline{\phi}_1}{\overline{\phi}_1 + \overline{\phi}_w} + \frac{\alpha_0(T)}{1 + q^2\xi^2} \tag{16}$$

(The constant term in Eq. (16) is there to account for the excluded volume decrease of lutidine concentration due to the presence of the silica, even in the absence of wetting, i.e., when $\alpha_0 = 0$).

Using Eqs. (14-16) in Eq. (11), we obtain

$$S(q) = \frac{1}{V}\left[(\rho_s - \rho_w) + (\rho_1 - \rho_w)\alpha(\bar{q}, T)\right]^2 \langle\delta\phi_s(\bar{q})\delta\phi_s^*(\bar{q})\rangle \tag{17}$$

But the scattering from the silica gel itself in pure $D_2O$ may be written as

$$S_{sg}(q) = \frac{1}{V}[\rho_s - \rho_w]^2 \langle\delta\phi_s(\bar{q})\delta\phi_s^*(\bar{q})\rangle \tag{18}$$

so that the scattering from the binary fluid mixture may be expressed in terms of $S_{sg}(q)$ (measured in a separate experiment) and $\alpha(\bar{q}, T)$ as

$$S(\bar{q}) = \left[1 + \frac{\rho_1 - \rho_w}{\rho_s - \rho_w}\alpha(\bar{q}, T)\right]^2 S_{sg}(\bar{q}) \tag{19}$$

To this must be added the pure critical fluctuations in the fluid given by

$$S_{crit}(\bar{q}) = (\rho_1 - \rho_w)^2 \chi/(1 + q^2\xi^2) \tag{20}$$

and a background term. Such an expression was found to provide an extremely good fit (Fig. 4) to the data for a wide range of temperatures and concentrations throughout the one-phase region of the pure system. The four fitting parameters used were $\chi, \xi, \alpha_o$ and a constant background term B.

The results showed that $\xi$ increased towards the critical temperature, as did $\alpha_o$ and $\chi$, although the accuracy was not sufficient to determine any critical exponents (in addition to the fact that the concentration of the "free" fluid, that which is not "frozen" in the wetting layers, is also changing with temperature). This behavior appears to be different to the behavior observed in Vycor. $\alpha_0$ was also found <u>not</u> to scale with $\chi$, which would be expected from a simple linear response theory indicating that the wetting response of the fluid near the silica surface is probably non-linear as might be expected. In the 2-phase region, $S(\bar{q})$ could be represented by

$$S(q) = \frac{C_1}{\left(1 + q^2 \xi_d^2\right)^2} + C_2 S_{sg}(q) \tag{21}$$

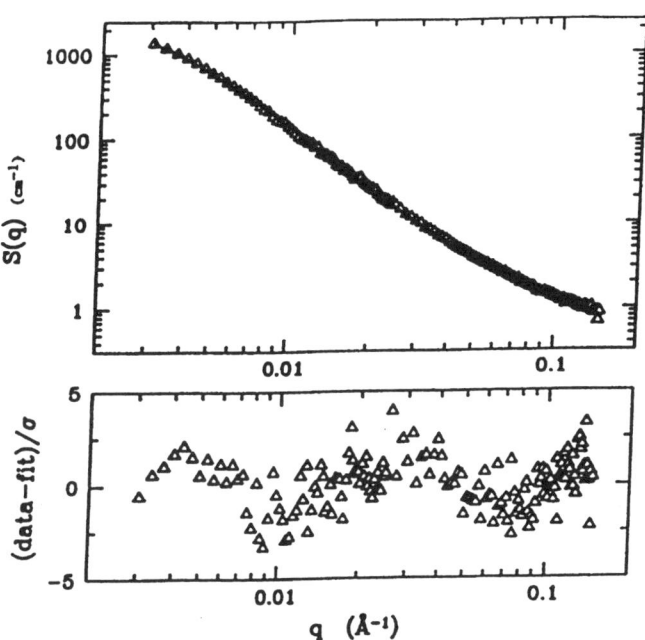

Figure 4. Result of fitting the theoretical expression given in Eqs. (18) and (19) to the measured S(q) (in absolute units) for a silica gel/water/lutidine sample containing 2.05% silica and 40 volume of lutidine at 26.80°C. The residuals in units of the standard deviation of the data are shown in the bottom portion of the figure. (From Ref. [14])

plus a constant background term, i.e., scattering due to the silica gel in an effective uniform fluid plus scattering from domains indicating that sharp interfaces between phase separated domains had formed. Eq. (21) is similar in spirit to the model of Debye et al. [17], and the models used by Wong et al. [18] and Lin et al. [11] to describe frozen domains. $\xi_d$ was too large to measure accurately. Within the instrumental resolution, the first term in Eq. (21) effectively looked like Porod scattering. These domains also appeared to coarsen with increasing time, as observed by an increase in the amplitude of the $q^{-4}$ term (proportional to the interfacial area by Porod's Law).

The conformation of polymers confined inside pore spaces is also of interest. At first thought it would appear difficult to get a polymer whose equilibrium radius of gyration $R_g$ in solution is greater than a typical pore size to enter the porous medium. However, it turns out, as predicted theoretically [19] that from a sufficiently concentrated (semi-dilute or more concentrated) polymer solution osmotic pressure will force the polymer chains to enter the pores, the criterion being roughly that the correlation length $\xi$ in the semi-dilute solution be comparable to the pore size. This was observed in SANS experiments by Lal et al [20] where polystyrene (PS) chains of equilibrium radius of gyration as large as 30 nm were imbibed into cleaned Vycor glass from semi-dilute solutions in times ranging from hours (for the smallest chains) to 65 days (for the largest chains). This was verified by first studying the SANS from the virgin Vycor sample with a contrast-matched solvent (deuterated dichloromethane) in the pores (which showed no characteristic "Vycor peak") and comparing it with the imbibed Vycor in which the Vycor peak had reappeared, indicating that PS chains entering the Vycor had destroyed the contrast-matching condition.

For the SANS experiments a mixture of hydrogenated and deuterated PS chains of equal molecular weights (MW) (h-PS and d-PS respectively), and hydrogenated and deuterated toluene solvent (h-toluene and d-toluene) was used. The SANS experiment can be made sensitive to the conformation of <u>individual</u> PS chains inside the Vycor if the following two conditions are simultaneously met:
1) The h-, d-toluene mixture exactly contrast matches the silica of the Vycor. (This was determined by careful subsidiary contrast matching experiments using varying h-, and d-toluene concentrations.)
2) The concentration of h-PS and d-PS chains is chosen such that the averaged SLD of the chains exactly matches that of the solvent.

Under these conditions it may be shown [20] that only single chain fluctuations will contribute to the observed S(q) which thus measures the form-factor or conformation of simple chains. Experiments were done for a variety of PS molecular weights. A typical S(q) curve for PS of MW 75K in toluene in Vycor is shown in Fig. 5. The results were compared with the S(q) for the same chains in dilute solution in toluene (i.e. "free" chains). A Debye-function fit (appropriate to free Gaussian chains in a good solvent) [20] was made to both sets of data. Such a procedure is valid for obtaining a radius of gyration from the small q region where the fit is reasonably good, although at large q both curves deviate. The radius of gyration of the confined chains were obtained

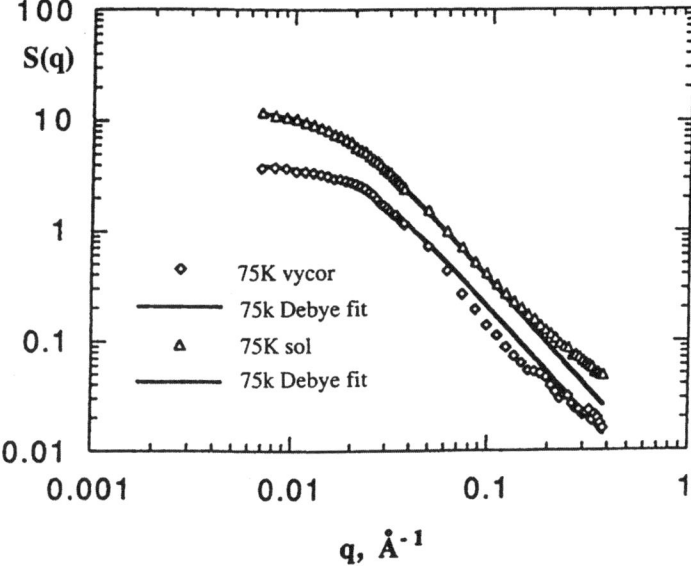

Figure 5. S(q) (arbitrary units) for 75K MW polystyrene chains in dilute solution ($\Delta$) in toluene and in toluene in Vycor ($\Diamond$). The solid lines are corresponding fits of the Debye-function for ideal polymer chains. (From Ref. [20])

to be always smaller than the radius of gyration of the free chains in the equivalent bulk solution. The free chains in dilute solution yield an S(q) that deviates from the $1/q^2$ Debye-like behavior due to excluded volume effects (yielding $q^{-1/\nu}$, where $\nu$ is the Flory exponent [21]). The chains in the Vycor deviate due to conformational changes arising from confinement in the tube-like pores. These were interpreted in terms of the theory developed by Daoud and de Gennes [22] and Brochard and de Gennes [23]. Briefly, the theory takes into account four main effects which govern the behavior of chains confined in non-adsorbing cylindrical pores. These are the confinement of the chains by the cylinder walls, which squeezes the chains laterally, the intrachain excluded volume effect which swells the chains and stretches them along the cylinder axis, the entropic elasticity of the chains which limits this stretching, and the interchain interactions which may also lead to segregation of the chains. Thus the polymer chains are stretched out into "cigars" along the tubes, which may at high concentrations segregate from each other (strong confinement limit) or overlap in an entangled manner (moderate confinement regime) and thus no longer be stretched along the cylinder axis. In the latter cases $R_{gz}^2 = R_g^2/3$, where $R_{gz}$ is the radius of gyration along the cylinder axis, and $R_g$ is that of the chain in bulk solution. One then has the relation

$$R_{gv}^2 = D^2/8 + R_{gz}^2 \qquad (22)$$

where $R_{gv}$ is the radius of gyration as measured in the Vycor pore space and D is the pore diameter of Vycor (known to be ~7 nm). Thus Eq.(22) provides a method for testing the relationship of the measured radii of gyration of the individual chains in Vycor and in the bulk solution, and was found to be satisfied extremely well for the different molecular weights studied. Thus one can conclude that the chains were in the conformation of ideal overlapping squeezed cigars. This conclusion was confirmed by also fitting the S(q) of the individual chains in the regime $D<q^{-1}<R_g$ to a "cylindrical" Guinier model.

## 3. Studies of Fluids at Interfaces and in Thin Films

At a liquid/vapor or liquid/solid interface, the liquid behavior may be modified in several ways. Near the melting temperature, the bulk may be crystalline in coexistence with a thin layer of liquid at the interface (surface melting) or the liquid phase may develop as crystalline layer at the interface (surface or interface freezing). There have been a fair number of studies of such phenomena over the last few years. Here, we shall discuss studies of this phenomenon on hydrocarbon liquid systems as probed by X-ray reflectivity.

Specular X-ray reflectivity yields the electron density profile normal to the liquid surface (averaged over the plane of the surface), and thus can reveal the existence of a surface layer with a slightly different density to that of the bulk. In the simplest theory, the specular reflectivity is given by the expression [24]

$$R = R_F \left| \frac{1}{\rho_0} \int_{-\infty}^{\infty} dz \, \frac{d\rho(z)}{dz} \, e^{iq_z z} \right|^2 \tag{23}$$

where $q_z$ is the wavevector transfer normal to the surface.

(This expression breaks down when the reflectivity approaches unity in the vicinity of total external reflection, at small values of $q_z$. In this region, there are standard methods, borrowed from optics, to calculate the reflectivity accurately by regarding the varying electron density normal to the surface as made up of a series of thin slabs with constant density, matching boundary conditions and using iterative procedures to solve for the fields [25,26]).

In the isotropic phase of a liquid crystal, the free surface can produce an alignment to a nematic phase which penetrates further into the bulk as the temperature is lowered towards the bulk isotropic-nematic transition temperature ($T_{IN}$) [27]. We may say that the nematic phase <u>wets</u> the surface slightly above $T_{IN}$. Similarly, the smectic phase can begin to wet the surface in the nematic phase as T tends to the nematic-smectic transition temperature and this effect has been studied by X-ray specular reflectivity techniques by

Pershan, Als-Nielsen and co-workers [28], using a liquid surface reflectometer. Fig. 6 shows the measured specular reflectivity from the surface of the liquid crystals Octyloxycyanobiphenyl (8OCB) as the temperature is decreased towards the nematic-smectic A transition temperature $T_{NA}$. Plotted are both $R(q_z)$ and $R(q_z)/R_F(q_z)$ where $R_F$ is the theoretical Fresnel reflectivity. For $q_z \ll Q_o$ (where $Q_o$ is $2\pi$ divided by the periodicity of the smectic layers), $R(q_z)$ follows the Fresnel reflectivity $R_F(q_z)$ fairly

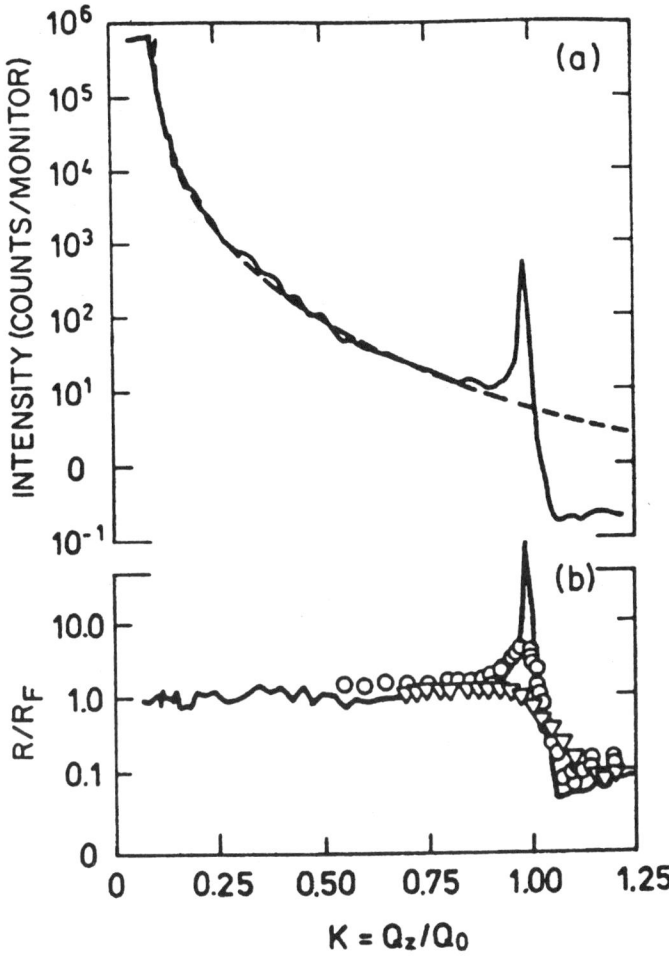

Figure 6. (a) Specular x-ray reflectivity from the surface of 8OCB at T-$T_{NA}$=0.05°C. The $Q_z$ scale is normalized to $Q_0$=0.199Å$^{-1}$. The dashed line is the calculated Fresnel reflectivity (b) R($Q_z$)/R$_F$($Q_z$) vs $Q_z$/$Q_0$. The solid line is for T-$T_{NA}$=0.05°C, open circles for T-$T_{NA}$=2.8°C and the triangles for T-$T_{NA}$=11.06°C. (after Pershan et al., Ref. [28])

well. For $q_z \sim Q_o$, the smectic peak appears with a width proportional to $\xi_{\|}$, the correlation length in the direction normal to the smectic layers. This peak sharpens as $T_{NA}$ is approached. For $q_z \gtrsim Q_o$, $R(q_z)$ dips sharply <u>below</u> $R_F(q_z)$. This is due to destructive interference between the surface reflectivity and that from the smectic layers. To fit the data, the authors used the formula given in Eq. (23) with

$$\rho(z) = \rho_0(z) + \rho_1(z) \tag{24}$$

where the first term was taken to be given by

$$\frac{1}{\rho_\infty}\frac{d\rho_0}{dz} = \frac{d}{dz}\,\Theta(z - z_o)\,B_s\,\exp\!\left[-\frac{z - z_o}{\xi_{\|}(T)}\right]Sin[Q_o(z - z_o)] \tag{25}$$

($\Theta(z - z_o)$ being the step-function which is 1 if $z > z_o$, and 0 otherwise) and the second term was taken to be

$$\frac{1}{\rho_\infty}\frac{d\rho_1}{dz} = C_1(\pi z)^{-1}\exp\!\left[\frac{-(\sigma z)^2}{2}\right]Sin(Q_1\,z) \tag{26}$$

where $C_1$ is defined to ensure $\int \rho_1^{-1}\,d\rho_1/dz = 1$.
The Fourier transforms then yields

$$R(q_z) = R_F(q_z)\,|\Phi(q_z)|^2 \tag{27}$$

where

$$\Phi(q_z) = \Phi_0(q_z) + \Phi_1(q_z) \tag{28}$$

and

$$\Phi_0(q_z) = i\,\frac{B_s}{2}\,\exp(-i\,q_z\,z_0)\left[\frac{\xi_{\|}Q_o - 1}{(q_z - Q_o)\xi_{\|} - i} - \frac{\xi_{\|}Q_o + 1}{(q_z + Q_o)\xi_{\|} + i}\right] \tag{29}$$

$$\Phi_1(q_z) = \left(\frac{C_1}{2}\right)\left[erf\left[\frac{q_z + Q_1}{\sqrt{2}\,\sigma}\right] - erf\left[\frac{q_z - Q_1}{\sqrt{2}\,\sigma}\right]\right] \tag{30}$$

The term $\rho_0$ represents a sinusoidal (smectic) density decaying into the bulk with correlation length $\xi_{\|}$, with a phase which is controlled by $z_0$, while $\rho_1$ represents a surface smeared with a roughness $\sigma$ but with also a sinusoidally varying component which is damped rapidly into the bulk. It is basically an empirical from chosen to represent the experimental data, and is presumed to correct for the fact that the form of Eq. (25) for $\rho_0$ does not accurately represent the smectic oscillations near the surface. Fitting of the data using Eq. (27) yields a value for $\xi_\parallel$ essentially identical to the bulk correlation length measured in X-ray scattering experiments on bulk samples in the region near the phase transition. It also yields a value for $z_0$ which is 0.25d, d being the layer spacing of the smectic-A phase $(= 2\pi/Q_o)$, yielding a maximum in $\rho_0$ exactly d/2 away from the surface. $Q_1$ turned out to be equal to $Q_0$ within a few percent. An interesting result of this experiment was the fact that the specular peak at $q_z = Q_o$ was extremely sharp in the transverse direction, while the bulk critical scattering showed a finite transverse correlation length $\xi_\perp$. Thus the smectic order at the surface had a much larger in-plane correlation than in the bulk.

In another experiment on the system dodecylcyanobiphenyl (12CB) which does not have a smectic phase, but a first-order transition from the bulk isotropic phase to the smectic-A phase at $T_{IA}$ = 57.7$^0$C, Ocko et al. [29] found a specular reflectivity that clearly showed interference effects developing due to smectic layers forming at the surface above $T_{IA}$ as the transition temperature was approached. The data showed quantization effects as a function of temperature due to discrete smectic layers forming. Up to 5 smectic layers were observed as the temperature was lowered. The data was again fitted with expression (23) with the form taken for $\rho(z)$ being given by

$$\rho(z) = \Theta(z - z_o) + H_n(z) B_s \, Sin(2\pi z/d) \tag{31}$$

where $\Theta(z)$ is the step-function defined in Eq. (25), d the smectic layer spacing and $H_n(z)$=1 for 0 < z < nd and zero elsewhere. The above expression was convoluted with a Gaussian profile to represent surface or interlayer roughness and the resulting reflectivity fitted to the experimental data by adjusting $z_0$, $B_s$ and the roughness parameter. The results showed an (incomplete) wetting of the free surface of the isotropic phase of 12CB by the smectic-A phase, but in discrete layers, rather than with an exponential decay into the bulk. Ocko has also used X-ray specular reflectivity experiments to demonstrate such layering transitions at smectic A-solid interfaces [30].

In view of the fact that the free surface has a disordering effect on a solid, and that surface melting appears to be common in ordinary solids [31], the ordering of liquid crystals at the free surface is somewhat unexpected. Holyst [32] has proposed a theory which may account for the layer-by-layer freezing in terms of the quenching of the out-of-plane surface fluctuations by surface tension effects.

Liquid crystals are not the only systems to show surface ordering effects. Even the simplest chain-like hydrocarbon molecules, namely the normal alkanes (which possess no nematic or smectic phases) show such behavior. Fig. 7 shows the specular X-ray

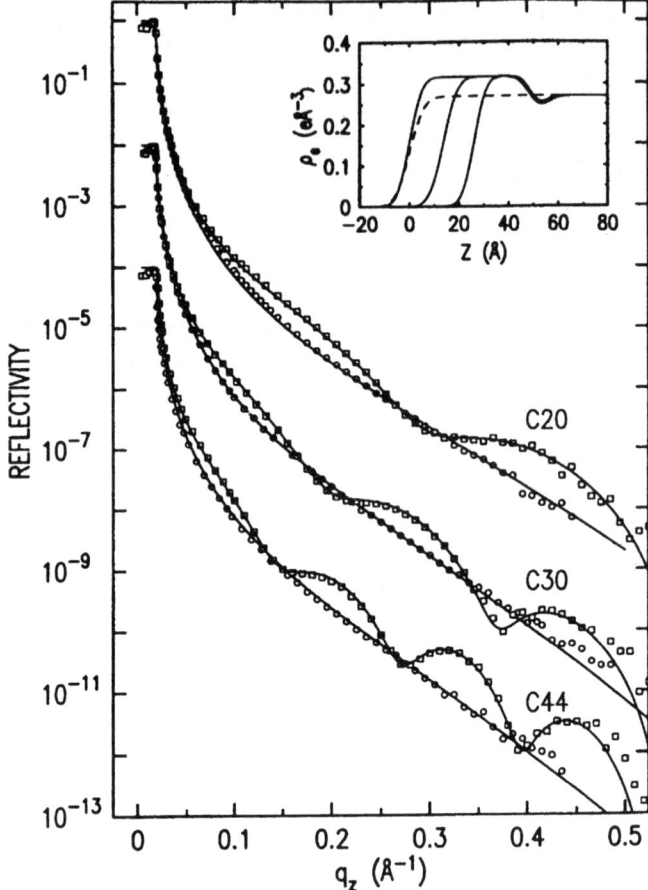

Figure.7. Specular x-ray reflectivity data for liquid alkanes with carbon numbers n=20, 30, and 44 respectively at $T_m$+4°C (open circles) and $T_m$+3°C (open squares), where $T_m$ is the bulk freezing temperature for each. The inset shows model fits in the crystalline (solid) and liqid (dashed) surface phases of the electron densities at the interfaces. (From Ref. [33])

reflectivity from the surface of 3 liquid alkanes (carbon numbers = 18, 20, 24 respectively) at roughly 4°C above their respective bulk crystallization temperatures $T_b$. [33]. At the higher temperature, the reflectivity is well fitted by the simple Nevot-Croce form as expected for a simple liquid surface with an effective roughness due to capillary wave fluctuations. However at about 3°C above $T_b$, oscillations in the reflectivity indicate the presence of a slightly denser layer on the liquid surface, with a thickness approximately equal to the chain length of the molecule. (The density profile fitted to the reflectivity for the C24 chain in this regime is shown in the inset to Fig. 7). The electron density of this layer is close to that of the bulk "rotator" crystalline phase, where the chains are oriented in a hexagonal structure normal to the layers, but the bonds along the chain are disordered. These reflectivity experiments thus suggest strongly that a single layer of this crystalline phase has formed on the surface of the bulk liquid slightly above $T_b$. This was confirmed by in-plane Grazing Incidence X-ray Diffraction (GIXD) experiments, and also by surface tension measurements [34]. Unlike the 12CB liquid crystal experiment discussed above, only one layer of the crystalline phase was seen to form before the bulk phase was reached via a strongly first order transition. Subsequent reflectivity and GIXD experiments showed that for long chain alcohols, a similar surface crystalline phase also formed slightly above $T_b$. This surface phase, however, consisted of a single bilayer of the hexagonally close-packed and tilted alcohol chains [35]. Reflectivity experiments have also shown surface crystallization in mixtures of alkane chains of different lengths [36]. For mixtures of chains of fairly comparable lengths, the thickness of the layer yields an average of the chain-length of the two species. For fairly disparate chain length mixtures on the other hand, one observes surface crystallization of either the majority component only or no surface crystallization at all when the concentrations are nearly equal. This has been explained [36] as being due to the extra energy cost of cocrystallizing chains of very different length adjacent to each other. GIXD experiments show that in these monolayer surface crystalline phases, there also exist tilted phases, i.e. where the alkane or alcohol molecules, instead of being oriented normal to the surface, develop a tilt angle towards nearest-neighbor or next-nearest-neighbor positions.

There have been measurements, testing the surface force apparatus (SFA) which indicate that many non-liquid-crystal fluids confined in thin films between two solid surfaces show ordering in layers parallel to the surface, not seen in the bulk, as evidenced by oscillatory forces between the surfaces as a function of separation [37]. (Oscillatory order of a bulk fluid in the vicinity of an interface with a solid has also recently been observed by X-ray reflectivity [38]). If the film is only a few molecular diameters thick, this layered structure exists across the film, and in addition the film may develop lateral order (epitaxially with the wall surfaces, or otherwise) forming a crystalline phase induced by confinement, as evidenced by computer simulations [39], and SFA experiments involving shear between the surfaces showing stick-slip behavior [40]. Obviously, apart from being of fundamental interest, such ordering effects are important in understanding lubrication or adhesion involving fluid films between solid surfaces.

We turn now to studies of fluctuations at liquid surfaces. The scattering may be calculated in terms of the height-height correlation function $\langle \delta z(\vec{r}) \, \delta z(\vec{r} + \vec{R}) \rangle$ where $\delta z(\vec{r})$ is the vertical height fluctuation (along the z axis normal to the surface) at lateral position $(\vec{r})$, and $(\vec{R})$ is the lateral separation between the two points. If we denote this function by $C(\vec{R})$, then the scattering cross section as a function of $\vec{q}$ is given by

$$S_{diff}(\bar{q}) = A\frac{(\Delta\rho)^2}{q_z^2}\iint dXdY\left[e^{q_z^2\,C(R)} - 1\right]e^{-i(q_xX+q_yY)}\qquad(32)$$

where $(\Delta\rho)$ is $(e^2/mc^2)$ times the electron density difference across the liquid surface, and A is the surface area. We have excluded the specular component of this expression, so that the above expression describes the diffuse or off-specular component of the scattering.

In the case of liquid surfaces, surface roughness is due to capillary wave fluctuations. There are some problems in connection with a truly first-principle calculation of such fluctuations (for a recent discussion, see Ref. 41), but one may write down at least a phenomenological expression for the surface free energy of the liquid and derive from it the spectral function for surface height fluctuations in the form:

$$\left\langle\left|\delta z(q)\right|^2\right\rangle = \frac{kT}{\gamma\left(q^2 + \kappa^2\right)}\qquad(33)$$

where $\gamma$ is the surface (or interface) tension, and $\kappa$ is the inverse of the capillary length defined by $\kappa^{-1} = (\Delta\rho_0)g/\gamma$, $\Delta\rho_0$ being the mass-density difference between the fluids on either side of the interface. For bulk liquids $\kappa$ is typically of the order of 10 cm$^{-1}$. Fourier transformation of Eq. (33) leads to a form for the height-height correlation function

$$C(r) = -\frac{1}{2}BK_0(\kappa r)\qquad(34)$$

where

$$B = kT/(\pi\gamma)\qquad(35)$$

and $K_0(x)$ is the modified Bessel function. At length scales $\ll\kappa^{-1}$ (which are, in practice, those relevant for scattering experiments), the Bessel function may be replaced by a logarithm, and, to prevent short (molecular) length scale problems, we may also introduce a lower length scale cutoff [42]. Thus, we finally write

$$C(r) = -\frac{1}{2}B\ln\left[\kappa\left(r^2 + r_0^2\right)^{\frac{1}{2}}\right]\qquad(36)$$

where $r_0$ is defined to give the correct lateral surface roughness given by the integral of Eq. (33) ($r_0$ turns out to be the inverse of the upper q cutoff for the capillary waves, $q_u$ defined below).

From Eq. (33) we see by integration over q that the true mean square roughness due to surface capillary waves is given by

$$\sigma^2 = \frac{1}{4}B\ln\left[\left(q_u^2 + \kappa^2\right)/\kappa^2\right] + \sigma_o^2 \tag{37}$$

where $\sigma_o$ is an "intrinsic roughness" due to the size of the molecules at the surface, and $q_u$ is an upper cut-off for the capillary wavevectors introduced to make the integral converge. It is $(1/r_o)$ where $r_o$ is the cut-off introduced in Eq. (36). Since $\kappa$ is in general $\ll q_u$, Eq (37) may be written as

$$\sigma^2 = \frac{1}{2}B\ln(q_u/\kappa) + \sigma_o^2 \tag{38}$$

Substituting this in Eq. (32), we may calculate the scattering after folding with the resolution function. If the latter is approximated by a Gaussian, Sanyal et al [43], have derived the form for the scattered intensity at $q_x$, $q_z$ (with $q_y$ integrated over assuming wide slits out of the plane of scattering)

$$\tag{39}$$

$$I = I_o \frac{q_c^4}{16} \frac{1}{q_z^3} \left(\frac{1}{2k_o\sin\alpha}\right)\exp\left(-q_z^2\sigma_{eff}^2\right)\frac{1}{\sqrt{\pi}}\Gamma\left[\frac{1-\eta}{2}\right] {}_1F_1\left[\frac{1-\eta}{2};\frac{1}{2};\frac{q_x^2 L^2}{4\pi^2}\right]\left|T(\alpha)\right|^2\left|T(\beta)^2\right|$$

where $\alpha$, $\beta$ are the grazing angles of incidence and scattering respectively, $I_o$ is the incident beam intensity, $k_o$ the incident wave vector, $q_c$ the wave vector corresponding to the critical angle of incidence, $\Gamma(x)$ is the gamma function, ${}_1F_1(x;y;z)$ is the Kummer function, and $T(\alpha)$ is the Fresnel transmission coefficient for incident angle $\alpha$,

$$\eta = \frac{1}{2}Bq_z^2 \tag{40}$$

and

$$\sigma_{eff}^2 = \sigma^2 + \frac{1}{2}(0.5772)B - \frac{1}{2}B\ln(2\pi/\kappa L) \tag{41}$$

L is the coherence length of the beam along the surface or the inverse of the resolution width in $q_x$ space. For $q_x <$ the resolution function width, this saturates and merges into the nominal "specular" reflectivity. For larger $q_x$, this has the asymptotic form

$$I(q_x,q_z) \sim q_x^{-(1-\eta(q_z))} \tag{42}$$

This is analogous to the algebraic decay $q^{(-2-\eta)}$ of $S(q)$ in a 2D crystal for which the displacement correlations possess logarithmic correlation functions (q being the lateral component of wavevector transfer, and the (1-$\eta$) rather than (2-$\eta$) arises in Eq. (42)

238

from integrating over $q_y$). In this sense the diffuse scattering around the "specular ridge" in the case of surface scattering is the analogue of the diffuse scattering around the Bragg rods in a 2D crystal. However, in the present case, the exponent $\eta$ is a continuous function of $q_z$, being given by Eq. (40), which can be calculated knowing the surface tension. Experiments carried out with X-ray synchrotron radiation on the surface of liquid ethanol show excellent agreement with the above predictions [43]. (See Fig. 8).

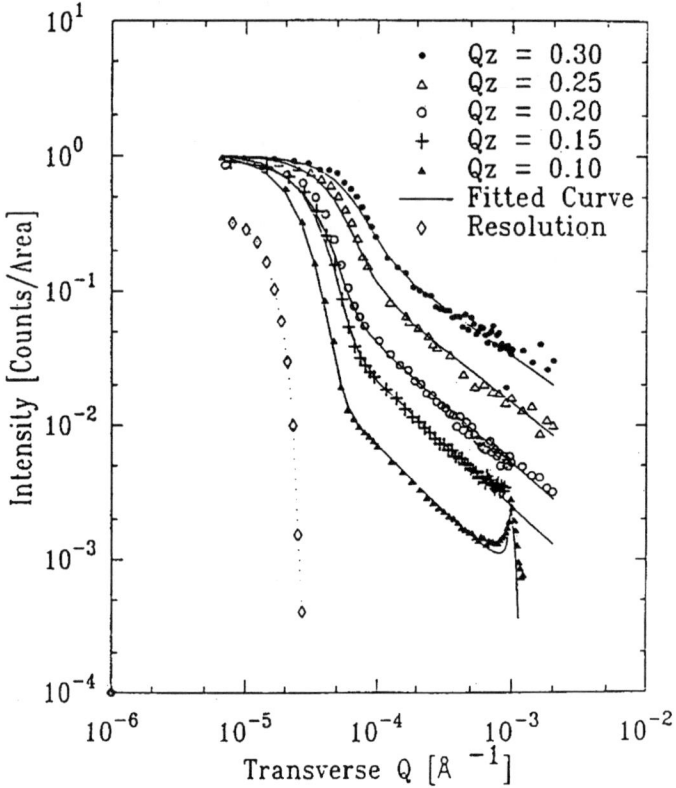

Figure 8. Log-log plot of the transverse diffuse scattering scans from the surface of liquid ethanol all normalized to unity at $q_x = 0$ for different values of $q_z$. Backgrounds have been subtracted and corrections for variation of illuminated area made. The solid curves represent the calculated scattering. The dashed curve represents the main beam profile converted to an effective transverse resolution at $q_z = 0.1 \, \text{Å}^{-1}$ (From Ref. [43]).

By Eqs. (38 and 41) the effective roughness $\sigma_{eff}$ measured in a reflectivity experiment is given by

$$\sigma_{eff}^2 = \frac{1}{2}B\ln(q_u / \Delta q) + \sigma_o^2$$

(43)

where $\Delta q$ is the instrumental resolution width ($2\pi/L$) [44]. The so-called "specular" reflectivity from a liquid will be governed by the Debye-Waller-like factor $\exp(-q_z^2 \sigma_{eff}^2)$ rather than $\exp(-q_z^2 \sigma^2)$. The fact that the effective roughness measured for a liquid surface by scattering is less than the true roughness has been known for some time [44] and is due to the unavoidable inclusion of some capillary-wave diffuse scattering inside the resolution broadening of the specular peak. A measurement of $\sigma^2_{eff}$ from specular X-ray reflectivity measurements on liquid alkanes of different chain lengths at different temperatures by Ocko et al [45] (see Fig. 9) yields consistent values for $q_u$, which appears to scale inversely with the chain length. This leads to the reasonable speculation that the short wavelength cut-off for capillary waves is at a length scale

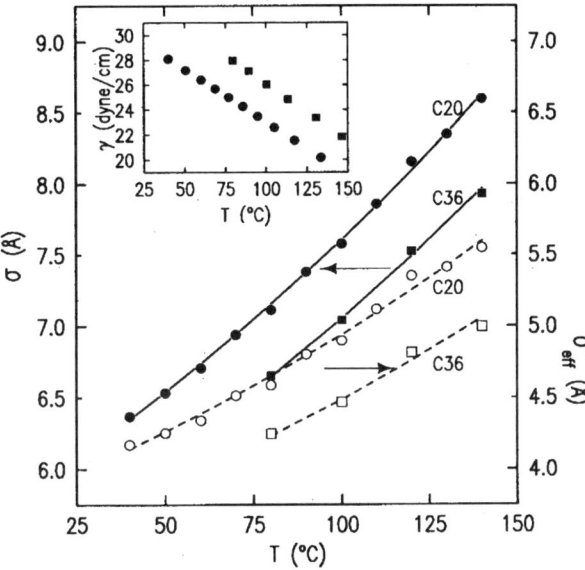

Figure 9. The "true" roughness $\sigma$ and the "effective roughness" $\sigma_{eff}$ for C20 and C36 alkane chains as a function of temperature. The lines are corresponding values calculated with $\sigma_o = 1.1$Å, $q_u = 0.44$Å$^{-1}$ for C20 and $q_u = 0.27$ Å$^{-1}$ for C36. This inset shows the measured surface tension $\gamma$ of C20 (circles) and C36 (squares). (From Ref. [45])

240

corresponding to the intermolecular spacing, as in the Debye cut-off for phonons in crystals, although a rigorous proof is lacking. Mode-coupling theory yields such a cut-off naturally by introducing a $q^4$ term in the denominator of Eq. (33), which yields an effective cut-off which is of the same order of magnitude as measured from experiments. We note that such a $q^4$ term in the denominator of Eq. (33) also occurs naturally from a curvature-resisting term in the surface free energy as for a surfactant-covered surface and has been used to fit X-ray scattering from such surfaces [46].

For thin liquid films, the Van-der-Waals interaction with the substrate can enormously increase the value of $\kappa$ defined in Eq. (33), which is now given by

$$\kappa^2 = A/4\pi\gamma d^4 \tag{44}$$

where A is the Hamaker constant for the Van der Waals interaction and d is the film thickness. In such cases, $\kappa$ may actually become larger then the resolution-width $\Delta q$ and a distinct shoulder is seen in the capillary wave diffuse scattering at a value of $q_x \sim \kappa$. This is seen in Fig. 10, which shows diffuse scattering (transverse diffuse scans as a function of $q_x$) from a thin polystyrene film on a silicon substrate [47].

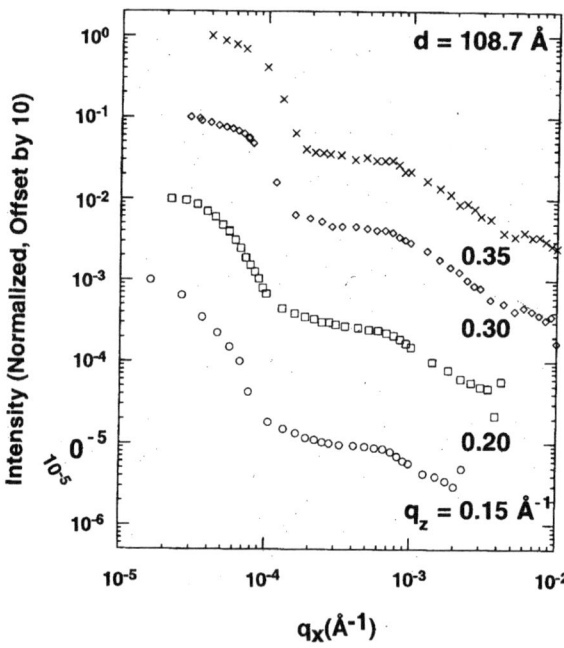

Figure 10. Transverse diffuse scans measured for a 108.7 Å film of polystyrene on a silicon substrate for different values of $q_z$ showing the finite thickness cut-off outside of the instrument resolution which is the width of the specular peak. (From Ref. [21])

Finally, we discuss briefly the effect of scattering with <u>coherent</u> X-ray beams. For conventional X-ray and neutron sources, the coherence length of the radiation falling on the sample is rather small (typically less than a micron in the directions transverse to the beam, and a few hundred angstroms in the longitudinal direction). This means that the sample scatters from a large number of incoherently scattering volumes, which results in the conventional statistical averages which appear in the correlation functions determining the scattering. However, if the beam is highly monochromatic and highly collimated and the sample is a long distance from the source (which is small in size), as is the situation at the present-day third generation synchrotron sources, then the coherence volumes can be of the order of tens of microns in size and the whole sample can scatter coherently. In general, due to interference effects, disorder or fluctuations in the atomic positions then results in the phenomenon of <u>speckle</u> (familiar to people working with optical lasers) in the scattered beam. As the atoms fluctuate in time, these speckles move in and out of the detector yielding a time-dependent scattered intensity. The auto-correlation function of this intensity can be used to infer the (slow) dynamics of the system. This is completely analogous to the technique of dynamical light scattering used for studying dynamical fluctuations in liquids, but the length scales studied can be much smaller (since $\bar{q}$ is larger) and optically opaque systems can be studied. For a system of independently diffusing scatters executing Brownian Motion, e.g., colloidal particles in solution, one obtains

$$\langle I(t) I(t+\tau)\rangle / \langle I^2\rangle = 1 + Ag_2(\tau)$$

where A is a prefactor that can be as large as unity for a <u>perfectly</u> coherent beam, but is often smaller, and $g_2(\tau)$ is given by exp $(-2Dq^2\tau)$, where D is the diffusion constant, which may be q-dependent for larger values of q. Such kinds of studies have been performed on colloidal, micellar and polymeric systems [48-51].

I wish to acknowledge my many collaborators in many of the above studies. They include L. Auvray, M. Deutsch, K.G. Huang, J. Lal, M.Y. Lin, B. Ocko, W. Press, M.H. Rafailovich, M.K. Sanyal, O. Seeck, E.B. Sirota, J. Sokolov, P. Thiyagarajan, M. Tolan, J. Wang, X.L. Wu, and X.Z. Wu.

## References

1. Guinier, A. and Fournet, D.C. (1955) *Small-angle Scattering of X-rays*, Wiley, New York.

2. Glatter, O. and Kratky, O. (1982) *Small-angle X-ray Scattering*, Academic Press, New York.

3. Russell, T.P. (1991) Small-angle Scattering at Synchrotron Radiation Sources, in *Handbook of Synchrotron Radiation*, Vol. 3, (G. Brown and D.E. Moncton, Eds.) North Holland, Amsterdam.

4. Feigin, L.A. and Svergun, D.I. (1987) *Structure Analysis by Small Angle X-ray and Neutron Scattering*, Plenum Press, New York.

5. Bacon, G.E. (1975) *Neutron Diffraction* (3rd edition) Clarendon Press, Oxford.

6. Squires, G.L. (1978) *Introduction to the Theory of Thermal Neutron Scattering*, Cambridge University Press.

7. Brochard, F. and de Gennes, P.G.,(1983) J.Phys.Lett. **44**, L-785.

8. Andelman, D. and Joanny, J.F. (1985) in *Scaling Phenomena in Disordered Systems*, (ed. R. Rynn and A. Skjeltorp) Plenum, New York, p.163.

9. Goldberg, W.I. (1985) in *Scaling Phenomena in Disordered Systems*, (ed. R. Pynn and A. Skjeltorp) Plenum, New York, p.151.

10. Goh, M.C., Goldburg, W.J., and Knobler, C.M. (1987) Phys.Rev.Lett. **58**, 1008.

11. Lin, M.Y., Sinha, S.K., Drake, J.M., Wu, X.L., Thiyagarajan, P., and Stanley, H.B. (1994) Phys.Rev.Lett. **72**, 2207.

12. Dierker, S.B. and Wiltzius, P. (1987) Phys.Rev.Lett. **58**, 1865 (1989); *ibid* **62**, 804 (1991); *ibid* **66**, 1185.

13. Frisken, B.J. and Cannell, D.S. (1992) Phys.Rev.Lett. **69**, 632.

14. Frisken, B.J., Cannell, D.S., Lin, M.Y., and Sinha, S.K. (1995) Phys.Rev.E **51**, 5866.

15. Liu, A.J. and Grest, G. (1991) Phys.Rev.A **44**, R7894; Monette, L., Liu, A.J., and Grest, G. (1992) Phys.Rev.A **46**, 7614; Liu, A.J. et al. (1990) Phys.Rev.Lett. **65**, 1897.

16. Maher, J.V., Goldberg, W.I., Pohl, D.W., and Lenz, M. (1984) Phys.Rev.Lett. **53**, 60.

17. Debye, P., Anderson, H.R., and Brumberger, H. (1957) J.Appl. Phys. **28**, 679.

18. Wong, P.-z., Cable, J.W., and Dimon, P. (1984) J. Appl. Phys. **55** , 2377.

19. de Gennes, P.G. (1985) *Scaling Concepts in Polymer Physics*, Cornell University Press, Ithaca, NY, 95.

20. Lal, J., Sinha, S.K., and Auvray, L. (1997) J. de Physique II,**7**, 1597.

21. Stauffer, D. (1985) *Introduction to Percolation Theory*, Taylor & Francis, London and Philadelphia.

22. Daoud, M. and de Gennes, P.G. (1977) J. de Physique **38**, 85.

23. Brochard, F. and de Gennes, P.G. (1979) J. de Physique **40**, L-399.

24. Als-Nielsen, J., Christensen, F. and Pershan, P.S. (1984) Phys. Rev. Lett. **48**, 1107; Wu, E.S. and Webb, W.W. (1973) Phys. Rev. A **8**, 2077; Als-Nielsen, J. (1985) Z. Phys. B **61**, 411.

25. Born, M. and Wolf, E. (1975) *Principles of Optics,* 5th Ed. (Pergamon, Oxford), p. 51.

26. Parratt, L.G. (1954) Phys. Rev. **95**, 359.

27. Sheng, P. (1976) Phys. Rev. Lett. **37,** 1059; Miyano, K. (1979) Phys Rev. Lett. **43,** 51; Allender, D. W., Henderson, G.L. and Johnson, D.L. (1981) Phys. Rev. A. **24,** 1086.

28. Pershan, P.S., Braslau, A., Weiss, A.H., and Als-Nielsen, J. (1987) Phys. Rev. A **35**, 4800.

29. Ocko, B.M., Braslau, A., Pershan, P.S., Als-Nielsen, J., and Deutsch, M. (1986) Phys. Rev. Lett. **57**, 94.

30. Ocko, B.M. (1990) Phys. Rev. Lett. **64**, 2160.

31. van der Veen, J.F. and Frenken, J.W.M. (1986) Surf. Sci. **178**, 382; Ohnesorge, R., Lowen, H., and Wagner, H. (1991) Phys. Rev. A **43**, 2870; Oxtoby, D.W. (1990) Nature **347**, 725.

32. Holyst, R. (1992) Phys. Rev. B **46**, 15542.

33. Wu, X.Z., Sirota, E.B., Sinha, S.K., Ocko, B.M., and Deutsch, M. (1993) Phys. Rev. Lett. **70**, 958; Ocko, B.M., Wu, X.Z., Sirota, E.B., Sinha, S.K., Gong, O., and Deutsch, M. (1997) Phys. Rev. E **55**, 3164.

34. Wu, X.Z., Ocko, B.M., Sirota, E.B., Sinha, S.K., Deutsch, M., Cao, B.H., and Kim, M.W. (1993) Science **261**, 1018.

35. Deutsch, M., Wu, X.Z., Sirota, E.B., Sinha, S.K., Ocko, B.M., and Magnussen, O.M. (1995) Europhys. Lett. **30**, 283.

36. Wu, X.Z., Ocko, B.M., Tang, N., Sirota, E.B., Sinha, S.K., and Deutsch, M. (1995) Phys. Rev. Lett. **75**, 1332.

37. Bhushan, B., Israelachvili, J.N., and Landman, U. (1995) Nature **374**, 607; Granick, S., (1991) Science **253**, 1373.

38. Yu, C, J., Richter, A.G., Datta, A., Durbin, M.K. and Dutta, P. (1999) Phys. Rev. Lett. **82**, 2326.

39. Cieplak, M., Smith, E.D., Robbins, M.O., (1994) Science **265**, 1209; Diestler, D.J. Schoen, M., and Cushman, J.H. (1993) Science **262**, 545.

40. Demirel, A.L., and Granick, S. (1996) Phys. Rev. Lett. **77**, 4330.

41. Napiorkowski, M. and Dietrich, S. (1993) Phys. Rev. E **47**, 1836.

42. McClain, B.R., Lee, D.D., Carvalho, B.L., Mochrie, S.G.J., Chen S.H., and Litster, J.D. (1994) Phys. Rev. Lett. **72**, 246.

43. Sanyal, M.K., Sinha, S.K., Huang, K.G., and Ocko, B.M. (1991) Phys. Rev. Lett. **66**, 628.

44. Braslau, A., Pershan, P.S., Swizlow, G., Ocko, B.M., and Als-Nielsen, J. (1988) Phys. Rev. A **38**, 2457; Meunier, J. and Langevin, D. (1982) J. Phys. (Paris) Lett. **43**, L185.

45. Ocko, B.M., Wu, X.Z., Sirota, E.B., Sinha, S.K., and Deutsch, M. (1994) Phys. Rev. Lett. **72**, 242.

46. Fradin, C., Daillant, J., Braslau, A., Luzet, D., Gourier, C., Alba, M., Grubel, G., Vignaud, G., Legrand, J.F., Lal, J., Petit, J.M., and Rieutard, F. (1997) Proceedings of 5th International Conference on Surface X-ray and Neutron Scattering (SXNS-5), J. Penfold and D. Norman, Eds., Physica. B (to be published).

47. Wang, J., Tolan, M., Sood, A.K., Wu, X-Z., Li, Z., Bahr, O., Rafailovich, M.H., Sokolov, J. and Sinha, S.K. (to be published).

48. Dierker, S,B., Pindak, R., Fleming, R.M., Robinson, I.K., and Berman , L. (1994) Phys. Rev. Lett. **73**, 82.

49. Thurn-Albrecht, T., Steffen, W., Patkowski, A., Meier, G., Fischer, E.W., Grubel, G., and Abernathy, D.L. (1996) Phys. Rev. Lett. **77**, 5437.

50. Mochrie, S.G.J., Mayes, A.M., Sandy, A.R., Sutton, M., Brauer, S., Stephenson, G.B., Abernathy, D.L. and Grubel, G. (1997), Phys. Rev. Lett. **78**, 1275.

51. Grubel, G., et al. (1994) ESRF Newsletter, Vol. 20, P. 1.

# SPECULATIONS AND CALCULATIONS IN THE PHYSICS OF FOAMS

WEAIRE, D., BRADLEY, G.

*Department of Physics.*
*Trinity College Dublin, Ireland.*
*e-mail: dweaire@tcd.ie, geoff@phy.tcd.ie*

AND

PHELAN, R.

*Shell Global Solutions.*
*Shell International Oil Products B.V.*
*Shell Research and Technology Centre, Amsterdam*
*PO Box 38000, 1030 BN Amsterdam, The Netherlands.*
*e-mail: r.phelan@research.kpn.com[†]*

**Abstract.** The physics of foam is reviewed with an emphasis on simulation. Some new results for the energies of ordered cylindrical foam structures are included.

## 1. Introduction

The properties of foams (Figure 1) present an interdisciplinary challenge to chemists, physicists, and engineers. Roughly speaking, their particular pre-occupations are with the *microscopic, mesoscopic* and *macroscopic* length scales, respectively. These relate to the properties of surfactant layers, thin films and their intersections, and average bulk properties.

*A.T. Skjeltorp and S.F. Edwards (eds.), Soft Condensed Matter: Configurations,*
*Dynamics and Functionality, 247-268.*
© 2000 *Kluwer Academic Publishers. Printed in the Netherlands.*

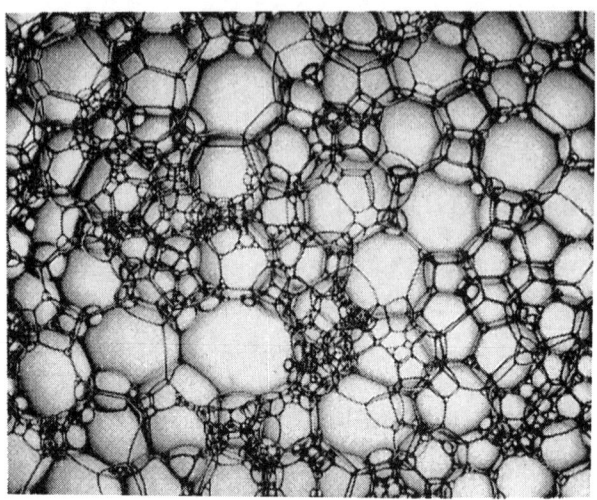

*Figure 1.* An aqueous foam.

Here we will concentrate on the favorite model used by the physicist to describe the structure of foam and its consequences, on the intermediate length scale. For more details, the reader may refer to the proceedings of a previous NATO school [20] largely devoted to this subject, and a recent monograph [2].

The model is extremely simple in its definition, but it has not always been easy to see its consequences. Simulations have been used to provide reliable results. This process has progressed to the point at which it may be said that the properties of static or quasi-static dry foams are well understood.

As Figure 2 indicates this still leaves much to be explored having to do with dynamical effects and wet foams.

The theoretical challenges posed by the new regime are formidable, so that it may be wise to pursue relevant experiments in the first instance. Two possibilities for such experiments are presented by Plateau frames ([13], [14]), see Figure 3, and cylindrical foam structures ([6], [24]). In both cases, dynamical effects can be explored by, for example, inducing a constant drainage of liquid by steadily introducing it at the top.

After reviewing the recent history of the subject with an emphasis on

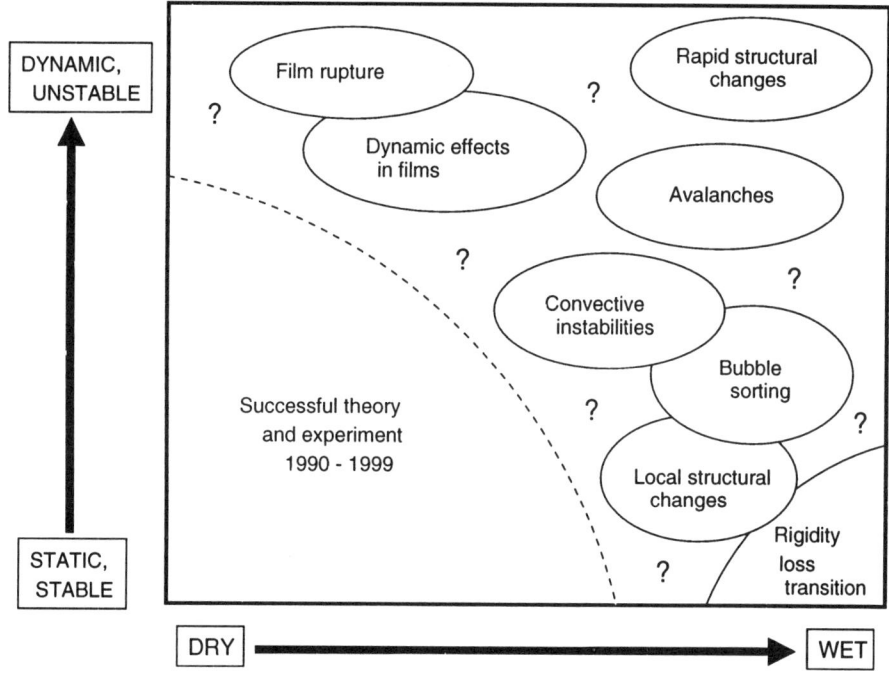

*Figure 2.* Many challenges remain in the physics of foams

simulation, we shall return to the two types of elementary experiment which we have recommended for the future.

## 2. The physics of foam

In the standard model of a foam, films of infinitesimal thickness meet in lines, which are joined at fourfold vertices. The energy of the surface is proportional to the total surface area, the gas enclosed in each cell (bubble) being treated as incompressible. For many purposes only a single material constant, the surface tension $\gamma$, is significant.

Another important parameter is the liquid fraction $\Phi_l$. When $\Phi_l$ is small (or ideally zero) we speak of a *dry* foam. When $\Phi_l$ approaches its maximum value for a stable foam (0.36 for monodisperse 3-d foam), the sample is described as *wet*.

The key elements of our understanding of dry foam in equilibrium date

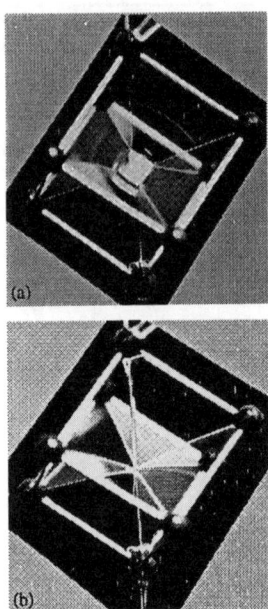

*Figure 3.* Soap films spanned in a cubic wire frame. Eight 4-fold vertices can be seen in (a). Introducing a flow of liquid from the top leads to the formation of a single 8-fold vertex (b)

back to the 19th century, when Plateau formulated the basic equilibrium rules: films meeting three-at-a-time at equal angles, fourfold vertices with equal angles, and the Laplace Law for each film. Of these rules, only the one which excludes more complicated vertices in very subtle. At this point, Plateau relied on a Belgian colleague, Ernest Lamarle, to back up his empirical findings with a long and complicated geometrical proof.

Between 1980 and 1995, this model was used to simulate *two-dimensional* foams (Figure 5), such as may be created by squeezing a blob of foam between two glass plates. This has proved to be a good guide to the properties of 3d foams, as was hoped. Experiments on 2d foam have also proved to be excellent for lecture demonstrations and teaching experiments.

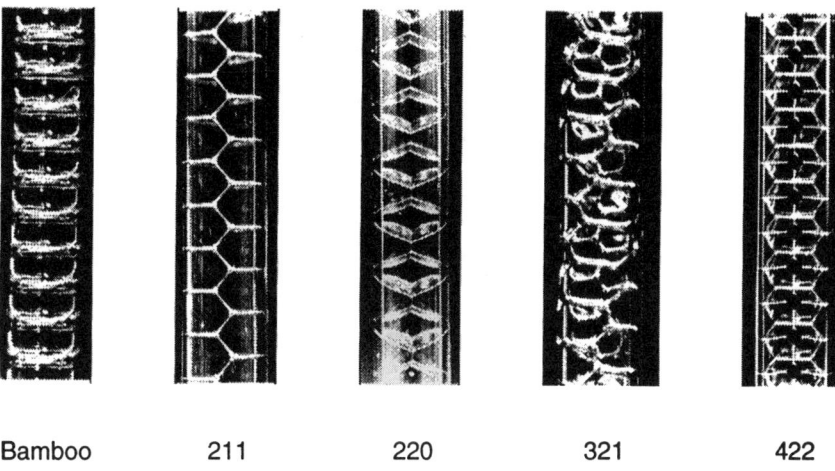

Bamboo     211     220     321     422

*Figure 4.* Equal sized bubbles introduced into a cylinder of comparable radius form ordered structures.

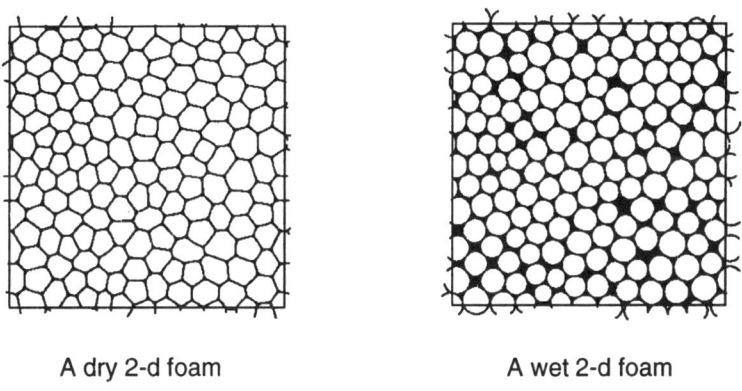

A dry 2-d foam          A wet 2-d foam

*Figure 5.* Output from simulations of 2-d foams

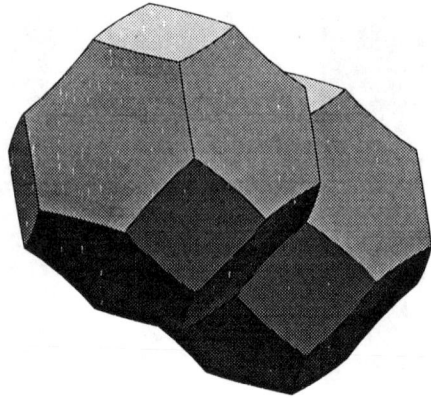

*Figure 6.*   The cubic unit cell of Kelvin's structure. This consists of two Kelvin tetrakaidecahedra

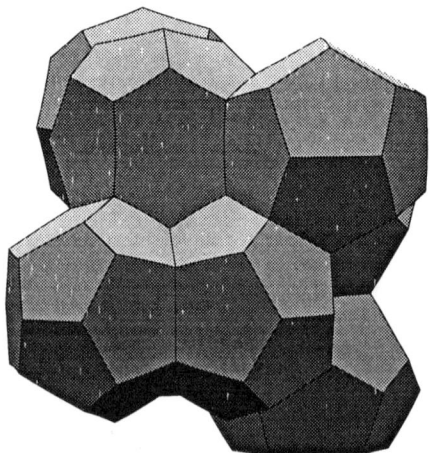

*Figure 7.*   The cubic unit cell of the Weaire-Phelan structure. This consists of six fourteen-sided polyhedra and two pentagonal dodecahedra.

## 3.  The Surface Evolver

In the early 90's it became possible to extend such simulations to 3d foams. Most of this work has relied on computer programs written by Ken Brakke for the Geometry Centre, under the heading of the *Surface Evolver* [1].

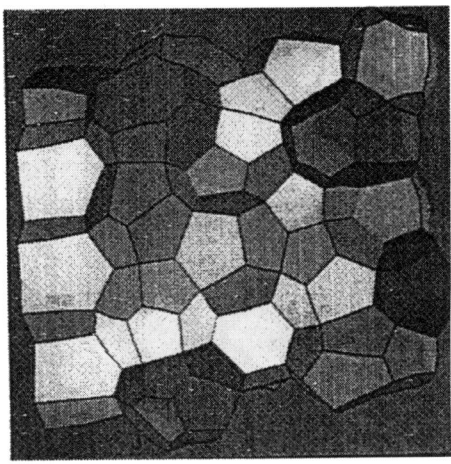

*Figure 8.* A reconstructed image of a dry foam using the Surface Evolver [26].

*Figure 9.* A simulated Kelvin's structure with 1% liquid fraction.

The program can create a specified sample of foam with $\sim 10^3$ cells, and equilibrate its structure. It can deal with dry or wet foams. Examples of output from the Surface Evolver are shown in Figures 6,7,8,9,10,11 and 12.

Previous 2-d simulations owe much of their efficiency to the direct use of the Plateau's equilibrium rules, summarised above. The Surface Evolver does not use such rules, but rather engages in the iteration of local variables

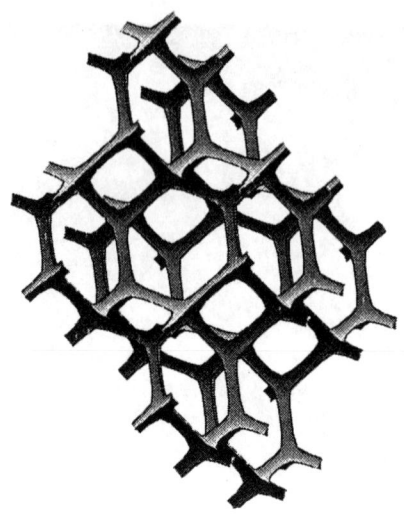

*Figure 10.* Part of a wet bulk Kelvin foam showing the network of Plateau borders.

*Figure 11.* Part of a bulk bamboo structure (cf. Figure 4). Here the Plateau borders around the edge of the cylinder can bee seen.

to minimise surface energy under given conditions. Of course, the rules must, in the end, be obeyed - but they are not put in at the beginning.

The Evolver works by taking some pre-defined structure which is topologically *similar* to the structure under consideration. This structure is

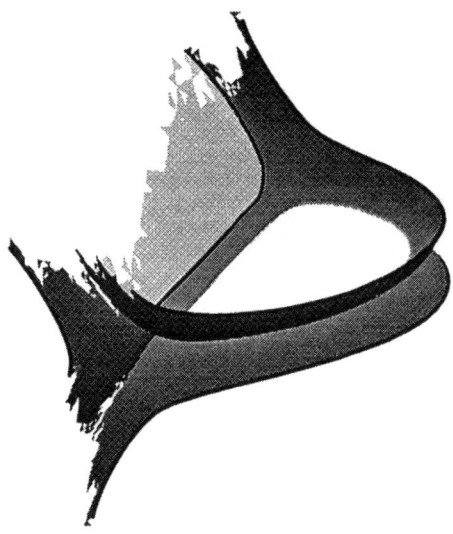

*Figure 12.* Part of a wet 211 cylindrical structure (cf. Figure 4). Here the films on the exterior have been removed and the films at the top connect into the films at the bottom after a rotation of 180 degrees, to complete the unit cell of the structure shown in the Figure 3.

made up of vertices (points) in 3-d space. Points are linked together using edges, and edges are linked to form faces. Faces may then be linked to form bodies (i.e. cells or bubbles). The Evolver minimises the energy of the area of this surface subject to certain constraints or additional energy terms. Constraints can be geometrical restrictions on vertices and edges, or specific integral quantities such as body volumes. Minimisation is done by evolving a surface down an energy gradient via any of a number of optional techniques. Typically gradient descent is used, alternative methods include conjugate gradient and Hessian methods (calculating the second derivative of energy and solving for the minimum energy).

The initial mesh is redefined internally in the Evolver to ensure that each face is made up of only three edges. Refinement of the mesh causes each parent facet to be sub-divided into four "child" facets subject to the same constraints as the parent. This can be see in Figure 13.

Minimisation is carried out at each level of refinement. Technically the only limitation on accuracy is computational power. Figure: 14 shows a plot

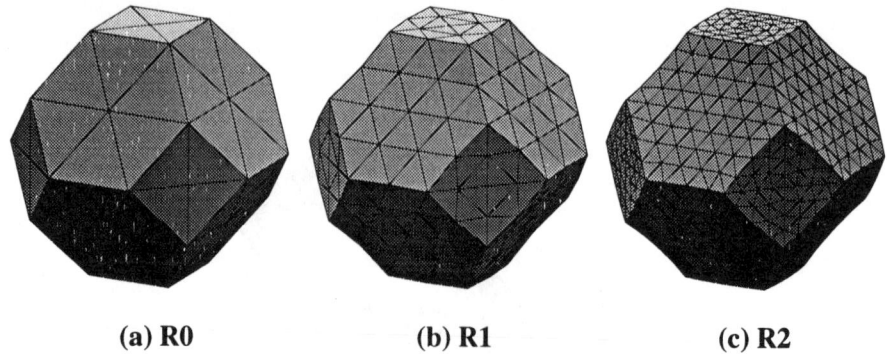

**(a) R0**   **(b) R1**   **(c) R2**

*Figure 13.* The stages of minimisation of a unit volume Kelvin cell: (a) shows the initial plane sided input structure while (b) and (c) show the first two refined structures with surface area minimised, and thus curved faces.

of the energy as a function of the total iteration number covering the first five refinements

It is important that a sufficient number of minimisation steps occurs at each level of refinement, and that the best possible mesh configuration is obtained for further minimisation at higher levels of refinement. Figure 15 shows the drop in the energy at a specific minimisation step.

Finally a power law is used to extrapolate a final energy for the structure based on the energy based on the current and previous energies.

However, the achievement of convergence and stability in structures is not always this simple to judge (see Brakke [19]). "Hand tuning" of the facets making up the surface is often required, particularly when using constraint functions and when modelling wet foams. Techniques for obtaining convergence include

- Equiangulation: juggling edge connectivity to give more equiangular triangles.
- Vertex averaging: replacing each vertex with the average of the centroids of its adjacent faces. This helps reduce the variation in face area and smoothes out spikes in the surface.
- Area-weeding: The removal of small facets and edges which may stall the minimisation.

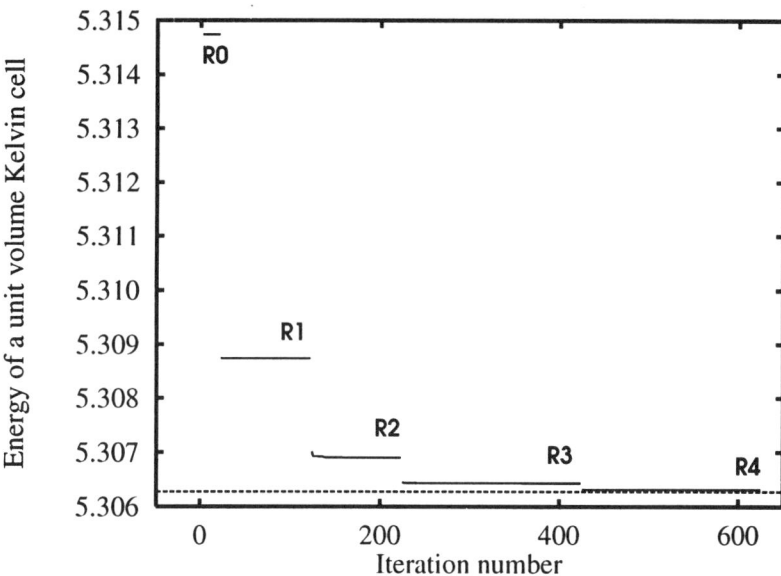

*Figure 14.* The minimisation of a unit volume Kelvin cell. Each step (R0 - R4) indicates an increasing refinement level (see Figure 13). The x-axis shows the total number of steps of conjugate gradient used in minimisation. The dashed horizontal line is the energy value obtained after further calculation and is accurate up to five decimal places.

- Minimisation techniques: Switching from gradient descent to conjugate gradient and back again to work around saddle points.
- Hessian techniques: Constraining Hessian iterations to move each vertex perpendicularly to the surface only removes problems with vertices moving off in strange directions that make convergence difficult.

In its present form the Evolver is has limited ability to handle topological changes. These must be induced by hand in a very time consuming process. It is possible to isolate vertices that do not have the topologies of one of the three minimal tangent cones. These can be adjusted to proper local topologies. It is also possible to isolate edges with more than three facets attached. These edges are split into triple facet edges and this splitting continues to propagate until some obstacle is reached. This process may leave stray facets around which also have to be removed by hand.

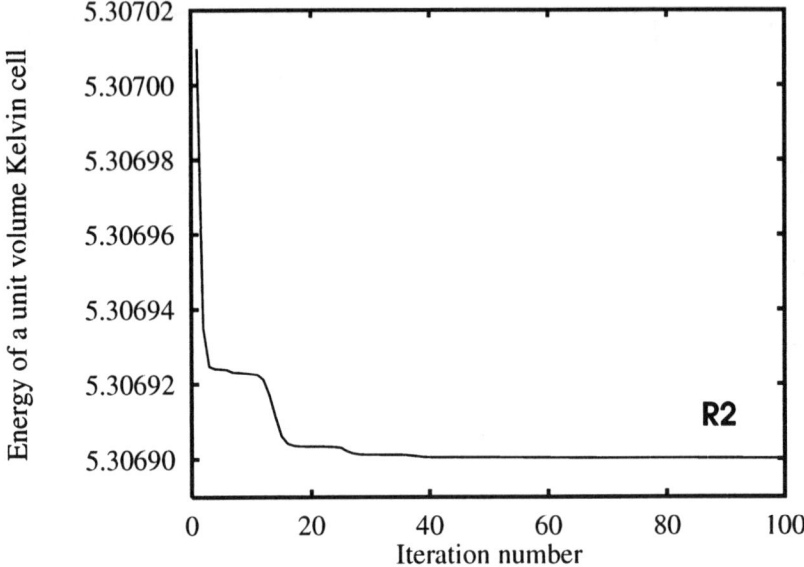

*Figure 15.* Energy versus total minimisation step at R2 refinement.

## 4. Alternative approaches

The results of the Evolver for the standard model are extremely accurate. Its only limitation (apart from possible difficulties with topological changes) is the availability of sufficient computer resources, which makes it practically impossible to pursue intensive calculations on very large systems. The work of Kraynik [15] represents the present state-of-the-art in such large calculations. An attractive alternative for wet foams is presented by a model used by Durian [17] and others, which treat bubbles as spherically symmetric objects with interactions which represent their deformation upon contact. This cannot, in practice, be made fully rigorous, but has proved valuable where the Evolver is inadequate.

## 5. Present understanding of quasi-static properties

Space does not allow more than a cursory summary of the present state of the art, so topics of interest will be itemised as follows:

## 5.1. COARSENING

The study of coarsening was the original motivation for observations on 2-d foams, by C. S. Smith in the 1950's. As he expected, it has been confirmed that soap froths show very similar scaling behavior to that of the ideal grain growth (that is, mean cell diameter $d \sim t^{\frac{1}{2}}$, although the two systems are not quite identical. This is well established in 2-d rather less so in 3-d, but there is little suspicion that it will be otherwise.

## 5.2. DEFECTS

In two dimensions, the ideal honeycomb structure is rendered unstable against coarsening by diffusion, wherever a topological defect is introduced. Some years ago, it was suggested that this progress shows anamolous features, but this no longer seems to be the case.

## 5.3. IDEAL STRUCTURE IN 3-D

In 1994, Weaire and Phelan overthrew the historical conjecture of Lord Kelvin, regarding the dry foam structure of lowest energy for equal-sized bubbles. Despite exhaustive searches of the family of structures of which the Weaire-Phelan (see Figure 7) structure belongs, it has not been bettered.

## 5.4. VERTEX STABILITY

While Plateau's 19th century rules on the instability of vertices in dry foams are elementary in 2d, this is not quite so in 3d. Weaire and Phelan were led by experiment and rough calculations to suggest that the symmetric junction of eight films is stable for *arbitrarily small* $\Phi_l$ [18]. However, Brakke [19] has found this not to be the case.

## 5.5. ELASTIC MODULI

A foam is a solid under small applied stress and so has a well-defined shear modulus. Extensive calculations by Kraynik [20] show that, while this may be highly anisotropic, its spherical average agrees well with a longstanding estimate by Stamenovic [21]

## 5.6. THE APPROACH TO THE WET LIMIT

In the wet limit, at which bubbles come apart, the elastic constant must go to zero. Whereas in an ordered foam this occurs by a finite jump, it is a smooth trend to zero for a disordered foam. The exact critical behavior, at this "rigidity loss" transition is still debated.

## 5.7. PLASTICITY

Under high stress a foam undergoes plastic deformation and has a finite yield stress above which it flows - see Figure 16. This region is coming to be well understood, in terms of topological changes. However there is evidence of *avalanches* of topological change in wet foams, which are currently under investigation.

## 5.8. DRAINAGE

Drainage of liquid at low flow rates takes place through the netwok of Plateau borders which are close to static equilibrium. The Foam Drainage Equation [22] , derived with a continuum approximation, gives a good account of a variety of interesting drainage effects. It has nevertheless been called into question recently [23]. In any case it will require significant modification at high flow rates, as described below.

## 6. Some challenges

Despite all this progress, a great deal remains to be achieved, as Figure 2 indicates. For example, whenever a drainage experiment is pushed to a high flow rate, the structure becomes unstable. What happens thereafter depends on the bubble size distribution. For a monodisperse foam, a *convective role* develops (Figure 16), with a velocity which increases as the flow rate is further increased. In a polydisperse foam, this and other dynamical effects eventually results in *bubble sorting* [24] and the structure is thereby restabilised.

This problem presents, in a manageable form, the kind of challenge which is posed by foams in many typical industrial contexts, in which there is a rapid mixing of liquid and gas, in a situation far from static equilibrium.

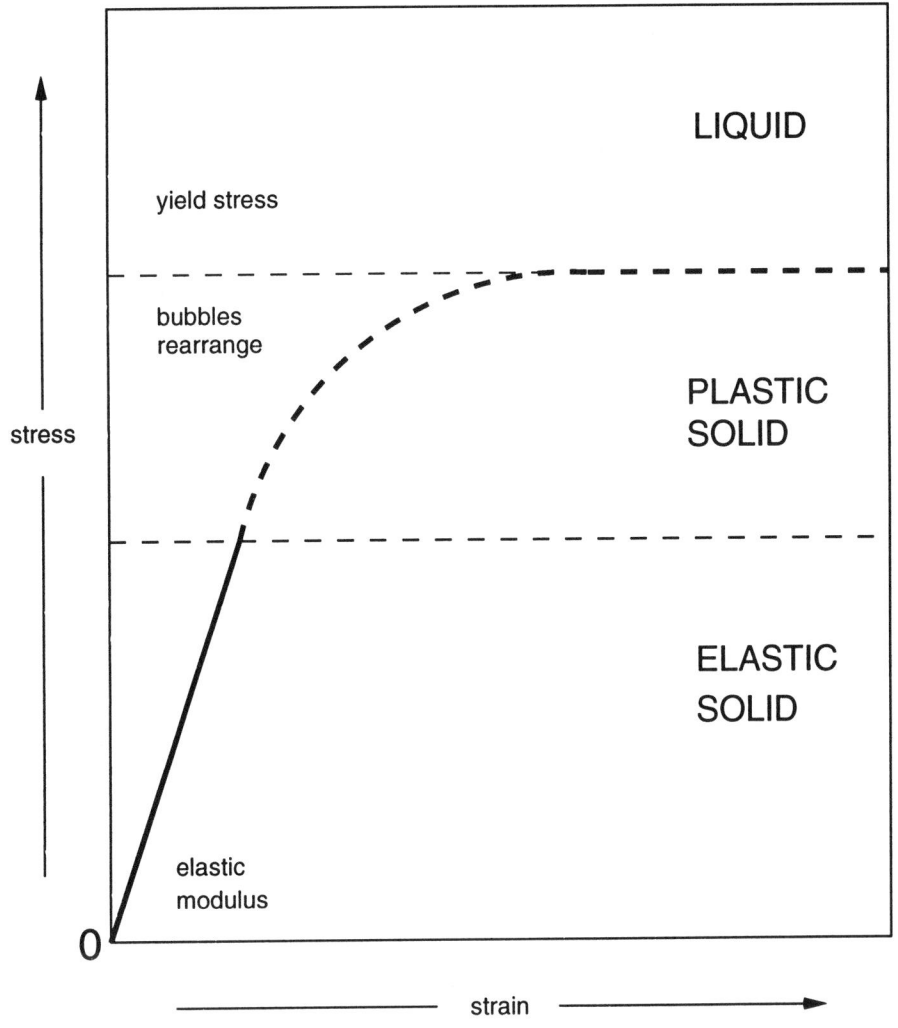

*Figure 16.*  A schematic diagram of stress against strain in a foam

At the same time it is still important to understand better the dynamical effects on a local scale, beyond the quasi-static region. Cylindrical foam experiments provide one of the best testbeds for this.

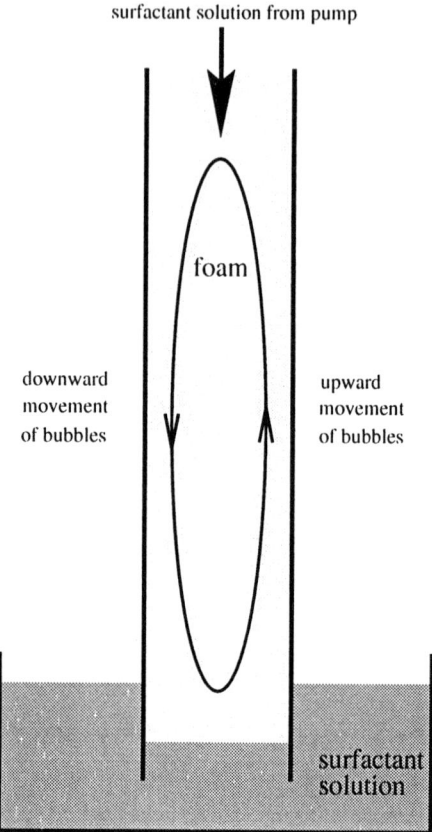

surfactant solution from pump

foam

downward
movement
of bubbles

upward
movement
of bubbles

surfactant
solution

*Figure 17.*   Convective motion of bubbles in a monodisperse foam.

## 7.  Cylindrical foams

The computational and experimental study of monodisperse foams confined within a cylindrical boundary has allowed us to better understand many of the properties of foam. Such a simple system - which will be described below - has helped us to understand the process of structural transitions both in wet and dry foams and will hopefully allow us to say something about their viscosity also. That said, the simple system has lead to its fair share of questions - some which still remain unanswered.

The experimental setup for such experiments is quite simple. A glass cylinder of radius $R$ is placed vertically into a solution of washing-up liquid.

Small $N_2$ bubbles are passed into the liquid via a pipette in this region. This causes bubbles of radius $r$ to be formed and pushed up into the glass cylinder. These bubbles then order themselves to have a hexagonal surface structure, which is perfect if the bubbles are sufficiently large. Indeed in this case the interior is also perfectly ordered, in any one of a great variety of structures (Figure 4). These hexagonal bubbles on the cylinder face form a triangular lattice wrapped on a cylinder with three spiraling families of $k$, $l$ and $m$ reticular lines [7]. The bubbles are labelled by an integer $i$ that increases as the bubble altitude increases in the cylinder. The neighbours of bubble $i$ are $i \pm k, i \pm l$ and $i \pm m$. Bubble $i + k$ can be reached from one step on the reticular line of family $k$ or by one step from $m$ followed by one step on $l$. Thus we have $k = l + m$. The dimensionless parameter $\lambda = \frac{R}{r}$ as our control parameter.

## 7.1. HISTORY OF CYLINDRICAL FOAM STRUCTURES

Since cylindrical foam structures are not well known, we will give a brief history of them

- It appears that Mann *et al.* [3] were the first to begin investigations of ordered cylindrical structures - though they seem to have stumbled upon them. In their experiment they passed air through oil which caused a series of equal volume bubbles to form and move up a glass tube. They immediately identified the structure as being highly ordered, but this seems to be the extent of their work.

- In 1955 Dodd [4] carried out a number of experiments in an attempt to obtain an experimental proof that Kelvin's proposed tetrakaidecahedron existed. He was able to place monodisperse bubbles, one at a time, into a cylinder, and create a structure which includes Kelvin cells.

- The first real attempt to categorise the ordered structures formed within a cylinder was performed by Weaire *et al.* [5]. It was hoped that would be a valuable test-bed for 3-d foam simulations. Following on from this work Pittet *et al.* [6] carried out a more detailed examination of these structures, pushing the categorisation further with an out look on examining the transitions from one structure to another. See next section.

- Hutzler *et al.* [8] carried out a number of experiments which yielded some rather interesting - and as of yet unexplained - results. In these experiments he observed a twist boundary slowly moving down the lenght of the cylinder.
- Recently Boltenhagen *et al.* [9] have been looking at deformations of cylindrical structures, and structural transitions into crystalline foams [10]
- A comprehensive computational cataloging of stability regimes for both dry and wet foams has been carried out by Bradley *et al.* [11]

## 7.2. STRUCTURAL TRANSITIONS IN CYLINDRICAL STRUCTURES

Pittet *et al.* [6] have carried out extensive work on structural transitions: here we attempt to give a brief overview of their work.

Starting with the basic experimental setup describe in section **7.**, and decreasing the bubble size in small increments every few minutes, they observed a reproducable *jump* from one ordered structure to the next more complicated ordered structure.

They observed the ordered structures formed and reversing the process, i.e. increasing the bubble size in small increments led them to discover hysteresis in the formation of the structures and it was possible to identify the transition points. Most of the ordered structures have quite a large quasi-stable range.

Slightly altering the experimental apparatus led to a different series of ordered structures being formed. This and other factors suggest there may be rules governing structural transitions.

Pittet *et al.* [7] discovered both experimentally and computationally that a single dislocation between two structures causes transitions. Such a dislocation is carried up by the rising froth. The motion of the dislocation may be carried in two ways. A climb - seen as cellular division (CD) or as a glide - due to edge flipping or neighbour switching.

Climbing leads to two possible structural transitions:

$$k = l+m, l, m \rightarrow k+1, l, m+1 \quad (CDI)$$
$$k = l+m, l, m \rightarrow k+1, l+1, m \quad (CDII)$$

Gliding also leads to two possible structural transitions

$$k = l + m, l, m \to k, l + 1, m - 1 \quad (GI)$$
$$k = l + m, l, m \to k, l - 1, m + 1 \quad (GII)$$

In more recent work Boltenhagen *et al.* [9] have catagorised the effects of compressing and dilating an ordered foam structure. They discovered that Pittets climb and glide dislocation rules applied to structural compression, and simply reversing the signs on the r.h.s. of the rules explain the laws of structural dilation.

## 7.3. RECENT CALCULATIONS

Recently Weaire and Phelan [25] and others have attempted to model ordered cylindrical structures. These attempts were hampered to some extent by the restrictions of the model used in the *Surface Evolver*. Brakke altered the program to loosen these restrictions. Here we present some of the output from the program and explain the relevance of such work.

There are two methods for obtaining ordered cylindrical structures: some may be generated using the bcc Kelvin unit cell, otherwise the must be set up individually. Such structures must be periodic in the *z-direction*, through a rotation of $\theta$ degrees.

Wet structures are much more difficult to model and generate (Figures 11 and 12) and are beyond the scope of this article.

### 7.3.1. *Theory of dry structure energy*
The energy of these structures may be understood as follows. Isolating several Kelvin cells in a cylinder we cause the cells to contort to fit the required shape. The Kelvin structure is *known* to have the property that it has the lowest surface area for a *single* space filling cell, thus it is then safe to assume that the contorted cells will have an energy which must be greater than that of an uncontorted Kelvin cell.

The energy of these structures can be broken into two parts: an interior film energy and a cylinder surface energy.

It is clear that for simple structure (where $\lambda < 3$ say) the energy contribution from the cylinder surface films will be reasonably large. However, for very large $\lambda$, this contribution will be very small. Also, the internal cell

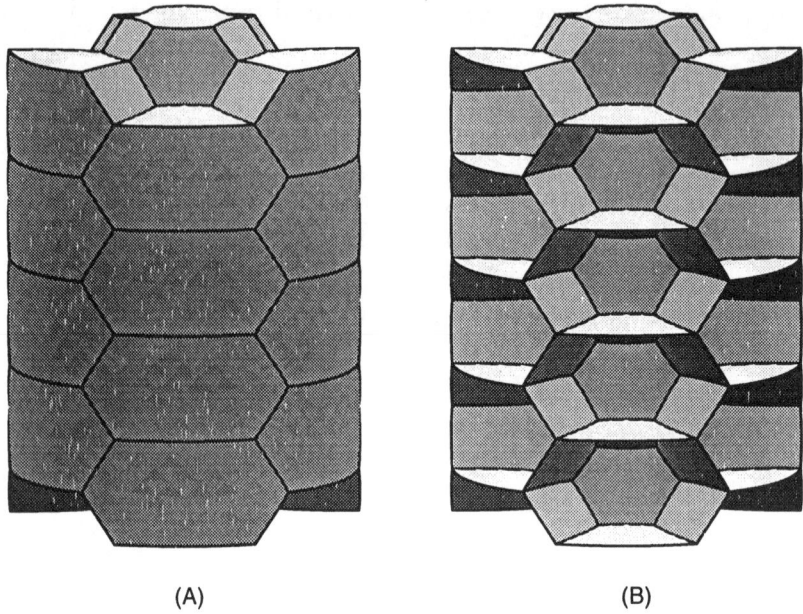

(A)                                    (B)

*Figure 18.*   Evolver output of a minimised 633-1 structure showing (A) the exteral hexagonal surface structure, and (B) Kelvin's tetrakaidecahedra as its internal structure.

structure for larger $\lambda$ will be that of Kelvins tetrakaidecahedra. Thus as $\lambda$ increases we would expect the total energy to converge towards the Kelvin cells energy.

Indeed, calculations show (Figure 19) that as the complexity of the structures increase (i.e. as they head towards bulk) the energy does indeed go down.

## 8.  Acknowledgements

Forbairt. Trinity College HPC scheme. ESA Topical Team. Stefan Hutzler. Ken Brakke. Claire Monnereau.

## References

1.  Brakke, K., **Experimental Math.** 1, 141.

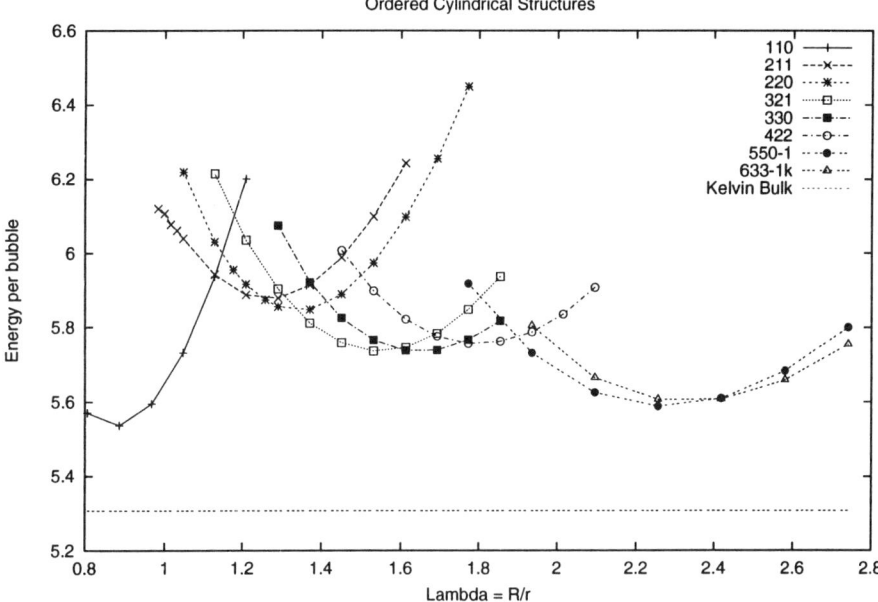

*Figure 19.* The total energy of various ordered cylindrical structures plotted against the control parameter lambda

2.  Weaire, D., Hutzler, S. **The Physics of Foam** Oxford University Press (1999) in press.
3.  Mann, W. B., Stephens, R. W. B. **Bubble formation in glass tubes** Phil Mag **15** Jan 1933.
4.  Dodd, J. D., **An approximation of the minimal tetrakaidecahedron** American Journal of Botany **Vol. 42**
5.  Weaire, D., Hutzler, S. and Pittet, N. (1992) **Cylindrical packings of foam cells** Forma, **7**, 259-263.
6.  Pittet, N., Rivier, N., Weaire, D. (1995) **Cylindrical packings of foam cells** Forma, **10**, 65-73.
7.  Pittet, N., Boltenhagen, P., Rivier, N. and Weaire, D., **Structural transitions in ordered, cylindrical foams** Europhys. Lett., **35** (7), pp, 547-552 (1996)
8.  Hutzler, S., Weaire, D., Crawford, R., **Moving boundaries in ordered cylindrical foam structures** Philosophical Magazine B, 1997, Vol. 75, No. 6, 845-857.
9.  Boltenhagen, P., Pittet, N. and Rivier, N. **Giant deformations and topological hysteresis of an ordered foam** Europhys. Lett., **43** (6), pp. 690-694 (1998)
10. Boltenhagen, P. and Pittet, N. **Structural transitions in crystalline foams** Europhys. Lett., **41** (5), pp. 571-576 (1998)

11.  Bradley, K. G., Brakke, K., Weaire, D. and Hutzler, S. **Computational modelling of ordered cellular structures** (preprint)

12.  Phelan, R. PhD. Thesis.

13.  M. in het Panhuis, S. Hutzler, D. Weaire and R. Phelan. **New variations on the soap film experiments of Plateau I. Experiments under forced drainage.** Philosophical Magazine B, 1998, Vol. 78, No. 1, 1-12.

14.  R. Phelan, D. Weaire, S. McMurry, S. J. Cox. **The deformation of soap film junctions by applied forces.** Philosophical Magazine Letters (in press)

15.  Editors: Denis Weaire and John Banhart. **Foams and films/International Workshop on Foams and Films, Leuven (Belgium), 5th and 6th March 1999.** ISBN 3-9085748-6-5

16.  Brakke, K. **The surface evolver and stability of liquid surfaces** Phil. Trans. R. Soc. Lond. A(1995).

17.  Durian D J 1997. **Phys. Rev. E 55** 1739

18.  D. Weaire and R. Phelan. **Vertex instabilities in foams and emulsions** J. Phys.: Condens. Matter 8 (1996) L37-L43.

19.  Kenneth Brakke. **Insatability of the wet cube cone soap film.** (preprint)

20.  Sadoc, S. F. and Rivier, N. (Eds.) (1998). **Foams and Emulsions,** Kluwer.

21.  D. Weaire. **On the Kelvin-Stamenovic estimate of the shear modulus of a foam** Philosophical Magazine Letters, 1998, Vol 77, No.

22.  D. Weaire, S. Hutzler, G. Verbist and E. Peters **A review of foam drainage** Advances in Chemical Physics, Volume 102. ISBN 0-471-19144-2

23.  Koehler, S. A., Hilgenfeldt. S and Stone H. A. **Liquid flow through aqueous foams: The node-dominated foam drainage equation** Preprint submitted to Phys. Rev. Lett.

24.  Hutzler, S., Weaire, D. and Shah. S. **Bubble sorting in a foam under forced drainage** (preprint)

25.  Weaire, D. and Phelan, R. **Cellular structures in three dimensions** Phil. Trans. R. Soc. Lond. A (1996) **354**, 1989-1997.

26.  C. Monnreeau and M. Vignes-Adler **Optimal tomography of real three-dimensional foams** Journal of colloid and interface science **202**, 45-53 (1998)

# COMPLEX PHYSICAL PHENOMENA IN CLAYS

*JON-OTTO FOSSUM**
NTNU Physics Department
Gloeshaugen, Sem Saelands vei 9
N-7491 Trondheim, NORWAY

**Abstract**
Recent experimental and theoretical research into the complexity of physical phenomena in clays is reviewed, and illustrated through the rich behavior displayed in physical model system clean chemistry customized synthetic clays, such as synthetic hectorites. Relations to naturally occuring clays and relevance to industrial applications are discussed.

Clays represent one of the traditional materials, whose applications have played important roles throughout traditional and modern human history [1]. A list of keywords within these contexts include building materials, ceramics, rheology modification, catalysis, paper filling, oil well -drilling and -stability, etc. [2]. Clays as geological materials have been widely studied and discussed by geologists, geo-chemists, geophysicists, etc, and modern industrial uses of clays are based on this history and knowledge [2]. However, fundamental studies of complex physical phenomena of clays, and resulting applications of clays beyond these traditional approaches and disciplines, are merely at its beginning. For example, one may observe that physicists in particular until recently apparently have not regarded clays as "interesting and accessible" experimental and theoretical model systems. With the increasing availability of clean chemistry customized synthetic clays, this is changing, and there is presently a growing scientific activity associated with including clays into modern materials science together with other and often generally better understood synthetic and complex adaptive materials such as colloids, polymers, liquid crystals, bio materials, etc. Clays may on one hand be considered as aqueous suspensions of physical colloids made up of plate like layered silicates as primary particles, and it is thus of importance to integrate clays into the specific and universal physics of hard colloid spherical, rod, platelet, etc suspensions [3]. On the other hand, dehydrated clays may be viewed as intercalation compounds [4], making studies of clays within the general context of "nano sandwiches" [5] of fundamental relevance. In the present review some of these points concerning clays' place within the general context of complex systems and materials [6] will be illustrated by means of recent experimental and theoretical findings and ideas on and about synthetic clays [7].

269

*A.T. Skjeltorp and S.F. Edwards (eds.), Soft Condensed Matter: Configurations, Dynamics and Functionality, 269-279.*

The most widely studied synthetic clay until now is laponite [8], which belongs to the family of swelling 2:1 clays (Some natural clays are so called 1:1, and not all 2:1 clays are swelling). Elementary introductions describing crystallographic structures and thus precise definitions of both natural and synthetic clays may be found in several text books, for example [2,9]. It should be noted that although the 2:1 swelling types maybe the most important class of clays in terms of industrial applications, it is the non-swelling clays one usually encounters in the "ground" and in ceramics. All dehydrated clays are based on a layered silicate meso structure. The 2:1 clays (or smectites) thus consist of 1 nano meter thick and charged (negative surface charge and a smaller positive edge charge) meso sheets, which in the dehydrated state stack (like decks of cards) by sharing charge-compensating cations. The synthetic clay Laponite is a particularly interesting model system due to the 25 nanometer diameter mono dispersity of the colloidal platelets This is different from natural and other synthetic clays which in general have a polydisperse distribution of micrometer sized platelet areas. The 1 nanometer thick platelets may individually be considered as single crystals. The crystal structure of such a smectite clay sheet is shown in Figure 1.

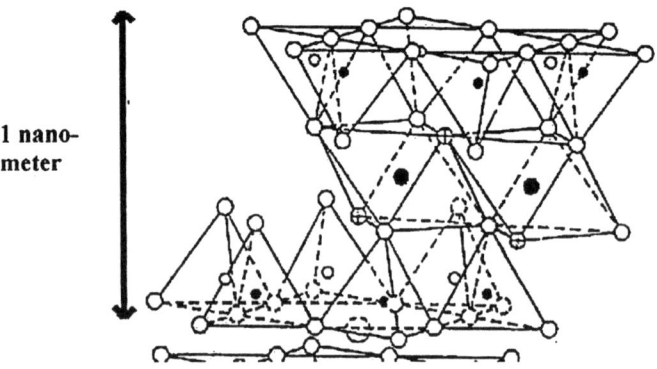

1 nano-
meter

*Figure 1:* Primary particle and crystal structure of a smectite clay. The one nanometer thick sheet maybe broken down into three atomic layers, i.e generally one metal hydroxide layer sandwiched in between two identical silicone tetrahedral oxide layers. Charge compensating interalyer cations are associated with the smectite primary particle surfaces. Details of these crystal structures may be found in [2,4,9] and at http://www.phys.ntnu.no/~fossumj/.

The addition of salt-water to these dehydrated mesoscopic layered systems, gives rise to interesting colloidal dispersion "phase" diagrams, which is illustrated in Figure 2 for the case of laponite [10]. Four separate regions of physical complexity have been suggested from experimental observations of clay-electrolyte-concentration diagrams of the type shown in Figure 2:

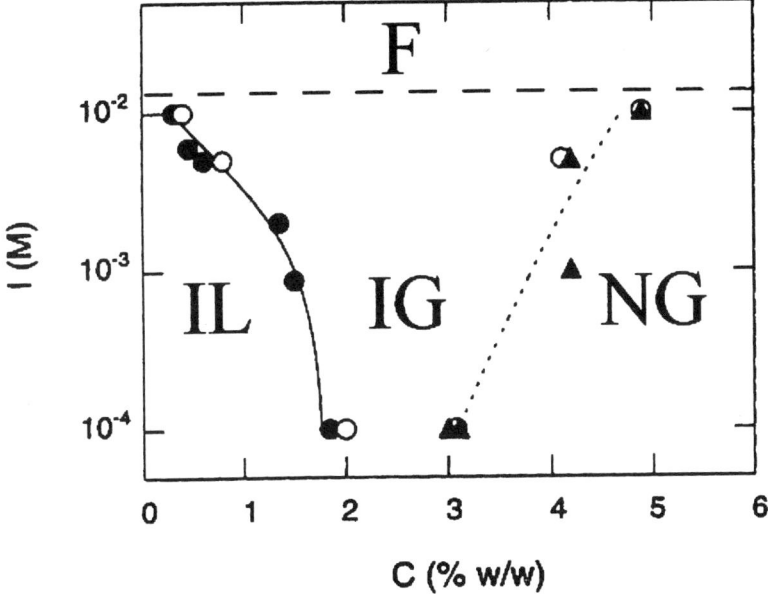

*Figure 2:* Clay-salt concentration diagram for synthetic laponite dispersed in water. C is the laponite weight percent and I is the electrolyte (NaCl in this case) concentration in mol/l. Adapted from [10]. A similar diagram has been established for the natural clay montmorillonite.

IL: Isotropic Liquid, IG: Isotropic Gel, NG: Nematic Gel, and F: Flocculation.

The simplest theoretical approach to discussing these complex behaviors is via traditional DLVO theory [11]. Transitions and aggregate structures within stable phases may thus be discussed in terms of an interaction potential between individual platelets, resulting from adding the electrolyte-independent van der Waals attraction and the double layer repulsion as characterized by an electrolyte concentration dependent Debye screening length [11]. The addition of these 2 forces may give possibilities of platelet-platelet interaction local potential minima, whose position and depth may be tuned by means of the chosen electrolyte concentration, enabling researchers to suggest the following scenarios:

IL region:

Brownian aggregates made up of several laponite platelets are suspended in water. The size and compactness of these aggregates may depend on electrolyte concentration via the Debye screening length. The aggregates in this phase are in general too small to scatter visible light, thus giving a "to the eye" transparent liquid with a viscosity which may be tuned "at will" by means of the salt-clay concentration. This phase may be made birefringent by application of high electric fields [12]. At low concentrations the liquid seems to be Newtonian, whereas upon approaching the IL-IG line the IL phase may display non-Newtonian and/or thixotropic [13] behavior.

IL-IG line:

The aggregate structure percolates when the clay concentration becomes sufficiently large. The precise natures of these transitions at various electrolyte concentrations are not known, although dynamic light scattering experiments [14,15] have been interpreted as being similar to scenarios given by mode coupling theories of structural glass transitions. Other recent experiments of the same type may suggest that the IL-IG line represents a purely dynamic transition [16]. The idea of a glass transitions in laponite suspensions was also recently discussed in terms of, and put into the fascinating context of a possible observation of aging of "Wigner" glasses in these systems [17].

IG region:

Percolating gels in which the laponite clay platelets have random orientations when macroscopically averaged, and where the details of the local platelet arrangements depend on the electrolyte concentration. At low electrolyte concentrations i.e. below 0.001-0.01 mol/l, this phase is transparent "to the eye", but as the F phase is approached, the local aggregate structure within the gel becomes micrometer sized and scatters visible light, and thus the gel shows macroscopic opaqueness. The gels are rather remarkable in that it only takes 1-2 percent weight of clay in order to make samples which "look and feel like stiff water". These gels are Bingham [13] solids with a salt-clay concentration dependent yield shear stress. Recent x-ray-, neutron- and light-scattering experiments on some of these isotropic gels [18] have been interpreted as suggestive of a percolating structure made up of bundled chain like aggregates consisting of sub units of small platelet stacks. Thus as for natural clays, one may imagine pore sizes on 3 different length scales: 1) sub 10 nanometer sized pores between platelets, 2) 10-100 nanometer sized pores between small platelet sub units, and 3) micrometer sized pores with "free" water in between sub unit aggregates. It has been suggested that these aggregates of sub units may be fractal at length scales between approximately 1 and 10 micrometers. These conclusions are different from interpretations of recent static light scattering experiments [16,19] in which micrometer sized impurities (or maybe possibly undissolved laponite micro aggregates) were filtered out of the laponite samples prior to salt induced colloidal reaggreagtion of dissolved laponite particles. The filtering procedures followed [16,19] were similar to the ones used earlier by Rosta and von Gunten [20]. One is at present thus faced with conflicting light scattering data and interpretations possibly resulting from differences in sample preparation procedures [19]. These discrepancies and debates also extend to suggested extremely slow laponite aggregation dynamics exceeding 1 year or more [18]. It has been pointed out that this slow dynamics may be a second order effect due to particle dissolution [21] thus giving rise to an effective platelet-platelet interaction changing with time. Clearly more experimental research may be needed also in these systems in order to distinguish real statistical physical aggregation and pattern formation at constant interaction, from chemical effects.

IG-NG line:

Laponite platelets or small stacked aggregates of platelets orient with respect to each other in domains over macroscopic length scales. The nature of this transition is even less clear than the IL-IG case. It has been suggested that this case could represent an example of an Onsager excluded volume Isotropic-Nematic type transition [22,23] in smectite clay gels, but this remains to be experimentally tested.

NG region:

Macroscopic nematic ordering of platelets as observed for example by macroscopic birefringence. There are to this date few detailed experimental studies of this phase, except at very high clay concentration near dehydration, where these clays should be discussed within the general context of intercalation compounds [4] rather than as colloidal suspensions.

F phase:

The salt concentration is so high that aggregates no longer are Brownian, and sediment, giving a clay-water "phase separation". The sedimentation level at given clay concentration decreases as the electrolyte concentration is increased, as the DLVO theory suggests [11].

Some of the ideas summarized above have been incorporated into simulations and theory recently. Monte Carlo (MC) simulations of infinitely thin nonintersecting discs each carrying a rigid point quadrupole, representing the laponite platelet charge distribution, has been performed [24,25]. This particular model suspension undergoes a reversible IL-IG transition above a critical quadrupolar coupling. It has been pointed out recently [26,27] that this model system in effect was inherently designed to aggregate reminiscent of a so called "house of card" structure in which the particles edge-to-face each other, a kind of structural arrangement which does not seem to emerge from experiments. At high particle concentrations this MC simulation is consistent with a NG excluded volume transition. This quadrupole MC model neglects several physical aspects of a real clay system, such as finite size of clay platelets dressed in their ionic double layers, van der Waals attraction, double layer fluctuations and interference effects, thus the model's relevance to the real clay systems seems to be doubtful or at best unresolved at this point in time. A more recent semi-analytical non-linear Poisson-Boltzmann theory for swollen clays gives an osmotic equation of state, which is in semi-quantitative agreement with experiments on laponite [26,28-30]. This excellent work demonstrates both the difficulties associated with, and the need for new theoretical work in this field.

Although the present review so far has focused on the synthetic clay laponite, it should be mentioned that the phase diagram in Figure 2 also is representative of the natural clay montmorillonite [22], and most likely also of all other swelling 2:1 smectite clays. We have recently undertaken experimental studies of the synthetic clay fluorohectorite [31], which have a distribution of sheets sizes up to several micrometers in effective diameter. It seems that in this pure system the polydispersity may represent "a third axis" to add to the phase diagram, in accord with recent theoretical work [32]. This is illustrated in Figure 3, where a series of constant clay concentration samples with increasing salt concentration from left to right, are shown. These gravity induced complex phase separations are presently under detailed study within our group.

The shear-thinning (Bingham solid) nature of the colloidal clay gels has enabled hydraulic fracturing experiments to be performed in these systems both in the IG and the NG phases [33]. Interestingly it was found that the fracture propagation velocity may be very close to the shear wave propagation sound velocity. Beautiful and complex fracture patterns emerged in these experiments from injection of water and gas into the gels. Other experiments looked simultaneously at the microscopic breakdown of the percolating aggregate structure and the macroscopic shear thinning properties in the IG [34]. In concert with a de Gennes scaling law for a semi-dilute polymer system, it has been suggested that there may exist a power law scaling between the yield stress and the clay volume fraction, with a power, which may be expressed in terms of the fractal dimension of the aggregate network. The same workers [34,35] have also performed static light scattering studies of the breakdown of the IG aggregate structure under shear, suggesting a successive breakdown scenario, depending on the rate of shear. However, the same question marks as mentioned above concerning filtering of samples prior to light scattering, should be addressed in these cases.

So far the focus of the present review has been on the soft condensed matter properties of clays, but will now turn towards the very interesting material aspects at the far other end of the phase diagram, namely dehydrated or only weakly hydrated clays. In this case one will rather classify these swelling 2:1 clays as intercalation compounds [4], and some important experimental techniques with which to study material properties in these cases are x-ray, neutron diffraction and NMR for mapping out water intercalation processes, as well as ultrasonic propagation etc for measuring related macroscopic elastic properties.

An example the dynamics of water intercalation into a synthetic fluorohectorite clay as measured by x-ray diffraction is shown in Figure 4. The (001) Bragg peak which is measuring the clay silicate interlayer stacking distance, moves a distance corresponding to 1 single monolayer of water, after a sudden step in temperature [36]. Other swelling clay systems also display this kind of behavior, and one, two, three, etc monolayers of water may be intercalated into the clay host system [4,36,37]. The dynamics of these phenomena are dependent upon the type of interlayer cation, which also together with the SiO clay surface determines the structural properties of the interacting intercalated water system [4,37]. The type of intercalated cation in this kind of system may be exchanged, and this enables manufacturing of so called pillared layered structures (PLS) in which large cations are intercalated [38]. The sample is then dehydrated leaving these cations as spacers in between face to face or edge to face connected clay sheets. These dry nanoporous systems may then be used industrially as molecular sieves, or as catalysts with large accessible surface areas. The PLS's have been and may still be utilized by physicists as model systems in which to study transport in well characterized customized nanoporous media [39]. Another interesting physical model system based on this type of cation exchange may be designed by exchanging into randomly

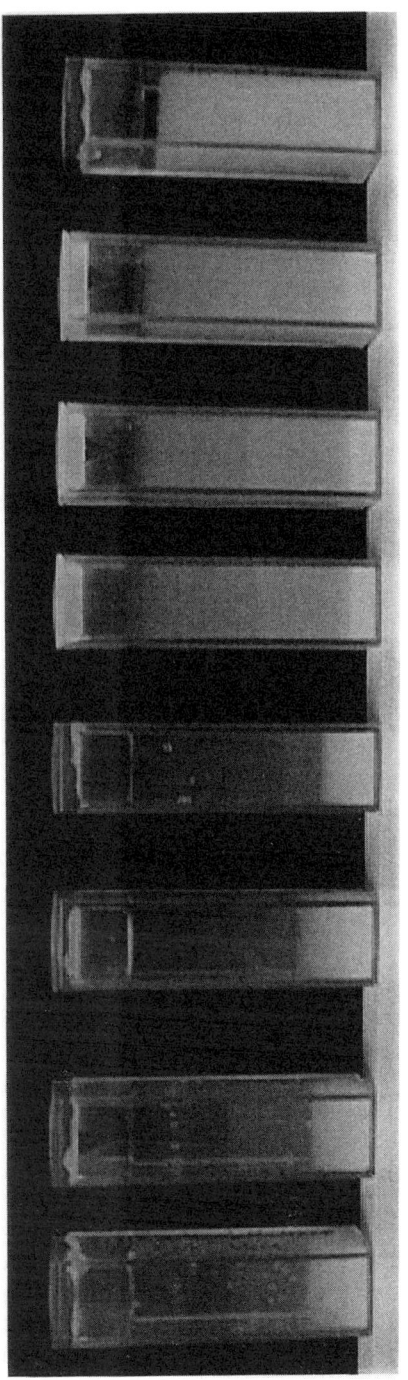

*Figure 3*: Gravity induced phase separation in the polydisperse synthetic clay fluorohectorite. Increasing NaCl concentration from left to right. The clay weight percent concentration is fixed.

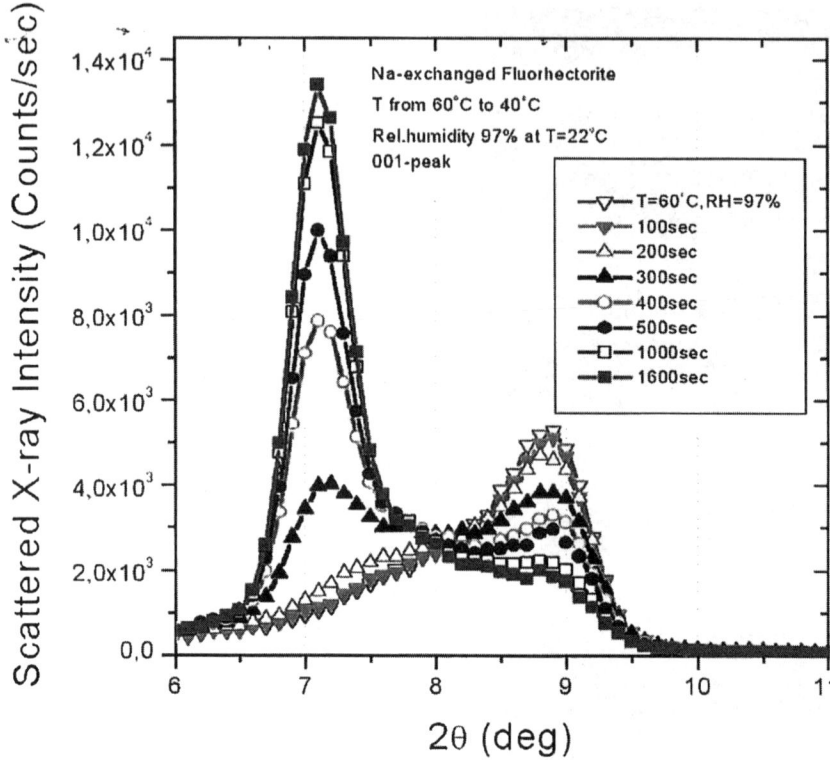

*Figure 4:* Dynamics of water intercalation into synthetic fluorohectorite. The interlayer distance (001) Bragg peak shifts dynamically after a sudden temperature step from a situation of zero monolayers of water intercalated (right peak) to a position of one monolayer of water intercalated (left peak) [36].

positioned magnetic interlayer cations, inducing magnetic spin-glass behavior at low temperatures [4,40]. Scale invariance in terms of fractal exponents of dehydrated and partly hydrated natural clays, has been reviewed previously [41,42]. Currently experimental work is ongoing in our group in order to investigate these phenomena in customized synthetic clays. Atomic Force Microscope (AFM) experiments are also ongoing at this time in order to map out and characterize the surface topography of customized synthetic clay films as was done for natural smectite clay films previously [43].

In concluding, clays represent rich physical model systems in which the broad spectrum of condensed matter physical phenomena may be observed in various parts of the salt-clay concentration phase diagram. The rich display of behaviors makes it crucial to utilize a variety of complementary experimental techniques in order to understand and develop these systems. Both basic studies of pure clay systems as discussed above, and mixtures of clays with surfactants [44], liquid crystals [45], colloids [46], bacteria [46], etc are of great interest, relevance and importance.

Clays as useful model systems are relatively new in the physics community, and a concerted effort of experimental and theoretical physicists and chemists together with researchers working in the more "traditional clay disciplines", is important for progress. There is lots of hard, interesting, challenging and good work to do for many years to come in this field.

# References

*Research supported by NTNU/NFR SUP 115185/420
*email: fossumj@phys.ntnu.no
*URL: http://www.phys.ntnu.no/~fossumj

1. R.E. Hummel, "Understanding Materials Science", *Springer-Verlag NY* (1998)
2. B. Velde, "Introduction to Clay Minerals", *Chapman and Hall, London* (1992)
3. F.M van der Kooj and H.N.W. Lekkerkerker, "Formation of Nematic Liquid Crystals in Suspensions of Hard Colloid Platelets", *J. Phys. Chem.* 102, 7829-7832 (1998)
4. S.A. Solin, "CLAYS AND CLAY INTERCALATION COMPOUNDS: Properties and Physical Phenomena", *Annu. Rev. Mater. Sci.* 27, 89-115 (1997)
5. C. Oriakhi, "Nanosandwhiches" in *Chemistry in Britain*, p 59-62, Nov (1998)
6. "Complex Systems and Materials" at *http://www.phys.ntnu.no/CPX* (1999)
7. J.O. Fossum, "Physical Phenomena in Clays", *Physica A* 270, 270-277 (1999)
8. Laporte Absorbents,**UK**; *Laporte Technical Reports*: P.K. Jennes Author
9. H. van Olphen, "Clay Colloid Chemistry", *Krieger Puplishing Company Florida*, 2nd ed.(1991)
10. A. Mourchid, E. Lecolier and H. van Damme, P. Levitz, "On Viscoelastic, Birefringent, and Swelling Properties of Laponite Clay Suspensions: Revisited Phase Diagram", *Langmuir* 14, 4718-4723 (1998)
11. J. Israelachvili, "Intermolecular and Surface Forces", *Academic Press London* (1992)
12. J.O. Fossum, A. Mikkelsen, A. Bakk, *Unpublished* (1999)
13. M. Reiner, "Deformation, Strain and Flow", *H.K. Lewis CO. LTD. London* (1969)
14. M. Kroon, G.H. Wegdam and R. Sprik, "Dynamic light scattering studies on the sol-gel transition of a suspension of anisotropic colloidal particles", *Phys. Rev. E* 54, 6541-6550 (1996)
15. M. Kroon, W.L. Vos and G.H. Wegdam, "Structure and formation of a gel of colloidal disks", *Phys. Rev. E* 57, 1962-1969 (1998)
16. S. Cocard, J.F. Tassin and T. Nicolai, "Aggregation and Gelation of Colloidal Discs", *Unpublished* poster at NATO ASI on Soft and Fragile Matter, St. Andrews, Scotland, July (1999)
17. D. Bonn, H. Tanaka, G. Wegdam, H. Kellay and J. Meunier, "Aging of a colloidal "Wigner" glass" *Europhys. Lett.*, 45, 52-57 (1999)
18. F. Pignon, A. Magnin, J-M. Piau, B. Cabane, P. Lindner and O. Diat, "Yield stress thixotropic clay suspension: Investigations of structure by light, neutron and x-ray scattering", *Phys. Rev. E* 56, 3281-3289 (1997)

19. T. Nicolai, *Private Communication* (1999)
20. L. Rosta and H.R. von Gunten, "Light Scattering Characterization of Laponite Sols", *J. Colloid and Interface Science* 134, 397-406 (1989)
21. A. Mourchid and P. Levitz, "Long-term gelation of laponite aqueuos dispersions", *Phys. Rev. E* 57, R4887-R4890 (1998)
22. J.P. Gabriel, C. Sanchez and P. Davidson, "Observation of Nematic Liquid-Crystal Textures in Aqueous Gels of Smectite Clays", *J. Phys. Chem.* 100, 11139-11143 (1996)
23. L. Onsager, *Ann. N.Y. Acad. Sci.* 51,627 (1949)
24. M.J. Dijkstra, J-P. Hansen and P.A. Madden, "Gelation of a Clay Colloid Suspension", *Phys. Rev. Lett.* 75,2236 (1995)
25. M.J. Dijkstra, J-P. Hansen and P.A. Madden, "Statistical model for the structure snd gelation of smectite clay suspensions", *Phys. Rev. E* 55, 3044-3053 (1997)
26. E. Trizac, "Quelques aspects statisiques de systemes complexes: coalescence balistique et suspensions d'argile", *These l'Ecole normale superieure de Lyon* (1997)
27. E. Trizac, *Private Communication* (1999)
28. E. Trizac and J-P. Hansen, "The Wigner-Seitz model for concentrated clay suspensions", *J.Phys.: Condens. Matter* 9, 2683-2692 (1997)
29. E. Trizac and J-P. Hansen, "Wigner-Seitz model of charged lamellar colloidal dispersions", *Phys. Rev. E* 56, 3137-3149 (1997)
30. R.J.F. Leote de Carvalho, E. Trizac and J-P. Hansen, "Non-linear Poisson-Boltzmann theory for swollen clays", *Europhys. Lett* 43, 369-375 (1998)
31. P.D. Kaviratna, T.J. Pinnavaia and P.A. Schroeder, "DIELECTRIC PROPERTIES OF SMECTITE CLAYS", *J. Phys. Chem. Sol.* 57, 1897-1906 (1996)
32. M.A. Bates and D. Frenkel, "Nematic-isotropic transition in polydisperse systems of infinitely thin hard platelets", *J. Chem. Phys*, 110, 6553-6559 (1999)
33. Y. Abdelhaye, G. Daccord, F. Duval, A. Louge and H. van Damme, "The hydraulic fracturation of colloidal gels", *C.R. Acad. Sci. Paris* t. 325, Serie IIb (1997)
34. F. Pignon, A. Magnin and J-M. Piau, "Structure and Pertinent Length Scale of a Discotic Clay Gel", *Phys. Rev. Lett.* 76, 4857-4860 (1996)
35. F. Pignon, A. Magnin and J-M. Piau, "Butterfly Light Scattering Pattern and Rheology of a Sheared Thixotropic Clay Gel", *Phys. Rev. Lett.* 79, 4689-4692 (1997)
36. J.O. Fossum, E.DiMasi, S.Berg-Lutnaes, et.al, *Unpublished work in progress* (1999)
37. S. Karaborni, B. Smit, W. Heidug, J. Urai and E. van Oort, "The Swelling of Clays: Molecular Simulations of the Hydration of Montmorillonite", *Science* 271, 1102-1104 (1996)
38. J.T. Klopprogge, "Synthetics of Smectites and Porous Pillared Clay Catalysts: A Review" *J. Porous Materials* 5, 5-41 (1998)
39. B.Y. Chen, H. Kim, S.D. Mahanti and T.J. Pinnavaia, Z.X. Cai, "Percolation and diffusion in two-dimensional microporous media: Pillared clays", *J. Chem. Phys.* 100, 3872-3880 (1994)

40. P. Zhou, J. Amarasekera, S.A. Solin, S.D. Mahanti and T.J. Pinnavaia, "Magnetic properties of vermiculite intercalation compounds", *Phys. Rev. B* 47, 16486-16493 (1993)

41. H. van Damme, "Scale invariance and hydric behavior of soils and clays", *C.R. Acad. Sci. Paris* t. 320 serie IIa, 665-681 (1995)

42. M. Sahimi, "Flow and Transport in Porous Media and Fractured Rock", *VCH Weinheim* (1995)

43. M. Zabat, M. Vayer-Besancon, R. Harba, S. Bonnamy and H. van Damme, "Surface topography and mechanical properties of smectite films", *Prog. Colloid Polym. Sci.* 105, 96-102 (1997)

44. I. Grillo, P. Levitz and Th. Zemb, "SANS structural determination of a nonionic surfactant layer adsorbed on clay particles" *Eur. Phys. J.* B 10, 29-34 (1999)

45. M. Kawasumi, N. Hasegawa, A. Usuki and A. Okada,"Nematic liquid crystal/clay mineral composits" *Mat. Science and Engineering* C6, 135-143 (1998)

46. J.O. Fossum et.al, *Work in progress* (1999)

47. F.P. Duval, P. Porion and H. Van Damme, "Microscale and Macroscale Diffusion of Water in Colloidal Gels. A Pulsed Field Gradient and NMR Imaging Investigation", *J. Phys. Chem. B* 103, 5730-5735 (1999)

# COMPLEX PARTICLE DYNAMICS DESCRIBED BY BRAID STATISTICS

ARNE T. SKJELTORP[1,2], SIGMUND CLAUSEN[1,2,3] AND GEIR HELGESEN[1]
[1]Institute for Energy Technology, Kjeller, Norway
[2]Physics Dept., University of Oslo
[3]Present address: SINTEF, Blindern N-0314, Oslo, Norway

## Abstract

Microscopic plastic spheres dispersed in a magnetic fluid are moved around by an external ac magnetic field. The spheres are confined to move in a thin layer between two plane parallel glass plates and may be observed in a light microscope. An n-strand braid records the particle trajectories. The fluctuations of these braids are investigated, and a wide range of dynamical behaviour is observed. For certain parameter values the fluctuations are highly intermittent and there is a hierarchical ordering of the dynamics in both space and time. In this case the motion is well modelled by a one-dimensional Lévy walk.

## 1. Introduction

Statistical mechanics is the standard tool to couple microscopic to macroscopic behaviour of a large collection of classical particles like atoms and molecules. The dynamics of a few particles can be handled by classical dynamics. Here, we want to present how a small discovery in physics involving small particles, led to a new use of an established mathematical field to offer an alternative description of particle dynamics. The discovery was how to marshal microscopic spheres into organised patterns using the "magnetic hole" effect [1]. The piece of mathematics needed was the braid theory [2-5].

The real world and mathematics, what is the relation between them? This is a general question, which has been posed many times, mostly by the mathematicians themselves, and requires a general answer. Probably the most general one is that mathematics is the "natural" language for a quantitative description of all natural phenomena. Mathematics can do without physics. Physics cannot do without mathematics. In creating a new branch of mathematics we do not have to prove its relevance to the real world. Whenever an essential progress was made in science, in particular in physics, various pieces of mathematical theories were either found waiting ready for their application or were created ad hoc. In any case, mathematics was indispensable. If the existing mathematics was not adequate enough - it was developed. Real important scientific discoveries cannot be described in a precise manner without mathematics.

*A.T. Skjeltorp and S.F. Edwards (eds.), Soft Condensed Matter: Configurations, Dynamics and Functionality, 281-291.*

Crystallography needed the group theory - it was ready. Relativity needed the non-Euclidean geometry - it was ready. Quantum mechanics needed the theory of the Hilbert spaces - it was ready. But the realm of mathematics is enormous. Let us be reminded of Stanislaw Ulam's quotation of Henri Poincare: „Mathematics is a language with which neither imprecise nor vague ideas can be expressed".

## 2. Experiment and model

The expression *"magnetic holes"* is used for a very simple, but unexpected physical phenomenon discovered by one of the authors some years ago [1]. The idea was to mix microscopic polystyrene spheres into a drop of a so-called magnetic fluid between two glass microscope slides. A magnetic fluid or ferrofluid is obtained by mixing nanometer sized magnetic particles like iron oxide into a nonmagnetic fluid, e.g. kerosene. In the absence of an external magnetic field the ferrofluid is not magnetised, see Fig. 1. However, if the sample is placed in a uniform magnetic field it becomes uniformly magnetised except within and the tiny plastic sphere. The lack of magnetic material within the volume of the sphere distorts the resulting magnetic field. In fact, the bubble would have the same field distorting effect as if one placed a small magnet in the centre of the sphere with a magnetic moment equal to the magnetic moment of the ferrofluid displaced by the sphere.

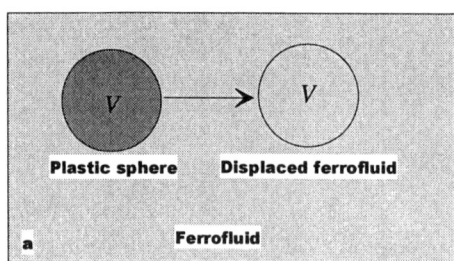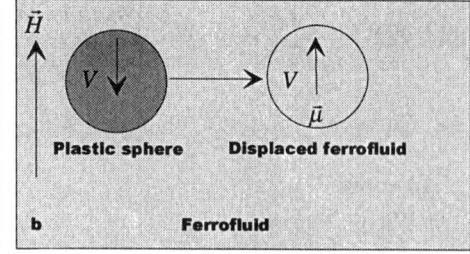

*Figure* 1. *Schematic view of the magnetic hole effect when an unmagnetised ferrofluid in (a) is subjected to an external magnetic field in (b). The magnetic moment carried by the sphere is equal in size but opposite in direction to the magnetic moment of the displaced ferrofluid by the sphere.*

But the magnetisation of this small magnet should point in the opposite direction to that of the whole sample. So, the sphere acquires an apparent magnetic dipole moments $\vec{\mu}$ antiparallel to the external field:

$$\vec{\mu} = -V\chi_{eff}\vec{H} .\tag{1}$$

Here, $V$ is the volume of a microsphere and $\chi_{eff}$ is the effective volume susceptibility of the ferrofluid In order to look at this particular experimental situation more formally; one has to apply the Maxwell equations. Imagine that two such plastic spheres with the same volume are found near each other. In the mathematical description of this situation each of the spheres can be pictured as a small magnet. Both are oriented in the same direction, opposite to the direction of the external magnetic field. But two magnets interact: e.g. if they are put side by side, they repel each other, Fig. 2. As the spheres carry the dipoles, the forces act on the spheres themselves. The plastic spheres can move in the ferrofluid and they will therefore move away from each other. Likewise, if the fields parallel to the axis connecting the centres of the spheres, the spheres will attract each other and be pulled together.

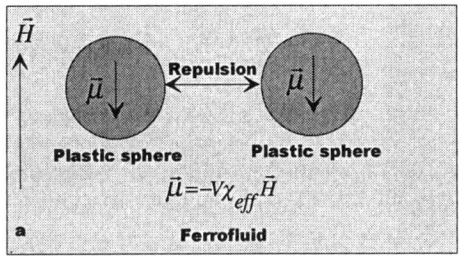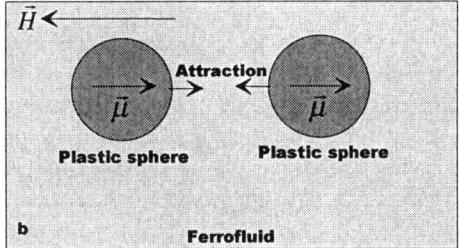

*Figure 2.* Schematic view of the magnetic hole effect producing (a) repulsive and (b) attractive interactions with different directions of the external magnetic field H.

The magnetic holes can be made almost identical, as the microspheres used to create them are manufactured in an ingenious polymerisation process developed by the late professor John Ugelstad, producing spheres of the same shape and size within accuracy better than 1 % [6]. In the experiments described below we used microspheres of a diameter 0.1 mm and below.

The ferrofluid [7] with microspheres dispersed in it was confined between two parallel glass plates, where the distance between the plates was typically twice the diameter $d$ of the microspheres, see Fig. 3. We placed the cell within a system of two pairs of coils producing a magnetic field rotating within the plane of the cell. The motion of the magnetic holes was also the same $xy$-plane.

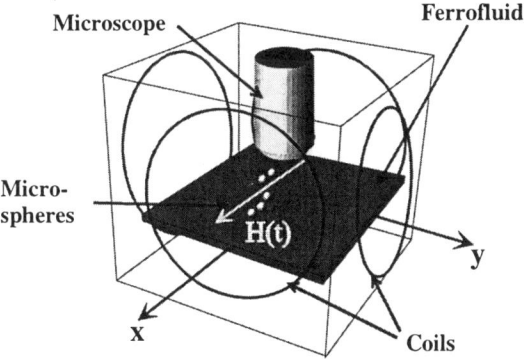

*Figure 3.* Schematic view of the experimental set-up where the sample contains plastic microspheres disper ferrofluid confined between two plane-parallel glass plates. Two pairs of coils are used to produce a magnetic (x, y)-plane.

We used circularly or elliptically polarised magnetic fields rotating within the sample $(x, y)$-plane with angular velocity $\omega_H$.

$$\vec{H}(t) = H\left[\cos(\omega_H t), \varepsilon \sin(\omega_H t)\right] \tag{2}$$

Here, $\varepsilon \leq 1$ is a measure of the field anisotropy.

The dipolar interaction energy of $n$ magnetic holes of diameter $d$ is given by:
$$U(\vec{r}_1, \vec{r}_2, ..., \vec{r}_N, t) =$$

$$\begin{cases} \displaystyle\sum_{i>j}^{N}\left[\frac{\mu^2(t)}{r_{ij}^3} - \frac{3\left(\vec{\mu}(t)\cdot\vec{r}_{ij}\right)^2}{r_{ij}^5}\right] & \text{if all } r_{ij} > \sigma \\ \\ \infty & \text{if any } r_{ij} < \sigma \end{cases} \tag{3}$$

Here $\vec{r}_{ij} = \vec{r}_j - \vec{r}_i$ is the vector joining the centres of the interacting microspheres. The components of the magnetic force acting on the $i$-th magnetic hole are given by

$$F_{i,\xi} = -\sum_{i>j}^{n}\frac{\partial}{\partial \xi_j}U(\vec{r}_1, \vec{r}_2, ..., \vec{r}_n, t) \tag{4}$$

where $\xi$ denotes either $x$ or $y$. The system is over-damped due to the large viscosity of the ferrofluid, and we may therefore neglect inertial forces. Thus, we assume that at any time the velocity of the $i$-th magnetic hole is proportional to the force given by Eq. (4):

$$\frac{dx_i}{dt} = \beta F_{i,x} \tag{5}$$

where $\beta=(3\pi\eta d)^{-1}$ and $\eta$ is the viscosity of the ferrofluid. Eqs. (3)–(5) can be transformed into a dimensionless form suitable for numerical integration by letting $H_x=1$ and $\beta=1/6$. By this choice of parameters, the threshold angular velocity for the stable rotation of a single pair of magnetic holes is equal to 1. The equations of motion were simulated using a fourth order Runge-Kutta algorithm and compared with our experimental results.

For a static magnetic field the minimum energy state is reached when all the $n$ microspheres are arranged in a linear chain oriented along the direction of the field. As soon as the field starts to rotate in the plane, the chain tries to follow the rotation of the field, but with a phase lag due to the viscous counterforce that slows down the motion. As long as the frequency is sufficiently low, the chain as a whole is able to follow the rotation of the field. This mode is trivial and independent of the number of microspheres $n$. The magnetic chain will follow the field with a certain phase lag when the driving frequency $\omega_H$ is less than a critical frequency, $\omega_c(n)$, which depends on the number of microspheres. For $\omega_H > \omega_c(n)$, the phase lag crosses a critical value and in each case the chain of microspheres is forced to split up into shorter pieces which are able to follow the rotation of the field.

With the existing experimental set-up it is possible to grab and digitise up to 25 images per second which gives us a continuous motion picture of the particle dynamics. Figure 4(a) shows a plot of the co-ordinates of the centres of five magnetic holes taken from the digitised images. The driving frequency is $f_H = 0.5$ Hz. In this case, $\omega_H > \omega_c(5)$, and it may be seen that the chain has split into two smaller pieces consisting of 2 and 3 microspheres. However, the microspheres trace out a very complex pattern and it seems difficult to analyse this. For comparison a picture of a *knot* is shown in the same figure. Could there be any connection? Only mathematics could provide us with an answer. Surely the theory of knots were already there, waiting to be used to describe real world phenomena. But could one do better? Obviously, the particle trajectories in Fig. 4(a) look rather messy and it is impossible to decide if the motion repeats itself or not. Things look much better if *time* is introduced.

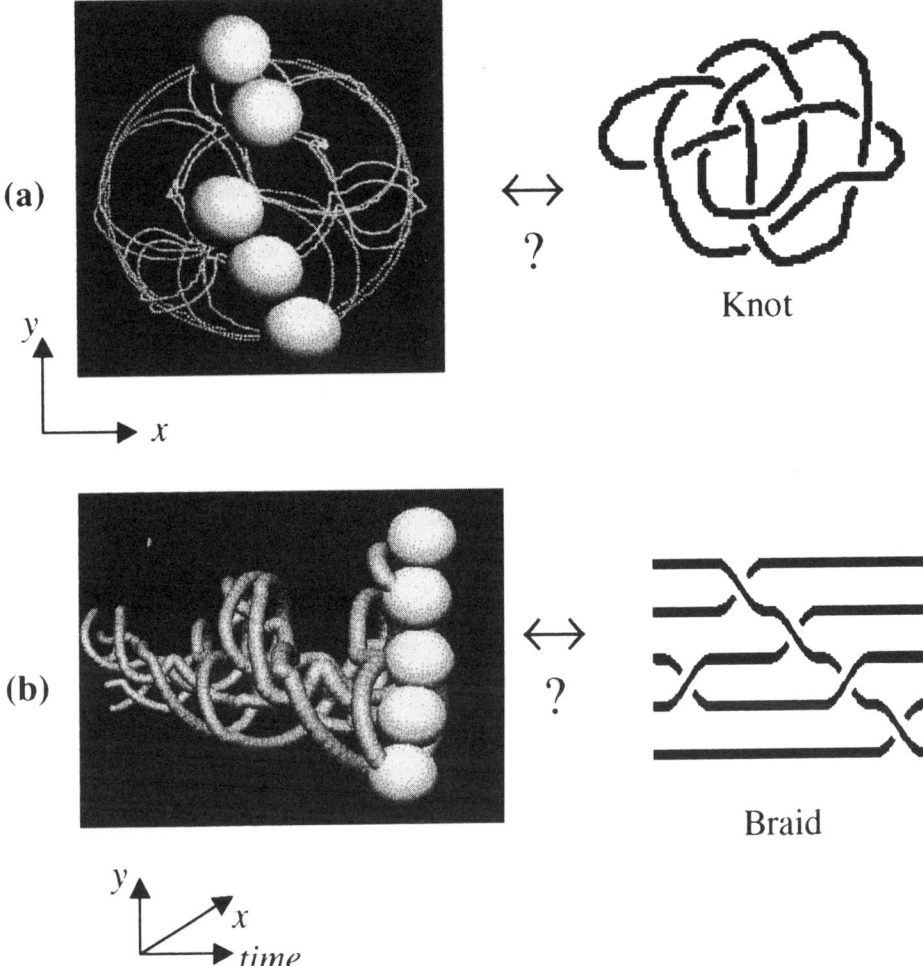

*Figure 4.* (a) Experimental observation of the trajectories of five magnetic holes subjected to a rotating magnetic field. The centres of the spheres were extracted from a series of digitised images. The total time span was about four full rotations of the magnetic field. Could there be any connection with the theory of knots? (b) The same as in (a) but the trajectories are now plotted vs. time t. Is it possible to analyse the resulting structure using braid theory?

Instead of plotting the co-ordinates of the magnetic holes in a $(x,y)$ co-ordinate system they are plotted in a three-dimensional space-time diagram $(x,y,t)$ with the time $t$ as the third co-ordinate as shown in Fig 4(b). One magnetic hole moving in a plane can be represented as a curve in a three-dimensional space-time diagram, and so several magnetic holes in motion produce a set of *braided* curves. This is a way of "freezing" the dynamics and the frozen structure, the *braid*, becomes the time history or world lines of the moving microspheres. The resulting braid structure of the motion in Fig. 4(b) looks rather complicated, but we will show how to describe the periodicity in a very compact manner. First, we will give a short introduction to braid theory.

## 3. Braids

Braid theory is a subfield of topology and an introduction may be found in the book by Birman [7]. Here, we will only give a brief description of the notions used in our analysis. Once a sequence of motions has been converted into braid notation, it can be further examined to obtain numbers characterising the structure and complexity of the braid and thereby the dynamics. In order to describe a braid without having to draw it, one may decompose it into a product of elementary braids; see Fig. 5a. For an $n$-strand system there exist $n$-1 elementary braids called generators and denoted by $\sigma_1$, $\sigma_2$,....,$\sigma_{n-1}$. In an elementary braid $\sigma_i$ the strand located at the $i$th position exchanges its place with the strand located at the $(i+1)$ th position.

In order to characterise the structure and complexity of a braid different numbers or *topological invariants* can be calculated. One such number is the *writhe* of the braid, $Wr$, which is simply the sum of the exponents of the braid word, a positive crossing adds +1 and a negative crossing adds -1. The writhe is therefore equal to the number of positive crossings minus the number of negative crossings. As an example consider the braid in Fig. 5(b), where $Wr=2$.

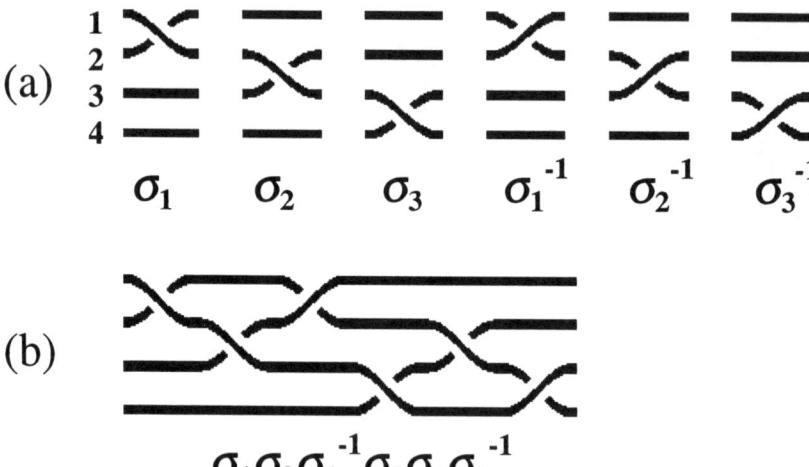

*Figure 5.* The generators of the 4-strand Artin braid group (a) and an example of a braid composed of them (b).

The recipe for the fluctuation analysis is now as follows: we make a video recording of the motion of $n$ magnetic holes where the output is a braid word describing the space-time braid of the motion. This braid is then divided into what may be denoted *half-period braids*. One half period braid is simply the space-time braid describing the motion of the microspheres during one half of the period $T$ of the rotating field, see Fig. 4. As the total braid grows with time $t$ the value of $Wr$ is extracted every time a new half period braid is plaited, i.e., whenever $t=mT/2$, where $m$ is the number of completed half periods. We set $t=m$ so that the unit of time is half a period of the rotating field. This approach results in a time series, $Wr(t)$. However, this time series does not give a complete topological description of the dynamics as there are several combinations of crossings of strands which results in the same value of $Wr(t)$. Nevertheless, our primary goal now is to study the aperiodic motion of the magnetic holes and the fluctuations in the patterns of the strands. In addition to the total writhe as a function of time, we are also interested in the successive half period variations defined by

$$\delta Wr(t)=Wr(t)-Wr(t-1)$$
$$(4)$$

which might be thought of as an average *writhe velocity*. $\delta Wr(t)$ equals the writhe $Wr$ of half-period braid number $t$, see Fig. 6.

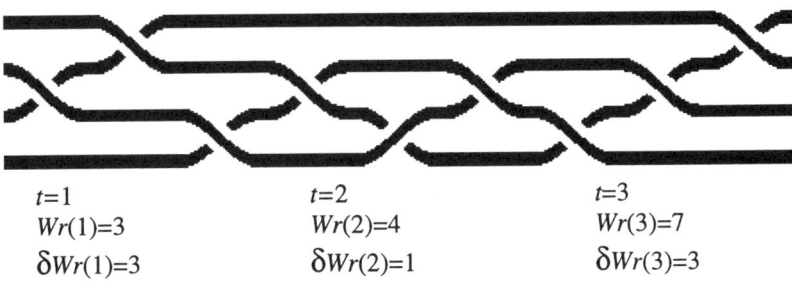

$t=1$           $t=2$           $t=3$
$Wr(1)=3$     $Wr(2)=4$     $Wr(3)=7$
$\delta Wr(1)=3$    $\delta Wr(2)=1$    $\delta Wr(3)=3$

*Figure 6.* Schematic illustration of the division of a braid into $t=3$ half-period braids. The accumulated value of the signed crossing number $Wr(t)$ and the half period differences $\delta Wr(t)$ are written below the braid.

## 4. Random fluctuations

In this review we will limit ourselves to describe the experimental result for the dynamics of $n=7$ magnetic holes at a fixed driving frequency $f_H=\omega_H/2\pi=0.25$ Hz and anisotropy parameter $\varepsilon=0.55$.

In the following we analyse the fluctuations in the writhe number, $Wr(t)$. This increases steadily with time $t$ due to the large majority of positive crossings of the space-time strands. In order to observe the fluctuations more easily the average increasing trend is subtracted from the original value of $Wr(t)$. The difference is denoted by $Wr(t)'$:

$$Wr(t)' = Wr(t) - t \cdot \overline{\delta Wr} .$$
$$(5)$$

Here, $\overline{\delta Wr}$ is the average value of $\delta Wr(t)$ averaged over the total number of half periods $N$:

$$\overline{\delta Wr} = \frac{1}{N} \sum_{t=1}^{N} \delta Wr(t)$$

(6)

The motion of the seven magnetic holes seems to be characterised by random fluctuations down to $\varepsilon \approx 0.62$ where a new type of behaviour is observed. At this value of the anisotropy parameter there is a finite probability for smaller chain pieces to stay separated from each other for long times. A continuous-time one-dimensional Lévy walk with a power law distribution of step lengths can model the overall motion. This model of a Lévy motion was proposed by Klafter, Blumen and Shleshinger [9]. Each time the chain of microspheres separates into smaller pieces a new step of the random walk is started. The separation times are defined as the times the chain pieces stay separated from each other and they are power-law distributed with an exponent equal to the exponent of the distribution of the step lengths.

Figure 7 shows $Wr(t)'$ for $\varepsilon=0.55$ with the associated half period variations $\delta Wr(t)$ plotted below. The driving frequency is still $f_H=0.25$ Hz. The motion of the magnetic holes was recorded for 10000 half periods of the rotating field, and the braids can thus be divided into a total of 10000 half period braids. The dynamical evolution is intermittent, and thus consists of both quiescent and more chaotic phases, which alternate temporally in an interspersed way.

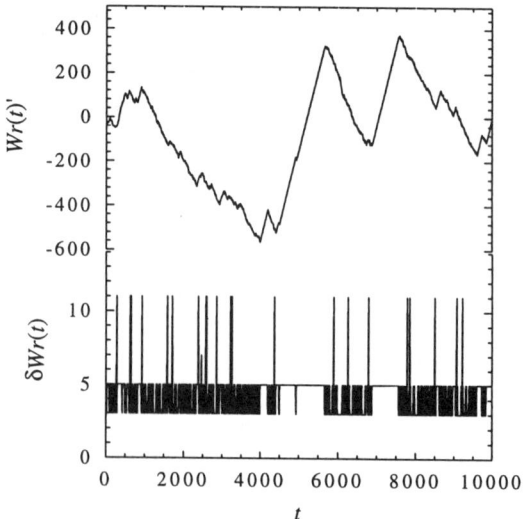

*Figure 7.* Time evolution of the signed crossing number $Wr(t)'$ of the space-time braid describing the motion of seven magnetic holes in an elliptically polarised field with $\varepsilon=0.55$ (upper curve). Time $t$ is measured in units of half a period of the driving field. The lowermost curve shows the associated half period differences $\delta Wr(t)$.

There is a *distribution* of step lengths, or equivalently a distribution of time intervals where long steps in $Wr(t)'$ are observed. During these long time intervals the microspheres move in a regular manner with the chain of magnetic holes divided into three smaller pieces which rotate with the field. Two of these chain pieces contain two microspheres whereas the third one contains three. The three chain pieces stay separated from each other for long times and the half-period variations $\delta Wr(t)=5$ during the steps, as seen in Fig. 7. This value equals the number of crossings of a half-period braid describing 2+2+3 magnetic holes rotating with the field. When the chain pieces are forced together again, the magnetic holes move in an aperiodic way for some time before they separate once more. During the separation times the signed crossing number $Wr(t)'$ increases at a constant rate causing the large-scale fluctuations seen in Fig. 7. We will show that the motion is well modelled by a one-dimensional Lévy walk with a distribution $\varphi(t)$ of quiescent time intervals or separation times $t$. The tail of this distribution follows a power law

$$\varphi(t) \propto t^{-(\alpha+1)} \tag{7}$$

with an exponent $\alpha$. The number $N$ of separation times $t_{sep}$ larger than $t$ goes as

$$N = N(t_{sep} \geq t) = \int_{t}^{\infty} \varphi(t_{sep})dt_{sep} \propto t^{-\alpha} \tag{8}$$

Figure 8 shows the distribution $N$ of separation times $t$ extracted from the time series shown in Fig. 7. It is possible to fit the data to a power law with an exponent $\alpha=1.27\pm0.13$ for $t>20$.

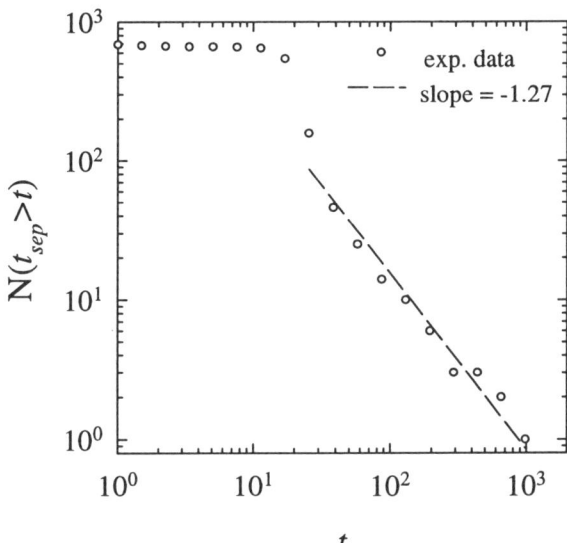

*Figure 8.*  Plots of the distribution $N(t_{sep} \geq t)$ of separation times $t$ for $\varepsilon=0.55$. The dashed line is the best fit to the experimental data in the region $t\in(20,1000)$.

A Lévy walk is a fractal generalisation of Brownian motion and it is possible to relate the diffusion exponent $\eta$ to the fractal dimension of space and time [10]. In this simple model of diffusion, space and time are coupled and a step in space is associated with a certain time span. The ensemble of time instants where jump events occur, form a fractal set with fractal dimension $d_t$:

$$d_t = \begin{cases} \alpha & \alpha \leq 1 \\ 1 & \alpha > 1 \end{cases} \qquad (9)$$

When do the stopover points form a fractal in space? An ordinary Brownian trajectory is so random that it is actually two-dimensional independent of the dimension of the embedding space. For a Lévy walk, where $1 \leq \alpha < 2$, the ensemble of stop-over points form a set with fractal dimension $d_r$ [10]:

$$d_r = \frac{2}{3-\alpha}. \qquad (10)$$

Thus, for the case studied here the ensemble of stopover points form a fractal set with $d_r=1.15\pm0.11$. The coupling of space and time through $d_r$ and $d_t$ is explicitly given by an expression for the diffusion exponent $\eta$:

$$\eta = \frac{2d_t}{d_r} = 3 - \alpha \qquad (11)$$

From this one gets $\eta=1.73\pm0.13$ using $\alpha=1.27\pm0.13$, in good agreement with the experimental result $\eta=1.74\pm0.03$ found in our experiments [3].

Recently, Wang and Tokuyama [11] has used a generalised Langevin equation to show theoretically that for the superdiffusive case, $\alpha>1.0$, $\eta=1.66\pm0.12$ and $d_r=1.21\pm0.09$, which is also in good agreement with our experimental result.

## 5. Conclusions

The random motion of microspheres in a plane has been studied. The two-dimensional trajectories of $n$ spheres generate a braid in a three-dimensional space-time. By studying the statistics of braid parameters a wide range of different dynamical behaviour is observed, ranging from periodic to random motion. For certain experimental situations the fluctuations were shown to be highly intermittent and a hierarchical ordering takes place in both space and time. In that case the motion is well modelled by a one-dimensional Lévy walk with a power law distribution of step lengths, which determines the fluctuation behaviour. The dynamical evolution consists of both regular and more chaotic phases which alternate temporally.

In conclusion, our experimental system is simple and can be well described using tools from braid theory. This coupled with computer simulations, which have not been discussed here, but which show good agreement with the experiments, allows us to look for general features of non-equilibrium phenomena and to make comparison with current theories.

### Acknowledgements

This work has been supported in part by the Research Council of Norway (S.C.). We want to thank A. Berge at NTNU in Trondheim, and Dyno Specialty Polymers for kindly providing the microspheres used in these experiments.

# References

1.   Skjeltorp, A.T. (1983) *Phys. Rev Lett.* **51**, 2306.
2.   Clausen, S., Helgesen, G. and Skjeltorp, A.T. "Braid Description of Few Body Dynamics", Int. J. of Bifurcation and Chaos, **8**, No. 7, 1383-1397 (1998). [http://www.wspc.com/journals/ijbc/87/13887.html]
3.   Clausen, S., Helgesen, G. and Skjeltorp, A.T. "Braid Description of Collective Fluctuations in a Few-Body System", Phys. Rev. **E58**, No. 4, 4229- 4237 (1998).
4.   Sigmund Clausen (Thesis, Physics Dept., Univ. of Oslo, 1998).
5.   For a popular science article about magnetic holes and braids written in French, see: Piotr Pieranski and Arne T. Skjeltorp, "Les tresses tissées par des trous magnétiques", *Pour la Science*, Avril 1997.
6.   The uniformly sized Ugelstad microspheres named after its inventor are commercially available from Dyno Specialty Polymers AS, Lillestrøm, Norway under the trade name Dynospheres [http://www.dyno.no/about.html]. They have a multitude of applications, ranging from laboratory scale experiments to industrial uses including biomagnetic separation, purification of biomolecules by liquid chromatography, magnetic resonance imaging diagnostics.
7.   Available from Ferrofluidics Corporation, 40 Simon St., Nashua NH, USA.
8.   Birman, J. S. (1974) *Braids, Links and mapping Class Group,* Annals of Math. Study, **82**, Princeton University Press.
9.   Klafter, J., Blumen, A. and Shlesinger, M. F. (1987) *Phys. Rev.* A **35**, 3081
10.  Wang, X. -J, (1992) *Phys. Rev. A* **45**, 8407.
11.  Wang, K.G. and Tokuyama, M. (1999) Nonequilibrioum statistical description of anomalous diffusion *Physica* A **265**, 341.

INDEX

Rayleigh instability, 113
renormalization group, 53
ribosome, 83

sausage-string pattern, 111
scaling, 4, 42, 45
segregation, 28
selection, 84
self-organization, 189
small angle scattering, 219
soap films, 18, 24
soft matter, 2, 8
specular reflectivity, 230
spin density wave, 48, 51
    fluctuation model, 44, 47, 61
    glasses, 71
    susceptibility, 44, 46, 52
superconducting gap function, 65
superconductivity, 38, 60
supramolecular architecture, 207

tissue culture, 101
transition metal oxide, 55
two dimensional froths, 17, 19

wetting, 230
Wigner crystal, 58